HUMAN EVOLUTION AND CHRISTIAN ETHICS

Can the origins of morality be explained entirely in evolutionary terms? If so, what are the implications for Christian moral theology and ethics? Is the latter redundant, as sociobiologists often assert? Stephen Pope argues that theologians need to engage with evolutionary theory rather than ignore it. He shows that our growing knowledge of human evolution is compatible with Christian faith and morality, provided that the former is not interpreted reductionistically and the latter is not understood in fundamentalist ways. Christian ethics ought to incorporate evolutionary approaches to human nature to the extent that they provide helpful knowledge of the conditions of human flourishing, both collective and individual. From this perspective, a strong affirmation of human dignity and appreciation for the theological virtues of faith, hope, and charity is consistent with a revised account of natural law and the cardinal virtues.

STEPHEN J. POPE is Professor of Theological Ethics at Boston College. He is editor of *Common Calling: The Laity and the Governance of the Church* (2004).

NEW STUDIES IN CHRISTIAN ETHICS

General Editor
Robin Gill

Editorial Board
Stephen R. L. Clark, Stanley Hauerwas, Robin W. Lovin

Christian ethics has increasingly assumed a central place within academic theology. At the same time the growing power and ambiguity of modern science and the rising dissatisfaction within the social sciences about claims to value-neutrality have prompted renewed interest in ethics within the secular academic world. There is, therefore, a need for studies in Christian ethics which, as well as being concerned with the relevance of Christian ethics to the present-day secular debate, are well informed about parallel discussions in recent philosophy, science or social science. *New Studies in Christian Ethics* aims to provide books that do this at the highest intellectual level and demonstrate that Christian ethics can make a distinctive contribution to this debate – either in moral substance or in terms of underlying moral justifications.

Titles published in the series:

1. *Rights and Christian Ethics* Kieran Cronin
2. *Biblical Interpretation and Christian Ethics* Ian McDonald
3. *Power and Christian Ethics* James Mackey
4. *Plurality and Christian Ethics* Ian S. Markham
5. *Moral Action and Christian Ethics* Jean Porter
6. *Responsibility and Christian Ethics* William Schweiker
7. *Justice and Christian Ethics* E. Clinton Gardner
8. *Feminism and Christian Ethics* Susan Parsons
9. *Sex, Gender and Christian Ethics* Lisa Sowle Cahill
10. *The Environment and Christian Ethics* Michael Northcott
11. *Concepts of Person and Christian Ethics* Stanley Rudman
12. *Priorities and Christian Ethics* Garth Hallett
13. *Community, Liberalism and Christian Ethics* David Fergusson
14. *The Market Economy and Christian Ethics* Peter Sedgwick
15. *Churchgoing and Christian Ethics* Robin Gill
16. *Inequality and Christian Ethics* Douglas Hicks
17. *Biology and Christian Ethics* Stephen Clark
18. *Altruism and Christian Ethics* Colin Grant
19. *The Public Forum and Christian Ethics* Robert Gascoigne
20. *Evil and Christian Ethics* Gordon Graham
21. *Living Together and Christian Ethics* Adrian Thatcher

22. *The Common Good and Christian Ethics* David Hollenbach
23. *Self Love and Christian Ethics* Darlene Fozard Weaver
24. *Economic Compulsion and Christian Ethics* Albino Barrera
25. *Genetics and Christian Ethics* Celia Deane-Drummond
26. *Health Care and Christian Ethics* Robin Gill
27. *Alcohol, Addiction and Christian Ethics* Christopher C. H. Cook
28. *Human Evolution and Christian Ethics* Stephen J. Pope

HUMAN EVOLUTION AND CHRISTIAN ETHICS

STEPHEN J. POPE
Boston College

CAMBRIDGE UNIVERSITY PRESS
Cambridge, New York, Melbourne, Madrid, Cape Town, Singapore, São Paulo

Cambridge University Press
The Edinburgh Building, Cambridge CB2 8RU, UK

Published in the United States of America by Cambridge University Press, New York

www.cambridge.org
Information on this title: www.cambridge.org/9780521863407

© Stephen J. Pope 2007

This publication is in copyright. Subject to statutory exception
and to the provisions of relevant collective licensing agreements,
no reproduction of any part may take place without
the written permission of Cambridge University Press.

First published 2007

Printed in the United Kingdom at the University Press, Cambridge

A catalogue record for this publication is available from the British Library

ISBN 978-0-521-86340-7 hardback

Cambridge University Press has no responsibility for
the persistence or accuracy of URLs for external or
third-party internet websites referred to in this publication,
and does not guarantee that any content on such
websites is, or will remain, accurate or appropriate.

James M. Gustafson
teacher, mentor, and friend

Contents

General editor's preface		*page* xi
Acknowledgments		xii
	Introduction	1
1	Evolution and religion	8
2	The indifference of Christian ethics to human evolution	32
3	Varieties of reductionism	56
4	Faith, creation, and evolution	76
5	Chance and purpose in evolution	111
6	Human nature and human flourishing	129
7	Freedom and responsibility	158
8	Human dignity and common descent	188
9	Christian love and evolutionary altruism	214
10	The natural roots of morality	250
11	Natural law in an evolutionary context	268
12	Sex, marriage, and family	297
Bibliography		320
Index of scriptural citations		346
Index of names and subjects		348

General editor's preface

This book is the twenty-eighth in the series *New Studies in Christian Ethics*. It contains an important dialogue with some of the earlier books in the series, notably Stephen Clark's *Biology and Christian Ethics*, Colin Grant's *Altruism and Christian Ethics* and Jean Porter's *Moral Action and Christian Ethics*. There are also points of mutual concern shared with Celia Deane-Drummond's recent *Genetics and Christian Ethics*. All of these books closely reflect the two key aims of the series – namely to promote monographs in Christian ethics that engage centrally with the present secular moral debate at the highest possible intellectual level and, secondly, to encourage contributors to demonstrate that Christian ethics can make a distinctive contribution to this debate.

Stephen Pope has already established a firm reputation as a creative Catholic theologian with his book *The Evolution of Altruism and the Ordering of Love* (1994). His particular contribution in *Human Evolution and Christian Ethics* is to engage critically and creatively as a natural-law theologian with sociobiologists. Quite a number of the latter have been highly critical of religion in general and of Christian theology in particular. He argues at length that they have often misunderstood (and oversimplified) what theologians today are attempting to do. But he is also aware that theologians themselves have all too often ignored evolutionary science. In contrast, he has read the science carefully and in the process developed a critical but sympathetic Christian ethical approach to sociobiological explanations of purpose and altruism that most other Christians simply ignore.

This is a careful and helpful book that offers an important bridge for those who wish to take *both* evolutionary theory and theology seriously.

ROBIN GILL

Acknowledgments

A number of the chapters of this book are developed on the basis of earlier published work. Some material has been based on the following publications: "Neither Enemy Nor Friend: Nature as Creation in the Theology of St. Thomas," *Zygon* 32 (spring 1997): 219–230; "The Biological 'Roots' of Personhood and Morality," *Josephinum Journal of Theology* 8.2 (summer/fall 2001): 91–101, and "Human Evolution and Moral Responsibility: Beyond the 'Free Will–Determinism' Conundrum," *Theoforum* 38 (2002): 365–389; "Agape and Human Nature: Contributions from Neo-Darwinism," *Social Science Information* 31 (1992): 509–529, anthologized in James B. Miller, ed., *An Evolving Dialogue: Scientific, Historical, Philosophical, and Theological Perspectives on Evolution* (Washington, DC: American Association for the Advancement of Science, 1998), pp. 421–440; "Familial Love and Human Nature: Thomas Aquinas and Neo-Darwinism," *American Catholic Philosophical Quarterly* 69 (summer 1995): 447–469; "The Ordering of Love: Self, Others, and Sacrifice," in Stephen G. Post, ed., *Empathy, Altruism, and Agape: Perspectives on Science and Religion* (New York: Oxford University Press, 2002); "The Evolutionary Roots of Morality in Theological Perspective," *Zygon* 33 (December 1998): 545–556; reprinted in James B. Miller, ed., *The Epic of Evolution: Science and Religion in Dialogue* (Upper Saddle River, NJ: Prentice Hall, 2004), pp. 189–198.

Some of the chapters of this book were written while I attended the John Templeton Oxford Summer Workshops at Wycliffe Hall, Oxford, in the summers of 1999–2002. Special thanks are due to fellow participants who made those three summers so intellectually stimulating and informative, and especially to Professor Alistair McGrath, the head of Wycliffe Hall, and Professor John Roche of Linacre College, Oxford, for their stewardship of this valuable program. Finally, I would like to thank Dean Joseph F. Quinn and Dean John J. Neuhauser of Boston College for granting me a sabbatical so that I could complete this book.

Acknowledgments

A number of friends have read the manuscript either in part or in whole. I am especially grateful to Anthony Annuziato, Don S. Browning, Lisa Sowle Cahill, Rebecca Fleitstra, Karl Giberson, Russell Hittinger, James F. Keenan, SJ, Gerald P. McKenny, and Terence L. Nichols. They helped me correct some of the significant defects of early drafts of this book. I am very grateful for their insight, generosity, and academic colleagueship. I would also like to thank four Boston College students, Ryan Ahearn, Rebecca Camacho, Lauren Escher, and Craig Gould, who provided excellent help in preparing this book for publication. My deepest debt of gratitude goes to my wife, Patti, and our three children, Mike, Katie, and Steve. I cannot imagine that any author could have a more supportive, understanding, and patient family.

A final mention must be made of my teacher, James M. Gustafson. Among his many accomplishments has been his ability to communicate to his students a sense of his deep devotion to learning, unwavering fidelity, and profound intellectual honesty. I dedicate this book, with admiration, gratitude, and loyalty, to Jim.

Introduction

For over a century Christian ethics has been deeply influenced by the social sciences and, in particular, by social theories of the kind developed by Marx, Weber, and Durkheim, but it has not engaged in an analogous enterprise when it comes to the natural sciences.[1] In this book I intend to explore the relevance of science, and specifically the information and insights of evolutionary theory, for Christian ethics.

The theory of evolution is now the primary explanatory context for understanding the origin of species.[2] Scientists and writers in the last thirty years have produced a significant body of literature dealing with "evolutionary ethics" and the "evolution of morality," but Christian ethics has for the most part ignored it. This inattentiveness takes place at a time when popular evolution-based writers represent the public face of science. The "sociobiology" proposed by Robert Trivers, E. O. Wilson, and Richard

---

[1] The term "science" will be taken to refer to the activities in which scientists seek to arrive at a relatively reliable understanding of the natural world. On the meaning of "science," see George F. R. Ellis, "The Thinking Underlying the New 'Scientific' World-Views," in Robert John Russell, William R. Stoeger, SJ, and Francisco J. Ayala, eds., *Evolutionary and Molecular Biology: Scientific Perspectives on Divine Action* (Vatican City State: Vatican Observatory Publications, and Berkeley, CA: Center for Theology and the Natural Sciences, 1998), pp. 251–280.

[2] This book provides neither a theoretical justification of the theory of evolution to convince Christian fundamentalists or other religiously based skeptics of its plausibility, nor an attempt to counter the popular misunderstanding and fear of the theory of evolution. Competent scientists have already dedicated many works to explaining the abundant evidence for evolution. For scientific arguments against "scientific creationism," see Tim M. Berra, *Evolution and the Myth of Creationism: A Basic Guide to the Facts in the Evolution Debate* (Stanford, CA: Stanford University Press, 1990); Kenneth Miller, *Finding Darwin's God: A Scientist's Search for Common Ground between God and Evolution* (New York: Cliff Street Books/HarperCollins, 1999); and Stephen Jay Gould, *Hens' Teeth and Horses' Toes: Reflections on Natural History* (New York: Norton, 1983), pp. 247–264. For a major, if somewhat dated, Catholic theological response to evolutionary theory, see Karl Rahner, *Hominization: The Evolutionary Origin of Man as a Theological Problem* (New York: Herder and Herder, 1965); Karl Rahner, "Natural Science and Reasonable Faith," trans. Hugh M. Riley, in *Theological Investigations*, vol. XXI: *Science and Christian Faith* (New York: Crossroad, 1988), pp. 16–55. A helpful survey has been provided by Don O'Leary, *Roman Catholicism and Modern Science* (New York: Continuum, 2006).

Dawkins attempts to provide the comprehensive explanation of social behavior in terms of evolutionary theory.[3] The slightly less overtly political "evolutionary psychology" developed in the 1980s by Leda Cosmides, John Tooby, and Donald Symons, and popularized by Steven Pinker and Robert Wright, strives to explain the deepest roots of human behavior in evolutionary terms, primarily through an understanding of the functioning of "evolved psychological mechanisms."[4]

In this book I argue that, despite various difficulties, Christian ethics and evolutionary theories are in principle consonant with one another. Distinct vantage points do not have to compete with one another if interpreted properly. If one accepts the axiom that, ultimately, "truth cannot conflict with truth,"[5] then one can argue that the knowledge provided by the natural sciences, including that pertaining to human evolution, is consistent with, and can help to shed light on, the truth affirmed in Christian faith.

Science of course does not provide Christian faith with direct and unambiguous intellectual justification, such that a person without faith would be convinced to adopt Christian belief solely or primarily on the basis of evidence given in the natural world. One cannot argue from evolutionary biology to Christianity, or vice versa. Since theology is an essentially interpretative enterprise, none of us can pretend to work from the vantage point of presuppositionless objectivity. Functioning within a tradition that is mediated historically, the study of theology involves both careful interpretation of magisterial texts and respectful dialogue with present forms of knowledge, including scientific findings about human evolution.

From a Christian standpoint, faith in the Creator requires theology to extend its range of sources to include science and other non-theological

---

[3] See Robert Trivers, *Social Evolution* (Menlo Park, CA: Benjamin/Cummings, 1985); E. O. Wilson, *Sociobiology: The New Synthesis* (Cambridge, MA: Harvard University Press, 1975); and Richard Dawkins, *The Selfish Gene* (New York: Oxford, 1976).

[4] See Jerome H. Barkow, Leda Cosmides, and John Tooby, *The Adapted Mind: Evolutionary Psychology and the Generation of Culture* (New York: Oxford University Press, 1992).

[5] John Paul II, "Message to the Pontifical Academy of Sciences on Evolution," *Origins* 26 (November 1996): 349, citing Leo XIII, *Providentissimus Deus*. See also Thomas Aquinas, *De Veritate* 1,8. As the pope put it in what was his best discussion of the science–religion relation, "Both religion and science must preserve their own autonomy and their distinctiveness. Religion is not founded on science nor is science an extension of religion. Each should possess its own principles, its pattern of procedures, its diversities of interpretation and its own conclusions ... While each can and should support the other as distinct dimensions of a common human culture, neither ought to assume that it forms a necessary premise for the other." John Paul II, "Letter to the Rev. George V. Coyne, S. J.," *Origins* 18 (November 1988): 377.

sources. Christian faith ought neither to interfere with the pursuit of scientific knowledge nor to require scientists to ignore relevant data, nor to encourage breaches of the procedures proper to scientific inquiry. As physicist Howard Van Till explains, "Linking a specific scientific theory with some religious belief system in such a way that one entails the other, for example, has a serious strategic disadvantage in that any discrediting of that scientific theory automatically tends to call into question the entire belief system attached to it."[6] The goal of science, he notes, is "to gain knowledge, not to reinforce preconceptions."[7]

The most popular term in the academy for the science–theology relation is "dialogue." Yet scientists and theologians do not learn from one another in the ways that microbiologists learn from biochemists or moral theologians learn from moral philosophers. In fact, scientists *qua* scientists have nothing to learn from theologians about how to conduct scientific research or about the scientific implications of their findings. Inserting theological questions into scientific inquiry is distracting as well as beside the point.

Scientists *qua* thoughtful human beings, on the other hand, are inclined to raise questions about the deeper meaning of their scientific work and to delve into matters that lie outside the domains with which the methods of science are suited to function. Some insights of science have important theological implications but, as wondering, imagining, feeling human beings, scientists raise kinds of questions that their professional training and specialization do not equip them to address. Theologians can alert scientists to ways in which they have attempted to exceed the proper limits of their disciplines and to the intellectual hazards of doing so. Christian ethicists can play a valuable role in disentangling evolutionary science from its ideological misuses, pointing out the shortcomings of distorted applications of evolutionary theory to various kinds of human behavior, and showing that moral and religious implications of evolutionary accounts of humanity can be interpreted nonreductionistically.

The unity of truth suggests that the findings of science and the insights of theology are ultimately compatible and, at certain points, mutually enlightening. Scientific perspectives on nature can clarify, enrich, and deepen the minds of those who view the natural world with the eyes of faith. Yet the wellspring of Christian convictions lies not in science but in

---

[6] Howard J. Van Till, Robert E. Snow, John H. Steck, and Davis A. Young, *Portraits of Creation: Biblical and Scientific Perspectives on the World's Formation* (Grand Rapids, MI: Eerdmans, 1990), p. 149.
[7] Ibid.

the personal religious experience made possible by living communities of faith. Approached in this way, knowledge of human evolution need not have the devastating impact on Christian ethics sometimes depicted by evolutionists such as Wilson and Dawkins. On the contrary, knowledge of evolution, and especially understood in terms of the notion of "emergent complexity," can make an important constructive contribution to Christian ethics, particularly with regard to our thinking about the natural law and the virtues. Science can help us understand the biological factors that allow for the human capacities that provide the basis for morality and religion.

AUDIENCE AND GOAL

We live in an increasingly secular culture in which many people find no grounds for taking seriously belief in God, never mind Christian faith. For some of those deeply influenced by evolutionary biology, Darwin's refutation of Paley's argument from design was the last nail in the coffin of theism.[8] Yet a number of scholars argue that knowledge of human evolution does not have to lead to this skeptical conclusion. What is sometimes characterized as a simple intellectual stand-off between science and religion is actually a much more complex and varied relationship. As historian of science John Brooke points out,

> There is no such thing as *the* relationship between science and religion. It is what different individuals and communities have made of it in a plethora of different contexts. Not only has the problematic interface between them shifted over time, but there is also a high degree of artificiality in abstracting from the science and religion of earlier centuries to see how they were related.[9]

What Brooke says here about the general categories of science and religion also applies to the categories of evolutionary theory and Christian ethics.

Coming from the opposite direction, some Christian ethicists insist so stridently that scientific (or other non-theological) modes of thought not be allowed to set the agenda for theology that they end up ignoring science altogether. But this stance obscures the fact that serious engagement with contemporary science need not diminish Christian identity. The Christian tradition itself generated a profound theological impetus for the

---

[8] See John Dupré, *Darwin's Legacy: What Evolution Means Today* (New York: Oxford University Press, 2003).
[9] John H. Brooke, *Science and Religion: Some Historical Perspectives* (Cambridge: Cambridge University Press, 1991), p. 321.

development of modern science.[10] Its colleges and universities were the places of many of the most ground-breaking scientific discoveries, and many of the greatest Christian theologians – from Augustine and Thomas Aquinas to Jonathan Edwards and Karl Rahner – developed their theologies in light of available knowledge regarding the natural world.[11] The policy of ignoring the natural sciences on grounds of Christian identity actually constitutes a break with the mainstream of the Christian tradition, not its continuation.[12]

While critical of evolutionary ideology, Christian ethics needs to engage evolutionary knowledge because it can help us better to understand important aspects of human nature and some of the enduring constituents of human flourishing. Christian ethics, especially as developed in the natural-law tradition engaged here, gives moral significance to the central constituents of human nature, so it must take seriously the massive body of literature and significant discoveries about where we come from, who we are, and what we need and desire as human beings. Knowledge of human evolution is a necessary source of insight for any contemporary Christian ethics that takes human nature seriously.

This book attempts to address fundamental questions of Christian ethics more than it considers practical or "applied" matters. One might think that a book on Christian ethics and human evolution would place these evolutionary writings in relation to Christian treatments of the same topics, for example to relate E. O. Wilson on the evolution of deception to Augustine's analysis of lying or contrast ethological treatments of aggression with the Sermon on the Mount. Yet this kind of analysis is neither particularly interesting nor intellectually fruitful. The most significant level of interchange concerns more fundamental questions about the nature of reality (metaphysics, and especially ontology) and God (theology), rather than practical moral questions. When a given evolutionist disagrees with a Christian moral teaching about sex or lying, for example, the point in

---

[10] See John H. Brooke, David C. Lindberg, and Ronald L. Numbers, eds., *God and Nature: Historical Essays on the Encounter between Christianity and Science* (Berkeley: University of California Press, 1986).

[11] The term "nature" can be used in many ways, including three major uses of the term found in this book: the "nature" or essence of an entity, the totality of the physical world, the world of creation as distinct from supernatural grace. Context will indicate which of these meanings of the term is intended.

[12] It might be added that while the Reformed theologian Karl Barth has often been regarded as indifferent to science, it is possible to develop his theology in a way that includes a more constructive relation to it. See Thomas Torrance: *Theological Science* (New York: Oxford, 1969; reissued in 1996 by T. & T. Clark); *Space, Time, and Resurrection* (Edinburgh: Handsel, 1976); and *Reality and Scientific Theology* (Edinburgh: Scottish Academic Press, 1985).

dispute is more often based on how he or she views human society or human nature and not only about the morality of sexual relations or speech.

The deepest moral disagreements are rooted in competing presuppositions about what is most real, how we can come to understand what is most real, and how this knowledge provides guidance for leading good lives and developing good communities. This book deals with the dispute between Christian moral realism – which holds that the world is intrinsically morally meaningful – and evolutionary ontological naturalism, which denies that it has any meaning other than what we human beings choose to make of it. It will devote some time to considering fundamental theological issues such as faith, creation, and providence, and metaphysical concerns regarding the place of teleology, directionality, and progress in the evolutionary process. Christian ethics cannot participate in dialogue with evolutionary theory without some, even if cursory, prior examination of these themes.

Theories of evolution do not make a direct contribution to Christian ethics. Evolutionary biology can provide neither a "foundation" for Christian ethics nor scientific "backing" to the contents of Christian ethics, even within the natural-law tradition. Our knowledge of nature, including evolution, cannot determine the content of theological or moral affirmations.

Knowledge of human evolution, however, can play a valuable role in helping us to understand important aspects of human nature and human flourishing. The natural-law tradition regards the moral life as the way to move toward the human good, and any account of the human good reflects some account of human nature and the conditions that make for its flourishing.

OVERVIEW

The basic structure of the book falls into three parts: the first part argues for the importance of current knowledge of evolution for Christian ethics in general (chs. 1–6), the second part examines ways in which evolution can enrich and inform our understanding of human nature and specifically regarding the themes of freedom, love, and human dignity (chs. 7–9), and the third part discusses the relevance of evolution to the natural-law tradition (chs. 10–12).

One of my central convictions is that Christian ethics can fruitfully employ evolutionary insights into human behavior as long as these are not distorted by unjustifiable kinds of reductionism. A nonreductionistic reading of evolution that recognizes its inherent directionality is consonant with Christian belief in creation and providence. The human race is the product of a process that has generated unprecedented forms of emergent complexity. Christian theologians have long maintained that God operates through the "secondary causes" made available by the evolutionary process. The account of human nature as constituted by emergent complexity helps us understand aspects of key notions in Christian ethics, particularly human freedom, love of neighbor, human dignity, morality, and natural law.

The twofold audience of this book causes a certain imbalance in the presentation of the material examined. It requires an explanation of some things that Christian ethicists already know but that scientifically inclined readers do not, and vice versa. A certain amount of introductory explanation is needed for each group, though not, it is hoped, to the point of tedium. Like most interdisciplinary projects, reading this book will require a certain amount of patience and intellectual generosity on the part of the expert reader.

The attempt to engage in interdisciplinary reflection that joins such diverse disciplines, or, more accurately, sets of disciplines, necessarily involves wading into discussions that lie outside any given author's expertise. This is particularly the case when a Christian wades into the study of human evolution, which, as Simon Conway Morris notes, is a field "riven with controversy."[13]

My own training is in Christian theological ethics rather than in the natural sciences. Anyone who is willing to engage in materials that so far outstrip his or her competence as I do here, as a Christian ethicist, has to compensate with a heavy reliance on respected authorities in various scientific fields. I realize that the issues broached in this discussion are of far greater complexity than I may appreciate, and that widely respected authorities frequently disagree with one another. As much as possible, I strive not to take a stand on major debates in the field of evolutionary biology, such as group selection, the extent of adaptation, the pace of evolution, and other issues. While attempting to avoid misrepresenting the authors whom I discuss, I no doubt make generalizations that are, from the point of view of scientific experts, coarse-grained, incomplete, and oversimplified. I believe nevertheless that the importance of the topic warrants the risk of gaffes, missteps, and even serious errors that others can correct.

---

[13] Simon Conway Morris, *Life's Solution: Inevitable Humans in a Lonely Universe* (Cambridge: Cambridge University Press, 2003), p. 270.

CHAPTER 1

Evolution and religion

This chapter examines four evolutionary theories regarding religion, offers a critique of them, and then argues that our knowledge of human evolution can be compatible with Christian ethics and the religious faith that it reflects. It begins with a discussion of the place of evil in nature because that presents the central objection to Christian faith.

RELIGION REJECTED BY EVOLUTION: THE "PROBLEM OF EVIL"

The challenge posed by the "problem of evil" was based not only in a growing awareness of the pervasiveness of pain, competition, and wastefulness in the natural world but also in the recognition that these are "built into" the very structure of nature itself. The advent of evolutionary theory brought with it the question of whether a good God could be the Creator and providential Governor of such a natural order.

Young Darwin assumed the truth of conventional Anglican Christianity, and as a college student he was impressed by the argument of design put forth in William Paley's *Natural Theology*. His reading of Lyell's *Principles of Geology* on the *Beagle*, however, convinced him that the earth changed gradually over a much longer period of time than either conventional science or religion had been aware. Malthus' *Essay on the Principle of Population* significantly shaped his view of human society as marked by the same ruthless "struggle for existence" that he found in the world of biological organisms.

Some of the seeds of Darwin's doubts about the Christian doctrine of God came from his increased awareness of both the inaccuracies of scriptural accounts of human origins and the philosophical weaknesses of natural theology. He gradually came to reject what he took to be the religious content of Scripture, particularly its attribution to God of the "feelings of a revengeful tyrant."[1] "Thus disbelief crept over me at a slow rate, but was at last complete," he confessed in his *Autobiography*.[2]

Darwin's views of science, his own life experience, and his philosophical proclivities all made it exceedingly difficult for him to reconcile divine benevolence with the harshness, randomness, and selfishness at the heart of the "struggle for existence." He experienced the heartlessness of nature and the human suffering it causes in a very personal way with the death of his beloved daughter Anne.[3] The experience of the fact that the world does not consistently reward virtue and punish vice led Darwin to reject the providential Creator of orthodox Christianity. Instead of benefiting the "greater good," nature rewards individuals who survive and their offspring.[4] The laws that govern the natural order, Darwin came to believe, could not have been created, or the course of evolution supervised, by a benevolent deity.[5]

Darwin's moral objections to major strains of biblical narratives were balanced by his admiration of some of its major ethical teachings, particularly those of Jesus in the Gospels. He gave no credence to the miracles and supernatural intervention into nature asserted by "revealed" theology. Some scholars believe that Darwin continued to use "God-language" to avoid scandal and outrage, despite the fact that he came to suspect that agnosticism (a term coined by his intellectual ally T. H. Huxley) was intellectually inescapable.[6] Yet others held that Darwin continued to use "God-language" as a way of expressing his sense of awe at the wonders of the natural world.[7] Thus he wrote of

> the extreme difficulty or rather impossibility of conceiving this immense and wonderful universe, including man with his capacity of looking far backwards and far into futurity, as a result of blind chance or necessity. When thus reflecting I feel compelled to look to a First Cause having an intelligent mind in some degree analogous to that of man, and I deserve to be called a theist.[8]

---

[1] *Autobiography*, in *Charles Darwin and Thomas Henry Huxley: Autobiographies*, ed. Gavin de Beer (London: Oxford University Press, 1974), p. 49.
[2] Ibid., pp. 86–87. See also A. Desmond and J. Moore, *Darwin* (New York: Penguin, 1992).
[3] See *Autobiography*, ed. de Beer, pp. 97–98.
[4] Neil Gillespie, however, argues that Darwin did not abandon theism. See *Charles Darwin and the Problem of Creation* (Chicago and London: University of Chicago Press, 1979). Gillespie holds that Darwin came to believe that the laws of nature work for the greater good of the whole of nature.
[5] See John Hedley Brooke, "The Relations between Darwin's Science and His Religion," in John Durant, ed., *Darwinism and Divinity: Essays on Evolution and Religious Belief* (New York and Oxford: Basil Blackwell, 1985), pp. 40–75.
[6] See Ernst Mayr, *One Long Argument: Charles Darwin and the Genesis of Modern Evolutionary Thought* (Cambridge, MA: Harvard University Press, 1991), p. 85.
[7] See Gillespie, *Charles Darwin and the Problem of Creation*, pp. 143–145.
[8] *Autobiography*, ed. de Beer, p. 54. See William E. Phipps, *Darwin's Religious Odyssey* (Harrisburg, PA: Trinity Press International, 2002).

At the same time, Darwin confessed an increase of "skepticism or rationalism"[9] in his adult years and growing reservation toward religion, and particularly regarding belief in a "personal God" and any "future existence with reward and retribution."[10]

Some of Darwin's theistic successors, given to a more benign interpretation of nature, argued that Darwin's science is fully compatible with theism as long as evolution is understood to be the natural means employed by God to create new species.[11] They regarded the evil present in the evolutionary process as a necessary component of a process that was generally good. Other followers of Darwin, however, argued that Darwinism implied the end of theism.[12] Psalm 19:2 announces, "The heavens are telling the glory of God; and the firmament proclaims his handiwork,"[13] but T. H. Huxley did not think so. The earth is anything but the peaceful garden of the Yahwist creation account in Genesis, and the "survival of the fittest" without the corrections of culture inevitably destroys the finest moral impulses of the human race. Moral virtue for Huxley, then, entailed a course of conduct that "in all respects" runs directly contrary to the "struggle for existence."[14]

The objection to Christian faith from the evil in nature was repeated with even greater intensity in the writings of some neo-Darwinians. Sociobiologists are essentially the latter-day heirs to Huxley in this regard. George C. Williams, author of *Adaptation and Natural Selection*, argues that genes are concerned only with self-replication, and that organic life follows suit by exploiting any opportunity for inclusive fitness maximization, whatever the cost in pain and suffering for other organisms: "Nothing resembling the Golden Rule or other widely preached ethical principle is operating in living nature."[15] Nature is simply a "process of maximizing short-sighted selfishness" that leads to results that are "grossly immoral"[16]

---

[9] *Autobiography*, ed. de Beer, p. 55.   [10] Ibid., p. 54.

[11] For example, Henry Drummond, *The Ascent of Man* (New York: James Pott and Company, 1894). See Stephen J. Pope. "Neither Enemy nor Friend: Nature as Creation in the Theology of Saint Thomas Aquinas," *Zygon* 32 (1997): 219–230.

[12] See James R. Moore, "Herbert Spencer's Henchmen: The Evolution of Protestant Liberals in Late Nineteenth-Century America," in John Durant, ed., *Darwinism and Divinity: Essays on Evolution and Religious Belief* (New York and Oxford: Basil Blackwell, 1985), pp. 76–100.

[13] Scriptural citations are taken from the New Revised Standard Version (New York and Oxford: Oxford University Press, 1989).

[14] *Evolution and Ethics and Other Essays*, reprinted in *Issues in Evolutionary Ethics*, ed. Paul Thompson (Albany, NY: SUNY Press, 1995), p. 133. See Moore, "Herbert Spencer's Henchmen," pp. 76–100.

[15] George C. Williams, "Huxley's Evolution and Ethics in Sociobiological Perspective," *Zygon* 23 (1988): 391.

[16] Ibid., 385.

and driven by the "inescapable arithmetic of predation and parasitism."[17] How, then, are we to respond to nature? We must use reason and culture to rebel against it, he proposes.[18]

Williams draws a theological conclusion from his view of nature. Commenting on the "harem polygyny" of Hanuman langurs in northern India, he notes that,

> Dominant males have exclusive sexual access to a group of adult females, as long as they keep other males away. Sooner or later, a stronger male usurps the harem and the defeated one must join the ranks of celibate outcasts. The new male shows his love for his new wives by trying to kill their unweaned infants. For each successful killing, a mother stops lactating and goes into estrous ... Deprived of her nursing baby, a female soon starts ovulating. She accepts the advances of her baby's murderer, and he becomes the father of her next child.
>
> Do you still think God is good?[19]

There are, however, two major problems with this argument. First, Williams' improper anthropomorphism involves an equivocal use of language that distorts his description of nature. The terms "harem," "celibate," "wives," and "love" do not apply literally to langurs, who obviously have no social institution of marriage that makes sense of the status of "wives" and "celibates."

Second, Williams' theological argument is akin to Darwin's rejection of design, but it does not pay the slightest attention to more complex theologies of creation. If God created the langur mating system, he reasons, then God cannot be good; no good God would produce such a cruel system. This critique does apply to a fundamentalist view of creation according to which God directly creates each species and the behavior patterns appropriate to it, but even fundamentalism holds that because of the "fall" there is a "gap" between the ordering of the natural world and the will of the Creator. Christian theologians do not hold that nature presents a perfect one-to-one expression of the plan of God. Williams' naïve assumption ignores the theological positions of those who believe that there is a distance between the ordering that has emerged in the natural world and the divine will of the Creator. The evolutionary process produced species under conditions marked by high degrees of contingency and chance. If this is the case, then one cannot attribute animal behavior patterns to the direct creative intention of God.

---

[17] Ibid., 398. [18] Ibid., 403.
[19] George C. Williams, *The Pony Fish's Glow* (New York: HarperCollins, 1997), pp. 156–157.

Sociobiologist Richard Dawkins develops this mistaken strain of thought even further. In fairness, it can be noted that the real sociobiological objection to divine benevolence comes not so much from the behavior of this or that species but rather from the very processes of evolution. Dawkins argues that a benevolent God cannot exist as the Creator of a world that is indifferent to moral goodness and evil, that is, one in which there is innocent suffering. The world, Dawkins writes,

> would be neither evil nor good in intention. It would manifest no intentions of any kind. In a universe of physical forces and genetic replication, some people are going to get hurt, other people are going to get lucky, and you won't find any rhyme or reason in it, nor any justice. The universe that we observe has precisely the properties we should expect if there is, at bottom, no design, no purpose, not evil and not good, nothing but blind, pitiless indifference.[20]

Trying to avoid moral anthropomorphism, Dawkins' argument nevertheless still slides from an indifferent universe to one that is "blind" and "pitiless," terms that only apply to beings actually capable of sight and mercy. This verbal vacillation mixes a sense of disappointment and resignation with outrage at the nature of the universe, but these are wholly inappropriate to a natural world in which nonhuman organisms have neither freedom nor moral intentions.

The universe is shaped by "blind" forces and genetic replication, to be sure, but it also provides a fitting habitat for human agents. As moral agents, we are embedded in nature but also influenced pervasively by cultural, social, and economic factors. There is no natural "balance sheet" according to which more virtuous people are automatically protected from cancer or AIDS. Dawkins and his evolutionary predecessors were not of course the first to discover that nature is not structured like a morally ideal legal system in which the just prosper and the wicked suffer. As theologian John Haught points out, "evolutionist complaints about the struggle, suffering, waste, and cruelty of natural processes add absolutely nothing new to the basic problem of evil of which religion has always been quite fully apprised."[21] The early Christians, for example, were well aware that God sends the sun to "rise on the evil and the good," and sends "rain on the righteous and on the unrighteous" (Matt. 5:45).

---

[20] Richard Dawkins, *River out of Eden* (New York: HarperCollins, 1995), pp. 132–133. See more recently but in the same vein, R. Dawkins, *The God Delusion* (Boston: Houghton Mifflin, 2006).

[21] John Haught, *Science and Religion: From Conflict to Conversation* (Mahwah, NY: Paulist Press, 1995), p. 59.

Concentrating on the long-term role of natural selection as well as contingent factors that contribute to human suffering, Dawkins ignores the fact that a great deal of suffering is not caused by either bad luck or nature but rather by human irresponsibility, selfishness, opportunism, and greed. The sweeping nature of his generalization that there is no "rhyme or reason" for human suffering ignores these factors; one can wonder whether most human suffering – even that occasioned by natural disasters such as Hurricane Katrina in New Orleans – is either caused or magnified by human moral failure and especially by injustice.[22]

Dawkins' objection to Christian theism lies in a questionable intellectual starting point: the premise that the proper understanding of all life comes from the natural sciences and that the most adequate explanation of human behavior is provided in evolutionary terms. This premise leads to the assumption that all references to the transcendent are illusions that must be rejected by rational people. Religious beliefs are so tenacious despite the counter-evidence because they speak to deep-set emotional needs. Thus religious practices – rituals, beliefs, taboos, sacred music, dance, moral codes – are so widespread across time and space because they are rooted in the evolved human emotional constitution.

Science proves that the attempt to meet human emotional needs by means of religion is illusory. The universe is purposeless, Dawkins holds, and those who affirm the existence of evolution are intellectually obliged to admit that religion is false. In nature, he writes, one sees, "the total prostitution of all animal life, including man and all his airs and graces, to the blind purposiveness of these minute virus-like substances [i.e. genes]."[23] Dawkins reveals his own confusion here. The term "prostituted" suggests payment for services rendered, and services of a particularly ignoble, demeaning kind. Yet nature cannot be "degraded" if it has neither inherent purpose nor innate moral status.

Dawkins states confidently that "we no longer have to resort to superstition when faced with the deep problems: Is there a meaning to life? What are we for? What is man?"[24] Here he presumes a radical but false dichotomy between religion and science because he equates religion with superstition. He mistakenly regards the notion of "God" as an explanatory hypothesis proposed by theology as an alternative to natural selection. Dawkins voices

---

[22] See Judith N. Shklar, *The Faces of Injustice* (New Haven and London: Yale University Press, 1990).
[23] Cited in Anthony O'Hear, *Beyond Evolution: Human Nature and the Limits of Evolutionary Explanation* (Oxford: Clarendon Press, 1997), p. 152.
[24] *The Selfish Gene*, p. 1.

a suspicion felt by other sociobiologists as well: "What has 'theology' ever said that is of the smallest use to anybody? When has 'theology' ever said anything that is demonstrably true and not obvious? . . . What makes you think that 'theology' is a subject at all?"[25] Dawkins by no means speaks for all scientists – many biologists are devout Christians – but he does articulate a view of theology as a pseudo-discipline that would be assumed by those who believe that science provides the only adequate avenue for coming to some genuine knowledge about religion. This improper reductionism will be examined at length in chapter three.

The central objection to Christian theism from Darwin to Dawkins has been the "problem of evil" – the belief that an absolutely benevolent, intelligent, and powerful God could not have created a natural world marked by so much pain and suffering. This objection does not offer an account of where the line is crossed into the exact "amount" of pain and suffering that makes belief in the Christian God implausible. In any case, Christians can reply to this criticism in several ways. Any response has to begin with the honest recognition of the fact that the presence of moral evil in the universe is ultimately a "mystery." Evil is a "mystery" not in the sense that it constitutes an intellectual problem that can be "solved" but because it is ultimately unintelligible, unreasonable, and counter-productive for everyone. It does not make sense, in other words, that creatures who possess the capacity for knowing and loving would deliberately turn away from God, who is absolute wisdom and love, and in so doing act in a way that is deeply harmful to themselves and others.

It is important to distinguish the willful decision to reject divine love from the broader notion of evil, which includes the disorder and harm that result from the workings of nature on finite beings. The fact that animals become sick and die, that they often kill to eat, that habitats can be marked by scarce resources and therefore are the scene of the "struggle for existence" and extinctions, that males compete with one another for access to females, and so on – all these are natural conditions that can bring "evil" on some animals and "good" for others. These are not "good and evil" in any moral or religious sense, but simply biological benefits and costs to various organisms. From a Christian perspective, the Creator has made a world in

---

[25] Cited from "Letters to the Editor," *The Independent* March 20, 1993, in Michael Poole, "A Critique of Aspects of the Philosophy and Theology of Richard Dawkins," *Science and Christian Belief* 6.1 (April 1994). See also Richard Dawkins, "A Reply to Poole," *Science and Christian Belief* 7.1 (April 1995): 45–50 and Michael Poole, "A Reply to Dawkins," *Science and Christian Belief* 7.1 (April 1995): 51–58.

which nature runs its course, lifetimes are relatively short or long, reproduction is achieved or not. The evil experienced by particular creatures takes place within a creation that overall is "very good" (Gen. 1:31).

Evolutionary science helps us to understand in a detailed way the fact that we are only creatures within a larger cosmos. It underscores our own dependence on forces much greater than ourselves, our vulnerability to harm, and our finitude. Christian faith affirms that God's providence works within the natural order as well as within history, but it does not presume that everything that happens will be for the immediate or even long-term benefit of human agents. The doctrines of creation and providence affirm that all things ultimately work for the glory of God, not that all things are for our benefit. Our sense of dependence and interdependence ought to underscore our gratitude for the goodness of the Creator and a corresponding sense of responsibility to use the power at our disposal for the benefit of creation and not for its destruction. Because Williams and Dawkins lack a doctrine of sin, they tend to ignore the extent to which human beings are responsible for the evils experienced by victims of our bad choices – and not only human victims but also victims in the animal world. Because they do not distinguish human or moral evil from physical evil, they project willful human vices and crimes (greed, rape, theft, etc.) onto the natural world of genes and organisms. If in the past Christian theologians have drawn too thick a line between human and nonhuman animals, Dawkins and the sociobiologists have erased it. The result has been a twofold confusion – one that tends to eliminate our accountability for wrongdoing and place the source of evil in our genes or other aspects of nature, and another that declares that God could not exist because divine existence would entail divine responsibility for the evil in the creation. In the end, for Dawkins, we are not responsible and God does not exist – and therefore nature itself is the cause and source of evil. This effectively proposes a new ontological dualism – evil nature versus human moral culture – without any account of how the "good" possibilities developed in culture could have emerged from such a morally dubious nature.

The traditional Christian understanding of the status of evil makes more sense of the facts of human experience than does neo-Darwinism. The Creator creates a world that is composed of finite creatures made to experience and manifest the goodness of creation and the glory of God. We human beings are given a special set of capacities that enable us to choose to pursue understanding and love. We possess the capacity to grow in understanding and love of ourselves, one another, and God – and in this lies true and complete human flourishing.

Creation in this view is deeply ambiguous. It is created good and we experience this goodness as nature's fertility, bounty, and beauty. The countless examples we find in the world of Darwinian adaptation speak to its intelligibility and order. At the same time, Christianity is aware of the indifference of nature and the power it has not only to support but also to thwart our well-being. Theologian Karl Rahner speaks in this regard of "a threatening, merciless, cruel, life-giving and life-destroying nature, a nature that humans experience as a multiplicity of impersonal and enslaving forces to which they seem helplessly delivered."[26] Alongside the elevating experiences of human dignity, freedom, and moral responsibility, nature forces on us the experience of "the power of death, the drive of instinct, the blind law of what is merely physical and chemical."[27] The conjoining of these realities underscores the ambiguity of nature as creation, the fact that nature appears to us as "simultaneously both ground and abyss, home and something foreign, bathed in splendor and yet sinister, heavenly and demonic, life and death, wise and blind."[28] Left with only nature, Rahner argues, we would be tempted to resign ourselves to nihilism or at best a Promethean struggle in which we can win a "thousand battles" but still face "final defeat" in the inevitable terminus of death. As a Christian, though, Rahner holds that the final unity of the ambiguous dimensions we experience in nature can only be achieved through understanding nature as the divine creation. Humanity and nature both have their unity in their origin: the one God who is "maker of heaven and earth." For Rahner, the negative experiences resulting from our place in the natural world underscore the "unmistakable dignity of the human person." The positive experiences we have as natural creatures remind us that we share with all animals a common divine origin: we are all "made by the one creative love of an eternal ultimate reality that lies beyond all duality and which we name God."[29] Christian theology thus offers an alternative to Huxley's and Dawkins' radical opposition between human beings and the evolutionary process because, for Christians, our ultimate meaning and final unity derives from our proper relation to our first cause and ultimate destiny, the Creator. Since nature and humanity are still in a process of developing, final harmony can only be spoken about as achieved in the future. Thus the Christian doctrine of creation must be complemented by "an eschatology of the eternal kingdom of God at the end of time."[30]

---

[26] Karl Rahner, SJ, "Nature as Creation," in Karl Lehmann and Albert Raffelt, eds., trans. ed. Harvey D. Egan, SJ, *The Content of Faith* (New York: Crossroad, 1999).
[27] Ibid. [28] Ibid. [29] Ibid., p. 84. [30] Ibid.

From the beginning of human history we have failed to accept God's offer of love. The central meaning of "guilt" in Christian theology is not the subjective conviction that we have done wrong ("guilt feelings") but the objective fact that we have repeatedly chosen to do wrong to both God and others. The cumulative effect of this refusal to accept God's love over the course of human history has been radically to limit our freedom to see and do good and even to recognize our failure in this regard. Rahner holds that because the choices we make become "objectified," the world we are born into is marked by "objectifications of guilt" and is "codetermined by guilt and by the guilty refusals of others."[31] Thus "the guilt of others is a permanent factor in the situation and realm of the individual's freedom, for the latter are determined by his personal world."[32]

Because of the deeply ambiguous history of human decisions, we inherit disordered dispositions to prefer self to others and to be unfairly biased in favor of our own "in-groups" and to the detriment of others. Humanity is pervaded by habits of ingratitude, apathy, moral blindness, and other vices, or, as Augustine understood it, "disordered love." The creation is good, but we have a disturbed relation to it because of our own disorder and as a result even creation itself has been damaged. Human nature is made in God's image, but we violate it when we exploit and ignore one another.

Christians have to admit that there is no tidy intellectual "solution" to the "problem of evil."[33] We affirm two things: that the world is the scene of innocent suffering (and we cannot make the deeply flawed assumption that all suffering is somehow "deserved") and that God is just and merciful (and hence not the source of innocent suffering). Creation possesses a relative independence from the Creator, in the sense that the course of nature runs according to the interaction of "natural laws" and contingent events (including the "chaotic" processes in nature). God "permits" or "allows" for the conditions that lead to human suffering but does not will that the innocent suffer.[34] Innocent suffering can be fought, resisted, avoided, mitigated, but it cannot be made "right" or "proper" by any theodicy. Trust in God can lead to quiet resignation in the face of unavoidable

---

[31] Ibid., p. 105.
[32] Karl Rahner, *Foundations of the Christian Faith: An Introduction to the Idea of Christianity*, trans. William V. Dych (New York: Crossroad, 1985), p. 107.
[33] See Karl Rahner, SJ, "Why Does God Allow Us to Suffer?," trans. Edward Quinn in *Theological Investigations*, vol. XIX: *Faith and Ministry* (New York: Crossroad, 1983), pp. 194–208.
[34] See Paul G. Crowley, SJ, *Unwanted Wisdom: Suffering, the Cross, and Hope* (New York and London: Continuum, 2005), pp. 81–84.

suffering but is accompanied by the conviction that evil does not have the final word in God's good creation. Chapter four below will develop these themes further.

RELIGION REPLACED BY EVOLUTION: E. O. WILSON

I now turn to four evolutionary approaches to the relation of science and religion. Given what sociobiologists take to be the obvious errors of Christianity, the existence and persistence of religion requires an explanation. Why would so many people give religion a place in their lives when it is so patently false? Where did it come from, how does it function, and what benefits do its adherents seek? Given its apparently deep roots in human nature, can it be simply abandoned as Dawkins urges?

E. O. Wilson admits that by "religion" he usually means the "blind faith"[35] taught to him as a young Christian fundamentalist.[36] Perhaps he was indeed asked to have "blind faith" in an almighty divine Father by ignorant and rigid religious authorities. Given the vehemence of creationism within this context, it is no coincidence that he came to regard religion as the archenemy of evolutionary theory.

Wilson believes that since the religious impulse is found throughout all cultures, one ought to inquire into its adaptive significance. "The highest forms of religious practice, when examined closely," Wilson argues, "can be seen to confer biological advantage. Above all they congeal identity."[37] Wilson argues that in our archaic past religion allowed small groups or bands of early humans to work together more effectively as hunters and gatherers than competing groups who were less religious. "When the gods are served, the Darwinian fitness of the members of the tribe is the ultimate unrecognized beneficiary."[38]

Human beings have a deeply implanted desire to belong to groups and a desire to feel a sense of purpose within the great scheme of things. Since we are a social species, religious needs cannot be confined to the private sphere of purely subjective feelings. They are addressed by the culture of every society in its myths, scriptures, rituals, priesthoods, doctrines, narratives, paradigmatic figures, and icons. E. O. Wilson maintains that the ideal of self-sacrifice functions to encourage loyalty to the group, thereby tacitly promoting the selfish interests of the individuals that have belonged to

---

[35] See E. O. Wilson, *On Human Nature* (Cambridge, MA: Harvard University Press, 2004), p. 265.
[36] Ibid., p. 249.   [37] Ibid., p. 188.   [38] Ibid., p. 184.

these groups: "membership in dominance orders pays off in survival and lifetime reproductive success."[39]

Religion is especially important because it supports morality, a human construct generated from biologically innate rules of mental development (or "epigenetic rules")[40] for the purpose of securing survival and reproduction. All human communities teach some moral code to encourage cooperation and to discourage cheating. Compliant individuals are usually given rewards by society such as power, status, and wealth that in turn contribute to greater longevity and more secure families; those who are not compliant are punished by being deprived of these goods.

Though Wilson is more willing than Dawkins to admit that religion has provided some benefits to the human race, he emphasizes two major moral criticisms of religion. First, religious tribalism has inspired wars of religions, persecutions, pogroms, and suchlike. Second, religious anthropocentrism – its view of human beings as superior to all other animals, as special, morally noble, and created in the image of God – has allowed human beings to wreak havoc on nature.

Complementing the ethical critique is the scientifically based argument that religion advances erroneous empirical views of nature that compete with what is presented by science. Pre-literate religion was the only way for people to comprehend the natural world and to cope with human suffering, but it has long been intellectually superseded by modern science. Religion persists now because of an emotional lag and because it continues to provide psychological comfort in private life and social cohesion in public life, especially through civil religion. Wilson is confident that in the long run institutional religion will diminish and finally disappear. He hopes that the deep emotional needs now addressed by religion can be satisfied by the new "sacred narrative" of the "evolutionary epic."[41]

Unlike Dawkins, who attempts to debunk religion in order to eliminate it altogether, Wilson wants to promote a new faith, complemented with a hope, based on a naturalistic epic. Scientific materialism will eventually overtake Christianity: "the final decisive edge enjoyed by scientific naturalism will come from its capacity to explain traditional religion, its chief competition, as a wholly material phenomenon. Theology will not likely succeed as an independent academic discipline."[42]

Make no mistake about the power of scientific materialism. It presents the human mind with an alternative mythology that until now has always, point for point in

---

[39] E. O. Wilson, *Consilience: The Unity of Knowledge* (New York: Alfred A. Knopf, 1998), p. 259.
[40] Ibid., pp. 246–247, also p. 257.   [41] Ibid., p. 265.   [42] Ibid., p. 192.

zones of conflict, defeated traditional religion. Its narrative form is the epic, the evolution of the universe from the big bang ... The true Promethean spirit of science ... constructs the mythology of scientific materialism, guided by the corrective devices of the scientific method, addressed with precise and deliberately affective appeal to the deepest needs of human nature, and kept strong by the blind hopes that the journey on which we are now embarked will be farther and better than the one just completed.[43]

Wilson's naturalism contains four components: a metaphysic (scientific materialism), a scripture or mythology (the epic), a method (the scientific method), and a set of cardinal virtues (faith [in science], ["blind"] hope, and ["Promethean"] courage). Each of these components is vulnerable to significant criticisms.[44]

First, scientific materialism is a metaphysical and not a scientific position. To be intellectually responsible, Wilson ought to explain his position as metaphysical, defend it against the relevant philosophical objections, and show its superiority to alternative metaphysical positions. Instead he treats it as a simple part of natural science.

Second, the "epic of evolution" presents the functional equivalent of Wilson's founding myth. It is noteworthy that Wilson does not, like Dawkins, use science only to debunk mythology. Yet Wilson's treatment of mythology as something constructed by theorists who need to invent a scientifically intelligible account of the world is based on a significant misunderstanding of the very nature of mythology, an extended symbolic story created by the human imagination. Religiously powerful myths are not constructed artificially for instrumental purposes. As Midgley observes, spiritual needs "are met through a slow and painful communal development, through the effort to find, in experience, new effective symbols, which must grow out of better ways of living and feeling."[45]

Third, Wilson speaks rather univocally of "the" scientific method, which somehow is corrective of the excesses that might be produced by imbalanced interpretations of the epic of evolution and its mythology. Perhaps this is shorthand for scientific methods in the plural, but there is no univocal scientific method.

Finally, the virtues that Wilson encourages – openness, tolerance, justice, respect for nature, and so on – are not derived from science. The

---

[43] E. O. Wilson, *On Human Nature*, pp. 196 and 209.
[44] See James M. Gustafson, "Sociobiology: A Secular Theology," *The Hastings Center Report* 9.1 (February 1979): 44–45.
[45] Mary Midgley, *Science as Salvation: A Modern Myth and Its Meaning* (Florence, KY: Routledge, 1994), p. 54.

suggestion that they are derived from his account of evolution is self-defeating. If cultivating virtue is worthwhile because it is a strategy for fitness, one has to infer that the best strategy would be to cultivate the appearance of virtue while acting opportunistically. If reputation is the key issue, appearance takes priority over the moral reality of virtue.

This sociobiological approach to religion and morality is driven by a forced dichotomy between religion as adaptive and religion as true. From a theological standpoint, there is no reason to argue that because the religious impulse has evolutionary roots it must be illusory. Evolution could be the means by which the Creator gave human beings a natural inclination to know and love God.

Critics sometimes miss the extent to which Wilson's vision is motivated by moral idealism, even if only in a rather modest form. He passionately advocates intellectual integrity, freedom of inquiry, religious tolerance, democracy, human rights, and ecological responsibility – and he argues against institutional religion because he regards it as an enemy to these values. He believes that the survival and well-being of our species depends upon establishing a reasonable consensus about our origins, our nature as human beings, our place in the natural world, and even our purpose, that is, about what makes human life worth living.

Wilson is particularly troubled by our inability to reach clear and stable moral agreement about how to solve the major social, political, and moral problems of our day. He believes that the close identification of ethics with religion lies behind the intractable nature of our moral conflicts, but he ignores the fact that religion also inspires peacemaking, forgiveness, and reconciliation as well as support for social justice and human rights. Wilson sees the need to eliminate one major source of a host of evils related to excessive in-group bias but he does not acknowledge that faith in a universal God of justice and love provides the most powerful basis for universal concern.

RELIGION SERVING EVOLUTION: D. S. WILSON

David Sloan Wilson argues that group selection provides the best perspective from which to understand religion as a social institution. He maintains that, under some circumstances, some human social groups function like single organisms and social insect colonies. Religion persists because it tends to be adaptive, at least for the group. Religious groups are "superorganisms" characterized by a composition of parts whose actions are coordinated with one another to enable the whole to function as an adaptive unit and who depend on social control mechanisms for their maintenance. Groups can function as adaptive units, "but only if special conditions are met. Ironically, in human groups it is often religion that provides the special conditions."[46] In contrast to the "selfish gene" of sociobiology, D. S. Wilson proposes a multilevel selection theory that acknowledges ways in which traits of groups as well as individuals can be adaptive. "Very simply, immoral individuals may best moral individuals within groups, but moral groups best immoral groups."[47] Group selection concerns traits that might have evolved to maximize the fitness of a group relative to other groups, just as individual selection concerns traits that might have evolved to give an advantage to an individual relative to other individuals within a particular group.

Religion functions as part of a community "superorganism." It allows for a high degree of collective organization and coordination, as in, he argues, the historic case of Calvin's Geneva.[48] Religion provides incentives for cooperation, a system for detecting and punishing cheats, and goals that effectively direct and motivate desired behavior. Groups strongly manifesting these traits will be more adaptive than groups less able to manifest them.

D. S. Wilson makes an effort to be as sympathetic and broad-minded as possible to religion. He does not think that religious beliefs can be true or that theology provides an intellectually coherent basis for understanding faith. He is just as dismissive of the truth-claims of religion as are other evolutionists. Yet unlike Dawkins and E. O. Wilson, he holds that religion can be good for societies when it functions as a group-level adaptation. Seeking to understand its positive function within society, D. S. Wilson echoes Emile Durkheim's practice of taking all appeals to "God" as nothing more than references to the social order. According to Steven Lukes, Durkheim regarded religion as "social" in three ways: "as socially determined, as embodying representations of social realities, and as having functional social consequences."[49] Wilson in effect "Darwinizes" Durkheim in arguing that the social function of religion brings adaptive

---

[46] D. S. Wilson, *Darwin's Cathedral* (Chicago: University of Chicago Press, 2001), p. 6.
[47] D. S. Wilson, "Evolution, Morality, and Human Potential," in Steven Scher and Frederick Rauscher, eds., *Evolutionary Psychology: Alternative Approaches* (Boston: Kluwer, 2003), p. 60. Wilson's view differs from that of Pascal Boyer, whose *Religion Explained: The Evolutionary Origins of Religious Thought* (New York: Basic Books, 2001) argues that religion was adaptive in earlier human environments in which communities were largely composed of kin groups, but that it has become maladaptive in conditions of large societies composed of unrelated individuals.
[48] D. S. Wilson, *Darwin's Cathedral*, pp. 89–91, 109–110, 123–124.
[49] Steven Lukes, *Emile Durkheim* (Stanford: Stanford University Press, 1985), p. 462.

benefits to groups. D. S. Wilson does not agree with Dawkins that religion is not "real," but its reality bears only on society rather than on some transcendent realm inaccessible to science. Like his sociobiological competitors, Wilson assumes that if religion serves a functional purpose it must be explained comprehensively in these terms. He does not acknowledge that religion can be the expression of a fundamental human recognition of our place in the creation and a means by which we are inclined to be in right relation to God.

RELIGION AS A "BY-PRODUCT" OF EVOLUTION: PINKER

Evolutionary psychology does not regard religion as an adaptation but rather as a "by-product" of the evolutionary process.[50] It identifies "information-processing modules" designed by natural selection to respond to certain aspects of the environment, process this information by means of particular algorithms, and produce behavior that solves specific adaptive problems faced by individuals. Psychological mechanisms are numerous and pertain to specific domains. Though psychological mechanisms themselves are adaptive, the many behaviors in which they are implicated are obviously not always adaptive. The mechanisms constitute our species-wide cognitive and emotional architecture; the behaviors themselves do not always promote fitness.

Denying that there is an adaptive "sense of the sacred" or "God module" specifically designed by nature for a particular adaptive function, evolutionary psychologists view religion as a powerful by-product of various psychological mechanisms that evolved for nonreligious purposes.[51] This position has the advantage of acknowledging that religion is not a unitary phenomenon. Some of the key mechanisms include "naïve" (or common sense) physics and biology, intrasexual competition, kin selection, reciprocity, attachment, intergroup bias, and coalitional psychology. Steven Pinker holds that evolution has given us "modules for objects and forces, for animate beings, for artifacts, for minds, and for natural kinds like animals, plants, and minerals."[52] The history of religion offers extensive evidence of the way that these "modules" can be tricked and misapplied to

---

[50] The by-product position is advanced by Boyer, *Religion Explained*.
[51] See D. M. Buss, M. G. Haselton, T. K. Shackelford, A. L. Bleske, and J. C. Wakefield, "Adaptations, Exaptations, and Spandrels," *American Psychologist* 53 (1998): 533–548. Gould coined "exaptations." See S. J. Gould, "Exaptation: A Crucial Tool for Evolutionary Psychology," *Journal of Social Issues* 47 (1991): 43–65.
[52] Steven Pinker, *How the Mind Works* (New York and London: W. W. Norton, 1997), p. 315.

issues that are inherently unintelligible. "For anyone with persistent intellectual curiosity, religious explanations are not worth knowing because they pile equally baffling enigmas on top of the original ones."[53]

If actions of individuals are known to reflect the intentions of agents, Pinker argues, then people make the inference that events in nature are caused by the intentions of unseen supernatural agents. If human agents can be appeased and bargained with, so might supernatural agents, and so forth. If the social group is organized according to various structures of reciprocity, then life as a whole can be considered so structured. Pinker believes that specific modules can be correlated with certain religious activities or practices: attachment by piety and religious affections, status hierarchy by ontology and institutional churches, reciprocity by cults of sacrifice and covenants, fear of death and desire for survival by promises of an afterlife, cooperation by religious loyalty, and indoctrinability by creeds and religious authorities.

Pinker complains that religion "exploits people's dependence on experts."[54] The powerful proneness to form alliances based on kinship is co-opted in a variety of ways by religion, especially by ancestor worship, by those who envision priesthood as "fatherhood," by the ethic of "brotherly love," and by ways in which religious groups attempt to function as "fictive kin."[55] Pinker claims that all religions strive to undermine families to build stronger loyalties to the larger religious group.

There are a variety of problems with Pinker's "explanation" of religion – he makes spurious claims to objectivity, attempts to force all knowledge to meet criteria appropriate only to scientific inquiry, employs philosophical and theological terms in nonsensical ways, vastly underestimates the extent to which culture influences human practices, and tends to ignore, or at least insufficiently recognize, the contingency and unpredictability of human behavior.

EVOLUTION SEPARATE FROM RELIGION: GOULD

The late Stephen Jay Gould challenged the evolutionary explanations of religion offered by sociobiology and evolutionary psychology. Whereas E. O. Wilson would explain religion by means of evolutionary theory, ultimately in order to eliminate or at least radically revise it, Gould wanted

---

[53] Ibid., p. 560. [54] Ibid., p. 557.
[55] T. Crippen and R. Machalak, "The Evolutionary Foundations of Religious Life," *International Review of Sociology* 3 (1989): 68.

to leave them independent of one another. His thesis was that, at their best, science and religion occupy separate intellectual spheres and have usually pursued a policy of peaceful coexistence summarized in the acronym, "NOMA," or "Non-Overlapping Magisteria." (By "magisterium" he means only something like a distinctive zone of reflection, discussion, and debate.)

This position flows from an apparently straightforward and clear claim: that science concerns itself with empirical realities, whereas religion addresses matters of meaning, ultimacy, and moral values. We could avoid all sorts of nasty fights, he argued, if only we would stop expecting science to provide validating evidence for religious dogmas or biblical events. So there should be no more natural theology, anthropic principle, or attempts to find scientific confirmation for religious beliefs. Nor, conversely, ought religion to be employed to resolve questions of a properly scientific nature. There should be no attempt to obtain satellite scans for evidence of the buried remnants of Noah's ark, or more geology based on a ten-thousand-year-old planet, or fundamentalist "creation science" – all of these claims violate NOMA. In short, "science gets the age of rocks, and religion the rock of ages; science studies how the heavens go, religion how to go to heaven."[56] Gould's main point was that this conflict is psychologically, ethically, scientifically, and religiously unnecessary and that the way to avoid it lies in a mutually agreed nonaggression pact based on uncompromising assent to NOMA.

Gould was suspicious of any ideological appeals to nature, especially in the guise of anything resembling social Darwinism. He insisted that we ought not to give ethical authority to science because the knowledge of nature produced by science does not generate moral wisdom: "we must simply admit that nature offers no moral instruction at all."[57] Yet if nature offers no moral instruction *per se*, is it not the case that our ideas and normative perspectives often function to direct our observation of nature and our ascertainment of relevant facts? If so, perhaps we need a more complex alternative than Gould's Weberian (and Nietzschean) dichotomy of value-free science versus value-invested religion.

The appeal of this position lies in its common-sense moderation. After Dawkins' vehement denunciations of the "God-meme pathology" and the "emptiness of theology," Gould seems refreshingly irenic, nuanced, and

---

[56] S.J. Gould, *Rock of Ages: Science and Religion and the Fullness of Life* (New York: Ballantine, 1999), p. 6.
[57] Ibid., p. 196.

broad-minded. Gould attempted to strike an admirable balance between his own self-professed agnosticism and skepticism, on the one hand, and respect for the consciences of religious people and an appreciation of the positive moral contributions made by religious communities, on the other. Rather than repeating the tired cliché that the Roman Catholic Church is simply a bastion of anti-intellectual dogmatism and anti-scientific authoritarianism, Gould saw that the church has been a cautious defender of science in general and evolutionary theory in particular.

Gould echoed what some Christian theologians have been saying for some time. Neo-orthodox theologian Langdon Gilkey, for example, suggests that we view science and religion as "two languages." Science concerns public, objective knowledge of proximate origins, and religion deals with personal awareness of ultimate origins. Religion is concerned about the "why?" questions, whereas science focuses on the "how?" questions.[58] Christians have to ask both kinds of questions but should avoid confusing or collapsing them. Answers to one type of question do not satisfy the other. They do not conflict because in principle they cannot conflict if properly understood.

The anti-Darwinian professor of law Phillip E. Johnson, author of *Darwin on Trial*,[59] attacked Gould's earlier presentations of NOMA for advocating an artificial separation of morality from reality. Moral claims, Johnson rightly points out, are highly dependent on descriptive beliefs and assumptions regarding human conduct. In fact, Gould's own book on the IQ controversy, *The Mismeasure of Man*,[60] provides ample demonstration of the dependence of moral positions on premises about what are purported to be descriptive realities. Johnson properly points out that Christian religious affirmations concerning God, Jesus Christ, and eternal life (to name just a few) are not about some vague "meaning" but purport to refer to *realities*, even if they are revealed and deeply mysterious.

Christianity describes the way the world *is* – created, fallen, and redeemed – and not only about how we ought to act within it. The theological description is prior to ethical prescriptions. Yet even though belief in creation, fall, and redemption concern what is most real, they cannot be classified as empirical questions, if that phrase means addressed through

---

[58] See Langdon Gilkey, *Creationism on Trial* (San Francisco: Harper and Row, 1985), pp. 49–52, 108–113. This position has been modified in favor of a more cooperative relationship in Gilkey's later book, *Nature, Reality, and the Sacred* (Minneapolis: Augsburg Fortress Press, 1993).
[59] Second edn., Downers Grove, IL: InterVarsity, 1993.
[60] New York: W. W. Norton, 1981.

laboratory work or field studies. More generally, Christians affirm the reality of the objects of their belief, but the truth and meaning of these beliefs cannot be established or even indirectly examined by means of scientific investigation. Yet they are not completely independent of what we know about humanity from scientific studies that pertain to human beings, any more than what we think about creation is utterly independent of what we know from natural science. This kind of complexity indicates the weakness of Gould's simple descriptive/prescriptive dichotomization.

Gould was not as strict an interpreter of NOMA as initially seems to be the case. He did believe that science can act as a check on religious claims, at least inasmuch as religion makes *empirical* claims about nature, human behavior, and the world. Gould should actually have called his principle "POMA," for "*Partially* Overlapping Magisteria" – a position which is both more interesting to think about and more difficult to explicate. POMA is exemplified in Gould's decision to rule out miracles and other forms of divine intervention, for instance, on the principle that since they are not accessible to science, they cannot be true. He aimed primarily at "creation science" appeals to revelation – for example, to uphold a ten-thousand-year-old view of the earth – but in so doing actually brought into his range any belief in a God who cares for and orders creation.

Gould, like Dawkins and E. O. Wilson, held as intellectually untenable any belief in the existence and activity of a personal, benevolent, and almighty God who orders history and nature. Gould not only registered the standard methodological restriction of science to nature but he also denied that belief in the transcendent has any intellectual support from logic or evidence. As one reviewer put it, Gould wished that "religion should make an orderly retreat from the assertion of truth to the propounding of moral values."[61]

Gould did not replicate Dawkins' aggressive stance toward religion because he also appreciated the positive value it has for millions of people. He was aware of the fact that though science has great explanatory power, it cannot provide answers for the big and inescapable existential questions that human beings encounter. Gould's modesty in this regard is to be admired, especially when contrasted with Wilson's overly confident substitution of evolutionary mythology and morality for religion or with Dawkins' dismissal of religious questions as silly adolescent anachronisms.

---

[61] Nigel Hawkes, review of Gould's *Rock of Ages*, in *The Tablet*, March 3, 2001, p. 309.

CRITIQUE OF EVOLUTIONARY ACCOUNTS OF RELIGION

All of these treatments of religion suffer from oversimplification. Dawkins does not recognize that what we call "religion" involves an extremely complex and sometimes very subtle collection of activities, practices, and beliefs. E. O. Wilson, too, collapses a rich, diverse, and complex array of human experiences, practices, and beliefs into one monolithic amalgam called "religion."

Loose, undisciplined evolutionary speculations about religion are all the more offensive because they are proposed under the banner of "scientific objectivity." In fact, however, this understanding of religion and science is far from being the product of detached (let alone genuinely scientific) reflection. Sociobiologists take for granted a way of thinking about religion that is a product of our own western, liberal effort, beginning in the late Middle Ages and escalating during the wars of religion, to establish a juridical and cultural separation of religion from the secular, public order, and the church from the state, in order to promote more securely, if not guarantee, civil order, domestic peace, and international security.

The change that comes about with the modern use of the word "religion" is underscored when viewed in contrast to an example of its medieval use. For Thomas Aquinas, faith concerned the whole of life – most of all, it connoted a sense of reverence for God and other human beings.[62] "Religion" regarded the particular duties of justice one owed to God. The term "secular" described mundane activities such as manual labor that, while ordinary, nevertheless received their deeper intelligibility from their place in the totality of life. Religion was by no means opposed to the secular.

The meaning of religion changed when it became a category used by modernity – from Machiavelli, Hobbes, and Locke through their contemporary disciples – in its attempt to establish and legitimate the modern state.[63] Religion came to be identified with activities, organizations, and beliefs that are purely private and have no contribution to make to politics or the political community, which is conceived as coming under the sole and undisputed authority of the state.[64] Evolutionary analysis of religion reflects the long-term effort of modernity to marginalize, control, and

---

[62] See *Summa Theologiae* (hereafter *ST*) II-II, 81 (Rome: Marietti, 1948). English translation by the English Dominican Province, 3 vols. (New York: Benziger Brothers, 1946).
[63] See N. Lash, *The Beginning and the End of "Religion"* (New York: Cambridge University Press, 1996).
[64] See Johann Baptist Metz, *Faith in History and Society*, trans. David Smith (New York: Crossroad, 1980), ch. 3.

finally eliminate religious communities – first as intellectually persuasive, then as socially viable. Sociobiology and evolutionary psychology provide an ideology intended to help along its demise.

The evolutionary agenda in effect echoes Hobbes' project of establishing "convenient articles of peace" to provide freedom from the fear of death. This stands in direct contrast to Augustine's classic definition of peace as the "tranquility of order."[65] In Hobbes' time, of course, the major competitor for authority in Christendom was the church, so one of his major concerns was radically to reduce and domesticate the power of the church and the normative order which it defended. As a seventeenth-century Anglican, Hobbes, unlike E. O. Wilson and Dawkins, continued to believe in the "true religion" as opposed to Catholicism (which preferred the pope to the sovereign) and Puritanism (which preferred the Christian conscience to the sovereign), but his philosophical subordination of church to state played a major role in the privatization of religion which Wilson applauds.

Paying attention to this intellectual and political lineage helps to bring into focus the *political* rather than the ostensibly purely scientific side of Wilson's critique of religion. In addition to his epistemological critique of religion – that religion cannot validate its truth-claims in the way that scientists can validate the truth-claims of science – he also objects to religion on the grounds that it violates the democratic ideal of political order. The democratic value of freedom is at odds with religion's allegedly irrational coercion, its slavish dependence on tradition, and its authoritarianism. For Wilson, the project of making a society democratic *requires* the weakening of religion.[66]

The modern creation of religion, however, brought with it a profoundly ironic result: an undercurrent of deep *intolerance* toward religion and a deeply *undemocratic* attempt to determine in advance, and not through extended and public conversation and debate, what kinds of ideas, standards, and ideals count as worthy for public discourse, and even deliberate social appropriation. One can advocate the separation of church and state without banishing all religious values and ideals from the public sphere. But as a matter of fact Wilson insists that anything having to do with religion – what he terms "transcendentalism" – is determined a priori to be inherently irrational, potentially divisive, and therefore first relegated to the private sphere and then eliminated altogether in the science-driven process of secularization.[67]

---

[65] See Thomas Hobbes, *Leviathan* (New York: Collier, 1962), ch. 17. Augustine, *De Civ. Dei* XIX, 13.
[66] See E. O. Wilson, *Consilience*, p. 256.   [67] See ibid., p. 265.

Sociobiologists and evolutionary psychologists fail to appreciate the fact that undermining the authority of religious communities as sources of moral guidance and wisdom has the effect of removing one of the major cultural sources of resistance to the dominance not only of the state but also of the ever growing bureaucratic and market forces over civil societies in the world today. The allegiance and strong sense of identity cultivated by vital religious communities provide an important alternative to the homogenizing forces of the market, the power of the state, and the ideologies that would deem them the only authorities that can legitimately use the power discovered by science and developed in technology. Evolutionists seek freedom rather than coercion.[68] But when they advocate the elimination of religious traditions or their isolation into a secluded moral "magisterium," they in effect help to undermine what has historically been a critically important counterbalance to the coercive power of the state and the market and thereby undercut one of the key roots of democracy itself.

From a Christian theological viewpoint, it is a mistake to force a choice between religion and evolution. Even if there are evolutionary conditions for why religion arose at a very early stage in human evolution, there is no reason why it should, for this fact alone, be an illusion or false. Religion and morality are social institutions that reflect central human needs and desires. Like other human institutions, they can be good or bad, healthy or sick. They are ways in which people are drawn to God and to the good. Scientific insights into the evolutionary roots of religion are compatible with the Christian view of the person as a self-transcending being made in the image of God. Philip Hefner is right to say that "religion generally, and Christian faith and theology, specifically, are systems of memes, cultural information, that have played a fundamental role in the emergence of persons, and continue to make significant proposals for understanding today what it means to be a person."[69]

The persistence of religions (despite the predictions of many observers of "secularization") across human cultures and through time testifies to the universal human desire to come to proper relation with what is ultimate. They can help us to underscore ways in which the human person is divinely created, habituated to the moral life, denigrated by sin, and healed by grace. A biologically informed view of human behavior, including the

---

[68] See, for example, ibid., p. 247.
[69] Philip Hefner, "Imago Dei: The Possibility and Necessity of the Human Person," in N. Gregerson, W. B. Drees, and U. Gorman, eds., *The Human Person in Science and Theology* (Edinburgh: T. & T. Clark, 2000), p. 80.

powerful desires that lead to religion, can deepen our appreciation of the many levels on which we are embedded in the organic world. It can also encourage a greater appreciation of our transcendence of the organic world in acts of knowing and loving God and neighbor. Pursuing this line of thought will enable us to appreciate both the biological roots and the moral fruits of the complex creature that is the human person.

CHAPTER 2

The indifference of Christian ethics to human evolution

Most approaches to Christian ethics assume that the natural sciences have little or nothing to contribute regarding the issues that they care about the most. This assumption runs against the theological bent of the tradition of Christian ethics, and especially against those streams that give a central place to human flourishing.

The variety of ways in which science and theology, and science and religion, have been related to one another is more or less replicated in discussions of the relation between evolution and Christian ethics. John Haught traces four modes of relating science and religion – "conflict," "contrast," "contact," and "confirmation."[1] This book recognizes that each of these responses is appropriate for various evolutionary positions. Christian ethicists tend to view evolution in one of three basic ways: as misleading and dangerous ("conflict"), as irrelevant ("contrast"), and as helpful ("contact" and "confirmation"). Christian ethicists generally tend to presume the irrelevance position: they accept the fact of evolution, show indifference to the theory of evolution, and reject naturalistic ontology. In this chapter I examine the two major approaches to Christian ethics that display the most consistent indifference to science in general and evolutionary theory in particular, namely, narrative theology and natural-law theory. The figures chosen for this chapter are representative of broad movements within Christian ethics; weaknesses in these positions reflect variations on weaknesses found in related schools of thought as well.

Disciplinary specialization presents a reason for theological indifference. A scientist can provide a biological account of the functioning of the major systems of the human brain without saying anything particularly relevant to ethics. Moreover, while some scientists have committed themselves to advanced theological education, few Christian ethicists are trained in the biological sciences. The demands of scientific education may account for why so many of the major figures involved in the science and religion

[1] See Haught, *Science and Religion*.

dialogue (such as Barbour, Peacocke) are scientists who become religious thinkers rather than vice versa. The stance of "leaving science to the scientists" can be an expression of appropriate disciplinary modesty. This same disciplinary self-restraint, however, does not make Christian ethicists eschew other highly technical disciplines, for example literary theory, cultural anthropology, or moral philosophy. Rather than completely avoiding descriptive accounts of human behavior, Christian ethics usually turns to social theory, typically Weberian, Durkheimian, or Marxist in character.

NARRATIVE THEOLOGY AND CHRISTIAN ETHICS: HAUERWAS

One version of the independence position comes from narrative theology and its interpretation of virtue ethics. This approach to theology takes its starting point from the fact that human life is historical, contingent, and finite and so needs to be described in terms of particular narratives recited within specific communities rather than by appealing to descriptions of an abstract and universal "human nature." We tell stories to bring to conscious expression and to share the experience of what it means to be human. Narrative is the "original form of theological expression,"[2] and the church is first of all a "story-telling community."[3] Biblical narratives can communicate religious experience in a particularly poignant way, form religious identity, and transform the perspective of listeners.

Since narrative theology is not a monolithic school of thought, one expects various narrative theologians to respond differently to claims about human evolution. Narrative theology is properly suspected of giving so much attention to biblical narratives that it tends to isolate theology from potentially valuable insights offered by science.

Theologian Stanley Hauerwas holds that Scripture is authoritative because of the way in which it nurtures and shapes the moral character and virtue of individual Christians and the Christian community as a whole. For Hauerwas, Christian ethics is a matter of learning to tell the story of Jesus in a way that allows Christians to make his story their story. The distinctive practices of the Christian community shape disciples to be able properly to hear and tell this story, and to witness to its truth by the truthfulness of their own lives as Christians. The primary task of the church

---

[2] Johann Baptist Metz, "A Short Apology of Narrative," trans. David Smith, in Stanley Hauerwas and L. Gregory Jones, eds., *Why Narrative? Readings in Narrative Theology* (Eugene: Wipe Stick Publications, 1997), p. 252. Nicholas Lash, "Ideology, Metaphor, and Analogy," in *Why Narrative?*, p. 136.
[3] Metz, "Short Apology," p. 255.

is not to influence the wider society or public policy: "The first task of the church is to exhibit in our common life the kind of community possible when trust, and not fear, rules our lives."[4] This is not to say that Christians should stand aloof from the public square, only that the center of gravity for the church lies in Jesus. "I have no interest," Hauerwas writes, "in legitimating and/or recommending a withdrawal of Christians or the church from social or political affairs. I simply want them to be there as Christians and as church."[5] This same condition applies to dialogue between Christian ethics and evolutionary science.

Hauerwas can acknowledge that the sciences give true knowledge within their own spheres of competence, but he denies that Christian faith depends on the sciences for its validation. The majority of Christian ethicists would agree. Yet he goes further than rejecting theological naturalism when he argues that there is *no* "independent criterion" of truth – no way of "making sense" of the world and one's own place in it – abstracted from the "concrete story of Jesus of Nazareth."[6] The truth about life is not "independent" of the truth revealed in Jesus. Since the gospel "promises to tell us everything we need to know about the world necessary for our salvation," there is no reason for Christian ethics to be concerned with the most adequate scientific depiction of the physical origin of the cosmos and the origin of species.

Hauerwas maintains that Christian ethics teaches a set of "skills" for living that allow one to act as a disciple of Jesus. It is impossible to separate questions regarding the truth of the narrative from its normative status: "At least part of what it means to call a significant narrative true," he holds, "is how that narrative claims and shapes our lives."[7] Narratives are to be evaluated in terms of the "richness of moral character and activity they generate,"[8] so the truthfulness of the narrative about Jesus is shown in the kind of community it forms.

---

[4] Stanley Hauerwas, *A Community of Character: Toward a Constructive Christian Social Ethic* (Notre Dame, IN: University of Notre Dame Press, 1981), p. 85.
[5] Stanley Hauerwas, *Against the Nations* (Minneapolis: Winston Press, 1985), p. 1.
[6] Stanley Hauerwas, "Why the Truth Demands Truthfulness," in Hauerwas and Jones, eds., *Why Narrative?*, pp. 303–310, at p. 305.
[7] Hauerwas, *Community of Character*, p. 97. Hauerwas' practical view of narrative opens itself to a diametrically opposite use than that intended. It is far from clear that people brought up with the Christian narrative are morally superior to, or are living more truthful lives, than people who are not. Conversely, the claim that a fundamentalist community produces good Christians by no means implies that their interpretation of Genesis is true. A skeptic could argue that because the Christian community is so often the scene of hypocrisy, ambition, and callous indifference to human suffering, the story of Jesus is not true.
[8] Ibid., p. 95.

The perceived value of evolutionary accounts of human origins in part depends on the status of human rationality. Christian ethics is concerned with human nature but as Hauerwas and his collaborator David Burrell note, "that 'nature' is not 'rationality,' itself, but the necessity of having a narrative to give our life coherence."[9] Hauerwas and Burrell oppose what they call the "standard account" of Anglo-American ethics, and particularly its assumption that an abstract and ahistorical rationality provides moral objectivity. "Human nature" should not be identified with a certain kind of "rationality." Narrativity is not just one of many forms of rationality and is, rather, paradigmatically human, constitutive of human beings in every culture. Christianity itself is "paradigmatically narrative."[10] The weakness of Hauerwas' theology of creation, however, inclines him to give minimal attention to human biological nature. There is more to "human nature" than the sheer capacity for narrativity – and this "more" includes our evolved species-wide biological constitution.

Rahner helpfully described our biological constitution as a "constitutive" aspect of our nature as material beings.[11] Since as finite beings we are material and spiritual, he writes, "one cannot conclude that what has been created, in its very first moment, divides into a separate material element and purely spiritual element, and not that both, albeit in a quite different way, are inner essential constitutive elements of every created being." Because our biological substrate and evolved nature is "constitutive" of who we are as human beings, we cannot reduce the human person to the capacity for narrativity or to the intersection of a set of particular narratives. God is the Creator of all things, and if the creation is ordered by its internal structures, patterns, and law-like regularities, then Christian ethics ought to take into account what can be known about the creation from the natural sciences. Natural sciences provide information about aspects of our human biological nature not accounted for by the biblical authors, their communities, and the early Christian community.

METHODOLOGICAL CRITIQUE OF NARRATIVE THEOLOGY: GUSTAFSON

James M. Gustafson is an exception to the general indifference of Christian ethics to evolutionary information. He has called persistently for Christian

---

[9] Stanley Hauerwas and David Burrell, "From System to Story: An Alternative Pattern for Rationality in Ethics," in Hauerwas and Jones, eds., *Why Narrative*, pp. 18–190, at p. 177.
[10] Lash, "Ideology," p. 136.
[11] Rahner, "Natural Science and Reasonable Faith," p. 29.

ethicists to pay more attention to the insights provided by the sciences, including evolutionary biology. He points out that contemporary Christians live with varying degrees of cognitive dissonance between the scientific assumptions they make about the world and the religious language they sometimes employ to describe it.[12] Few people, even devoutly religious people, employ Christian symbolism as their "first language" when speaking about their daily experience. Christian theologians must address this dissonance in a conscientious and systematic way.

Gustafson is highly critical of Christian ethicists who attempt to avoid the challenge presented by the sciences by engaging in a "retreat into sectarian religion."[13] "The first article of the Apostles' Creed," writes Gustafson, "necessitates understanding heaven and earth as carefully as possible."[14] He finds the strategy of "retreat" inadequate on several grounds: it underestimates the cultural pervasiveness of scientific ways of understanding the world, isolates Christian theological claims from critical analysis of the sciences, makes Christian discourse unpersuasive and sometimes even unintelligible to interlocutors who do not share Christian presuppositions, promotes intellectual complacency, and presumes an excessively narrow and self-enclosed view of the church.[15] Not everyone agrees with the accuracy of Gustafson's characterization of "sectarianism" or with his application of this category to particular theologians, but there is no doubt that the field of Christian ethics does not pay much attention to scientific treatments of human evolution.

Sharply critical of Hauerwas' narrative ethics, Gustafson argues that science provides an epistemologically optimal vantage point from which to understand the world in which we live. His theological ethics is deeply shaped by the Christian tradition, and especially its Reformed branch, but he believes as a matter of intellectual integrity that one cannot be confined to Christian accounts of truth. The truthfulness of the Christian narrative includes its practical import as displayed in the transformed lives of individuals and communities, yet the truthfulness of Christianity is not *only* practical and the structure of human rationality is not *exclusively* of a narrative kind. Theological truth-claims must be "checked against facts

---

[12] See Gustafson, *An Examined Faith: The Grace of Self-Doubt* (Minneapolis: Fortress Press, 2004), ch. 2, "Confronting Cognitive Dissonance."
[13] Ibid., p. 44. See also James M. Gustafson, "The Sectarian Temptation: Reflections on Theology, the Church, and the University," *Proceedings of the Catholic Theological Society of America* 40 (1985): 83–94.
[14] Gustafson, *An Examined Faith*, p. 44.   [15] See ibid.

and figures and political analysis,"[16] and reliance on biblical stories must be supplemented by modes of critical moral discourse.

Knowledge of the Christian narrative, according to Gustafson, is a necessary but not sufficient condition for Christian ethics. Gustafson stands in sharpest contrast to Hauerwas when he gives the highest epistemological priority to knowledge of the natural sciences, and especially to well-established theories regarding the material world in which we live. Gustafson likens his role as theologian to be something of a "traffic cop" working at the intersection of theology with scientific and other non-theological modes of discourse.[17] His openness to the kinds of knowledge provided by scientific inquiry is complemented by a refusal to shy away from the theologically uncomfortable implications of scientific findings, for example about the place of humanity within the cosmos, or the eventual demise of the solar system and our planet.

Theological plausibility runs along a spectrum, Gustafson maintains. Minimal plausibility requires that theological claims not be in contradiction to the findings of the natural sciences: "The substantial content of theology, if it is not in perfect harmony with scientific knowledge, cannot be in sharp incongruity with it, and what we say about God must be congruent in some way with what we know about human experience and its objects through the sciences."[18] At the other end of the spectrum, the most persuasive theological claims are those given positive support by the natural sciences: "I believe it is proper and necessary to test what is said theologically in light of the relevant sciences, and to use indicators from the sciences in what one says substantively about God."[19] In another place he states even more strongly that "the 'substantial content' of ideas of God cannot be incongruous (rather than must be 'in harmony') with well-established data and explanatory principles established by relevant sciences, and must 'be in some way *indicated* by these.' "[20]

Gustafson refuses to accept any move that would shield theology from the challenges presented by the sciences, and his work displays the courage of his convictions. The world depicted by the natural sciences is one that deflates the pervasive human and especially Christian pretension to be the

---

[16] James M. Gustafson, *Varieties of Moral Discourse: Prophetic, Narrative, Ethical, and Policy* (Grand Rapids, MI: Calvin College and Seminary, 1988), p. 26.
[17] See James M. Gustafson, *Intersections: Science, Theology, and Ethics* (Cleveland, Ohio: Pilgrim Press, 1996).
[18] James M. Gustafson, *Ethics from a Theocentric Perspective* (Chicago: University of Chicago Press, 1983 [1981]), vol. I, pp. 251–252.
[19] Ibid., p. 258.   [20] Ibid., p. 252.

center and goal of the entire order of creation. Christian ethicists today agree with Gustafson's rejection of the assumption that God exists "*simply for the sake of human beings*"[21] but shy away from his prophetic critique of idolatrous anthropocentrism and his reminder that "God will be God."[22]

NARRATIVE THEOLOGY: PLACHER

Narrative theologian William Placher, a former student of the distinguished Yale theologian Hans Frei, offers another set of considerations regarding narrative and science. Placher, one of the best-known exponents of "post-liberal theology," learned from Frei to read the Bible in light of the "shape of Biblical narratives" rather than first and foremost as isolated texts, and to give primacy of place to the biblical message communicated in the narratives rather than to cultural or philosophical frameworks that are extraneous to the narratives themselves.

Rather than start with the modern world and its framework, Placher (following Frei and George Lindbeck) argues that Christians ought to "start with the biblical world, and let those narratives define what is real, so that *our* lives have meaning to the extent that we fit them into *that* framework."[23] Placher describes the Christian viewpoint as one in which "the biblical world absorbs our world."[24] This means, he explains, that Christians, "will try (a) not to hold views incompatible with what we take to be its central claims, and (b) regularly to consider whether its categories might be unexpectedly helpful in understanding any aspect of our lives."[25] One would be hard pressed to find a position more opposed to the attempt to give theological claims mediation through scientifically grounded knowledge.

Placher targets the tendency of mainline Protestant churches to use "pop psychology" or other secular modes of thought in a way that trivializes Christian narrative and doctrine. He shares with Gustafson a worry over the thinning of the theological content of pastoral life, and he would agree with Gustafson's sharp criticism of what he calls "liberal Protestant sentimentality backed by current psychobabble."[26] Yet Placher is sharply

---

[21] Ibid., vol. II, p. 322; emphasis added.   [22] Ibid.
[23] William Placher, *Unapologetic Theology: A Christian Voice in a Pluralistic Conversation* (Louisville, KY: Westminster John Knox, 1989), p. 161; emphasis in the original.
[24] William Placher, "Being Postliberal: A Response to James Gustafson," *Christian Century* (April 7, 1999): 391. See Placher, *Unapologetic Theology*, p. 127. Here he draws from George Lindbeck, *The Nature of Doctrine: Religion and Theology in a Postliberal Age* (Philadelphia: Westminster Press, 1984), p. 117.
[25] Placher, "Being Postliberal": 391.
[26] Gustafson, *An Examined Faith*, p. 36.

criticized by Gustafson for proposing that Christians ought to "begin" in the "biblical world," since, Gustafson argues, every reader is already pervasively shaped by contemporary culture (including modern scientific culture) and cannot "return" to, or adopt, a pre-modern point of view.[27]

Though not reluctant to engage non-Christian ways of thought, Placher's primary identity is defined by Christian faith. Placher, like Hauerwas, shows a proclivity for focusing theology primarily on the internal structure of the Christian faith, and on its realization in practices of the Christian community, rather than in external engagement and dialogue. At the same time, Placher is engaged with contemporary philosophy, anthropology, and sociology. Theology ought to begin with and give special weight to the Christian narrative, but it cannot stay there. Christians must take seriously other forms of discourse but they ought to relate to these in a way that does not diminish the character of specifically Christian discourse.

Christian ethics in Placher's perspective would be allowed to enter into dialogue with evolutionary theorists as long as neither the intellectual agenda nor the framework of credibility is controlled by epistemological and ontological naturalism. Instead of Gustafson's "traffic cop" who stands among and between different disciplines and perspectives, Placher chooses to engage in the dialogue from within a decidedly Christian perspective. He understands that the metaphor of "absorbing" the world into the text captures only part of what happens in Christian ethics. A contemporary person cannot simply adopt a first-century identity, and the interchange between a contemporary Christian and an ancient text involves something like a "fusion of horizons"[28] that transforms both the agent and his or her reception of the text. Unlike Gustafson, Placher insists that the interpretative process be guided primarily by the Christian narrative rather than by non-Christian sources such as science or philosophy. He does not, however, provide a way of identifying the line one crosses when moving from Christian-based interpretation into an unacceptable reliance on non-Christian sources.

Gustafson criticizes narrative theologians for thinking that this line can be easily identified and sharply drawn. He argues that they naïvely underestimate the extent to which all of us are influenced by the cultural world in which we are immersed.[29] Narrative theology can acknowledge the

---

[27] See Gustafson, *Ethics from a Theocentric Perspective*, vol. 1, and *An Examined Faith*.
[28] Hans Georg Gadamer, *Truth and Method*, trans. J. Weinsheimer and D. Marshall, second rev. edn. (New York: Continuum, 1989 [1972]).
[29] See *An Examined Faith*, pp. 85–86.

influence of scientific culture on all of us, yet it might respond that Christians must be committed to upholding the distinctive message of the gospel precisely because of the powerful influence of science on contemporary culture. Yet Gustafson is not willing to concede that the Scriptures constitute a single coherent narrative, or that they are best understood in exclusively narrative terms, or that they can be used to judge scientific culture while themselves remaining immune from correction by scientific sources.[30]

Gustafson makes a valuable point in arguing that since narratives and more discursive forms of rationality are subject to bias and ideological distortion, the "traffic" has to move in all directions and in complex ways. As theologian Nicholas Lash, also a friend of narrative theology, points out, "Christian religious discourse is subject to purificatory and corrective criticism from the historian and the literary critic, as well as from the sociologist and psychologist of religion."[31] If this is the case, and it surely is, then Christian religious discourse should be subject to the purificatory and corrective criticism of the natural sciences as well. Gustafson is not alone in making this argument. Pope John Paul II adds his own recognition of the scientific correction of theology when he writes: "Science can purify religion from error and superstition; religion can purify science from idolatry and false absolutes. Each can draw the other into a wider world, a world in which both can flourish."[32]

The claim that the Christian will try "not to hold views incompatible with what we take to be [the Bible's] central claims" suffers from vagueness. Many and perhaps even most Christians around the world take the Genesis account of creation to be literally accurate. They may believe that faith requires them to affirm some version of "scientific creationism" or "intelligent design" on the grounds that it saves the plausibility of the Bible's "central claims." This principle seems to have deeply anti-intellectual implications that Placher himself would eschew. The central flaw with Placher's hermeneutic is its failure to acknowledge that there are many "central claims" in the Bible and many "worlds" of the Bible, not just one, and that each of them has to be interpreted, understood, evaluated, and critically appropriated in light of contemporary knowledge, including that concerning evolution. The same is true of the "many worlds" of the history of Christian doctrine, the larger Christian tradition, and its institutional life.

---

[30] See ibid., pp. 38–39.   [31] Lash, "Ideology," p. 122.
[32] John Paul II, "Letter to the Rev. George V. Coyne, S. J.": 378.

Placher believes with the historian of science Thomas Kuhn that "the methods of science do not provide a uniquely rational and objective way of discovering truth."[33] Kuhn famously attended to the "theory-laden" character of all descriptions of the world, including scientific descriptions. Evolutionary theory generates explanations that seem to be irrelevant to the descriptions shaped by "paradigms" used within other scientific models or from the humanities. All such models possess certain anomalies and have their own intellectual shortcomings. Placher finds in Kuhn a perspective that helpfully uncovers some of the problems of the ahistorical presuppositions of scientific positivism. Kuhn's theory not only underscores the limits of scientific forms of rationality but also points to an important dimension of how theology proceeds as an intellectual enterprise. Placher believes theology functions within religious communities in ways that are similar to how scientific paradigms function within scientific communities. Everyone thinks from within a paradigm that has been adopted by a community, and those who belong to the Christian community will understand its convictions as they live within the way of life made possible by its distinctive practices, narratives, and traditions.

Placher believes there is room in Christian theology for the kind of critical scrutiny valued by Gustafson. The Christian tradition itself offers resources for self-criticism and, as we have seen, this includes internal criticism. Theories are evaluated from within the context of a tradition and its story. If the best theory is the one that allows members of a community to tell the best story, then, presumably, the best theology is the one that allows Christians to tell the best story about their community and its tradition.[34] Placher appreciates the narrative basis of theology but does not allow an excessively pragmatic view of truth to suggest that the apparent virtues of a community make its stories true.[35]

Narrative theology as practiced by Placher (as distinguished from Hauerwas) does not necessarily inhibit Christian ethicists from entering into dialogue with nontheological views of humanity, including those concerning evolution. He concedes that the internal logic of faith "drives

---

[33] Placher, *Unapologetic Theology*, pp. 47–51. See Thomas S. Kuhn, *The Structure of Scientific Revolutions* (Chicago: University of Chicago Press, 1970).

[34] Placher, *Unapologetic Theology*, pp. 108–110. Placher here interprets MacIntyre. See Alasdair MacIntyre, *After Virtue: A Study in Moral Theory* (Notre Dame, IN: University of Notre Dame Press, 1980), and *Whose Justice? Which Rationality?* (Notre Dame, IN: University of Notre Dame Press, 1988), esp. pp. 143–144.

[35] Placher, *Unapologetic Theology*, p. 164.

us to conversation beyond the borders of Christian faith."[36] He would also properly insist that evolutionary biologists recognize their dependence on their own particular tradition, one that encourages certain kinds of questions and discourages others. This requires a self-critical awareness of the intellectual assumptions of modern science.

Narrative theology, properly interpreted, would thus seem able to accept the fact of evolution and even the theory of evolution, but not the epistemological and ontological reductionism presupposed by the "evolutionist" agenda. In theory, the Christian community can be open to criticism both on grounds internal to its own tradition and according to criteria generated by other traditions as well. Yet in practice, the emphasis on being "absorbed" into the world of the Bible and on fidelity to the Christian story can mute the relevance of sources external to the Christian tradition and even effectively immunize it from critical assessment by scientific perspectives.

I differ from Gustafson to the extent that my center of theological gravity resides in Trinitarian Christian faith and a theology of creation shaped by redemption. I do not adopt the distance from the Christian tradition represented in the "traffic cop" image, and I do not think it helpful to abandon Christian claims that are not directly supported by scientific evidence. Distinctive competencies of science and theology suggest the dialogue need not lead to an erasure of proper boundaries between these two ways of understanding the world. Christian ethics must take the findings of the sciences seriously, but there is no reason, as Rahner puts it, that the disciplines of theology and metaphysics should "justify themselves before the natural sciences."[37]

Yet I agree with Gustafson that at times the interpretation of central Christian beliefs must be subject to evaluation in light of discoveries made by the natural sciences, including evolutionary biology. Beginning with Christian faith can still allow Christians to be clear about the conditions under which the interpretation of faith would be revised by the insights given by the sciences. Placher leaves open the question of what kind of revision of theology might be prompted by knowledge of evolution, or even whether knowledge of evolution can have any influence on the interpretation of biblical narratives. Such a project runs against the grain of Placher's methodology. I differ from Placher in pursuing a theology of creation that allows evolution to tell us something about God's purposes

---

[36] Ibid., pp. 167–168.
[37] Rahner, "Natural Science and Reasonable Faith," p. 21.

and an account of Christian ethics that finds its rationale in human flourishing, properly construed.

It should also be noted that many scientists would find problematic, at least on face value, Placher's claim that "the jury is still out on the question of whether science engages in objective descriptions of an objective world."[38] This claim is perfectly reasonable if one assumes that "objective" means something like the strong meaning associated with terms such as "unrevisable," "incorrigible," or "infallible" or if it implies that human beings are best understood by methods that regard us as "objects" for whom self-consciousness and reflection are trivially significant. Human beings are objects, of course, but we are not "nothing but" objects. As Charles Taylor points out, naïve views of objectivity ignore the fact that "our interpretation of ourselves and our experience is constitutive of what we are, and therefore cannot be considered a view of reality, separable from reality, nor as an epiphenomenon, which can be by-passed in our understanding of reality."[39]

Science is a human activity and there is no purely rational examination of the truth untainted by the limitations of the particular historical perspectives of their agents. Yet this qualification does not require us to abandon the more modest expectation that science can provide a relatively reliable knowledge of the structures and processes of nature. Most evolutionary biologists would have difficulty with Placher's generalization that it is "very difficult to claim that ... [scientific] 'progress' consists in getting closer to describing the way the world actually is."[40] This claim is simply wrong if it suggests that Darwinism represents no advance over the theory of special creation, or that neo-Darwinism does not provide relatively more precise and empirically accurate accounts of animal behavior than its predecessors provided. Evolutionary biology involves some "story telling," but it does more than relay narratives. It employs models to solve problems, but it also makes discoveries about the organic world. Acknowledging scientific progress implies neither "scientism," the claim that there are no other forms of rationality than provided by science, nor "reductive naturalism," the assumption that science constitutes, in Placher's words, the "supermethod" by which all other forms of human inquiry can be judged.[41] Placher's criticism is aptly applied to Dawkins, E. O. Wilson, and other

---

[38] Placher, *Unapologetic Theology*, p. 141.
[39] See Charles Taylor, "Self-Interpreting Animals," *Human Agency and Language: Philosophical Papers*, vol. 1 (New York: Cambridge University Press, 1985), p. 47.
[40] Placher, *Unapologetic Theology*, pp. 139–140.   [41] Ibid., p. 141.

evolutionists at their reductionistic worst, but this abuse does not justify ignoring the findings of evolutionary biology altogether.

The use of narrative within Christian ethics needs to be complemented with critical argumentation and systematic theological reflection. Christian ethics is obviously deeply dependent on the biblical narrative. A healthy respect for intellectual pluralism can allow for dialogue between Christian ethics and evolutionary theory. There is no reason why according central importance to narrative must issue in a "blanket dismissal" of scientific attempts to account for aspects of human nature. While it is important to honor the distinctiveness of the Christian story and to appreciate the centrality of discipleship for Christian ethics, we need to acknowledge that discipleship is a way of tutoring, habituating, and training the *humanum* of flesh and blood human beings. Seen in this light, Christian ethics ought to welcome any insights that might be forthcoming from knowledge of our evolved humanity.

NATURAL-LAW THEORY

We turn now to a stream of Christian ethics that one would expect to be more open to appropriating evolutionary knowledge of human nature for Christian ethics, namely, the natural-law tradition. Yet somewhat surprisingly a number of the most prominent approaches to natural law regard scientific knowledge of human evolution as essentially irrelevant to both Christian ethics and natural law. They recognize the need to criticize the ideological, materialistic, and reductionistic presuppositions of evolutionism, but once evolutionary ideology has been debunked, many natural lawyers assume, there is no further need for considering the significance of human evolution for our interpretation of human nature.

Human evolution can be acknowledged without being considered to have any great moral significance. Knowledge of human evolution gives us a greater understanding of what constitutes our phylogenetic background and it underscores the fact that morality is natural to us as cognitively advanced social animals. Knowledge of evolution shows that we are social on account of our animality as well as rationality (the higher "political" specification of sociality still pertains only to human beings). In the end, this view argues, evolutionary sciences do not tell us anything important about human nature that we do not already know from philosophical observation, literature, the social sciences, and religious tradition.

As one would expect, this characterization does not fit all streams of natural-law thinking, as we shall see in the discussion of Jean Porter and

Larry Arnhart in chapter eleven, but it is true of three other versions of natural law current today: the personalist natural law of Pope John Paul II, the strictly Thomistic version of natural law developed by Servais Pinckaers, OP, and the "new" natural-law theory of Germaine Grisez and John Finnis.

PERSONALIST NATURAL LAW: JOHN PAUL II

The most influential recent reading of the natural law comes from Pope John Paul II, the tradition's most representative figure in living memory. A staunch opponent of widespread moral relativism, the pope denounced the gradual erosion of Christian influence on modern cultures and their increasing indifference to the objective moral law.

John Paul II spoke first and foremost from an explicitly scriptural perspective. This pleased Protestants who have been critical of natural law, which they have taken to be grounded in the erroneous assumption that reason ought to have more moral authority than revelation.[42] The pope's account of natural law drew from Scripture but also built on Thomistic precedents and incorporated modern notions of human rights. It was developed on the distinctively personalist and Christocentric anthropology represented in a key theme employed by the Second Vatican Council's "Pastoral Constitution on the Church in the Modern World": "only in the mystery of the Word made flesh does the mystery of humanity truly become clear."[43]

The encyclical *Veritatis Splendor* enunciated familiar themes from the natural-law tradition. It held that natural law is "inscribed" in the heart of every person, grounded in the human good, and prohibits "intrinsically evil acts,"[44] and it regarded reason as a gift of God that takes its proper orientation from faith — especially today, when knowledge of the natural

---

[42] See Karl Barth, *Church Dogmatics* III/4, trans. A. T. Mackay et al. (Edinburgh: T. & T. Clark, 1961), ch. 12: "The Command of God the Creator," pp. 3–31; Reinhold Niebuhr, *An Interpretation of Christian Ethics* (New York: Crossroad, 1979 [1935]), ch. 5, but modified in *The Nature and Destiny of Man: A Christian Interpretation*, 2 vols. (New York: Charles Scribner's Sons, 1941), vol. I, p. 281; but also Reinhard Hütter and Theodor Dieter, eds., *Ecumenical Ventures in Ethics: Protestants Engage Pope John Paul's Moral Encyclicals* (Grand Rapids, MI: Eerdmans, 1998).

[43] *Gaudium et Spes*, par. 22, in Austin Flannery, OP, *Vatican Council II*, rev. trans. (Northport, NY and Dublin: Costello Publishing Company, 1996), p. 185.

It should also be noted that "personalism" has been used by progressive as well as traditionalist interpretations of Catholic ethics. A revisionist form of personalism is found in Louis Janssens' classic article, "Artificial Insemination: Ethical Considerations," *Louvain Studies* 5 (1980): 3–29.

[44] John Paul II, *The Splendor of Truth: Veritatis Splendor* (Washington, DC: United States Catholic Conference, 1993), par. 81.

law has been blurred in the "modern conscience." The pope repeatedly insisted that an adequate grasp of the natural law depends on revelation, faith, and adherence to the teachings of the magisterium. As an "expert in humanity," the church claims to have the most profound grasp of the principles of the natural law as well as the best vantage point from which to understand its secondary principles and their application.

This is not to say that John Paul II denigrated reason or regarded natural law as intrinsically unintelligible. He would have in principle agreed with Richard McCormick's denial that there are any "mysterious ethical norms which are simply impervious to human insight."[45] He continued to reaffirm the ancient view of moral standards as inherently intelligible and accessible to all, and he pointed for evidence to the increasing popular and legal appeal of human rights around the world. Natural law *qua* human rights provides the basis for the infusion of ethical principles into the political arena of pluralistic democracies. It also provides criteria for holding accountable criminal states or transnational actors who engage in destructive practices such as genocide, slavery, and human trafficking.[46] Natural law holds out the best resource for countering both amoral relativism and the tyrannical misuse of power.[47]

John Paul II's personalist natural-law theory gave an important place to fixed, knowable, and clear moral rules that apply to communities and individuals. Its treatment of reproductive issues is illustrative. States as well as couples, no matter what difficulties and hardships they face, "must abide by the divine plan for responsible procreation."[48] Sounding a theme from Popes Pius XI and Paul VI, John Paul II warned his listeners: "The moral law obliges them in every case to control the impulse of instinct and passion, and to respect the biological laws inscribed in their person."[49] Papal personalism does not revise the ethics earlier backed by the "physicalism" of the moral manuals that regarded our reproductive biology as directly designed by God, rather than as resulting from the interaction of natural selection and random variations.[50] Moving even further back historically, John Paul II retained the Thomistic assumption that the

---

[45] Richard McCormick, SJ, *The Critical Calling: Reflections on Moral Dilemmas since Vatican II* (Washington, DC: Georgetown University Press, 1989), p. 204.
[46] See John Paul II, *Veritatis Splendor*, par. 80, citing *Gaudium et Spes* par. 27.
[47] See John Paul II, *"Evangelium Vitae," The Gospel of Life: On the Value and Inviolability of Human Life* (Washington, DC: United States Catholic Conference, 1995), par. 70.
[48] Ibid., par. 97.   [49] Ibid.
[50] See Charles E. Curran, "Natural Law in Moral Theology," in Charles E. Curran and Richard A. McCormick, SJ, eds., *Readings in Moral Theology No. 7: Natural Law and Theology* (Mahwah, NY: Paulist Press, 1991), pp. 254–263, 280–284.

The indifference of Christian ethics

"order of nature" is a direct reflection of the will of God, the "Author of nature."[51] Natural law proscribes not only artificial contraception, abortion, infanticide, and euthanasia, but also newer biomedical procedures regarding experimentation with human embryos and human cloning.

This approach to natural law was theological and ontological: nature is creation, the human creature must be understood in personal terms, and the person must always act in accord with the plan of the Creator. This theological understanding of creation, however, is never informed by, coordinated with, or even placed in contact with scientific ways of viewing nature. This is not because the pope denied the proper autonomy of science. On the contrary, his treatment of the Galileo case, evolutionary science, and the dialogue between science and theology all indicate a great respect for the proper authority of science.[52]

Yet the pope's moral agenda strongly shaped his reading of evolution, and particularly his rejection of the view that the human soul can emerge from matter and his insistence that each individual soul is directly created by God. His categorization of evolutionary science as pertaining to the body but not to the soul implies that the core of the person lies outside the province of science. His approach to natural law proceeded on the basis of phenomenological, metaphysical, and theological reflection on the human person while ignoring scientific data and theories about human behavior. The insulation of this theological perspective is understandable in the writings of a figure whose concern is primarily pastoral, but the same is not true for others of like mind in the academy who have the time and talent to address this important but neglected issue.

NATURAL LAW PRACTICALLY INDIFFERENT TO EVOLUTION: PINCKAERS

Servais Pinckaers, OP, makes a strong appeal for retrieving the "authentic" Thomas Aquinas of happiness and virtues against a modern moral theology that has been excessively preoccupied with obligations, duties, and laws. On this point, he echoes the work of secular philosophers who have complained about the impoverished nature of modern tendencies to

---

[51] God is referred to as the "Author of nature" in a variety of places, including *ST* I, 92,1 ad 1; I-II, 60,1 ad 2; I-II, 97,1 ad 1; II-II, 154,12 ad 1.

[52] See John Paul II, "Lessons of the Galileo Case," his "Address to the Pontifical Academy of Sciences," October 31, 1992, *Origins* 22 (1992): 371–373. See William R. Shea and Mariono Artigas, *Galileo in Rome: The Rise and Fall of a Troublesome Genius* (New York: Oxford University Press, 2003).

identify the entire scope of morality with obligations.[53] Pinckaers believes that natural law has an important role to play in moral theology, but one that needs to be seen within the specifically theological context of freedom and life in the Holy Spirit.[54] He calls into question the practice of moral theology that followed from the Council of Trent to the Second Vatican Council. The official "manuals" of modern Catholic moral instruction drew from the natural law to provide moral imperatives binding on all human beings. They regarded the golden rule and the Decalogue as expressing the core content of the natural law and assumed that distinctively Christian themes supplement this moral doctrine with religious motivation. Morality in this perspective is fundamentally a matter of elaborating the requirements of the natural law in various areas of special ethics, for example in banking, war, or biomedical practices. The "manuals" thus assigned biblical texts about Christian life to the realm of spirituality or ascetical theology rather than to moral theology.

Pinckaers has been dedicated to correcting these defects. He does this in part by returning to the theology of Thomas Aquinas, for whom ethics began not in law, duties, and obligations, but in the desire for happiness. Thomas maintained that happiness is found in the life of virtue and especially in the spiritual and moral development of the agent's interior life, the font of all of his or her actions. The cardinal virtues of prudence, justice, temperance, and fortitude are brought to perfection when they are influenced by the grace-infused theological virtues of faith, hope, and charity. The Christian moral life is inspired by the love of God and neighbor, which is given its most vivid and challenging depiction in the Sermon on the Mount. Christ alone offers the full revelation of true and ultimate happiness, and the Beatitudes describe the way of life that is appropriate to it.

Thomas critically appropriated the philosophy of the Greeks and Romans, including their doctrine of the natural law, within the theological context of Christian faith. Natural law was not taken, as it was later in the modern period, as an autonomous theory derived by reason independently of any theological claims. Natural law concerns genuine human happiness, and since happiness is found ultimately in union with God, natural law without theology is inconceivable (at least in a

---

[53] See Bernard Williams, *Ethics and the Limits of Philosophy* (Cambridge, MA: Harvard University Press, 1985), ch. 10.
[54] See *La vie selon l'Esprit: essai de théologie spirituelle selon saint Paul et saint Thomas d'Aquin* (Luxembourg: Editions Saint Paul, 1966).

genuinely Thomistic context). We are not genetic "survival machines," as Dawkins puts it, but made in the image of God, and therefore we can only be fully satisfied by God. To be made in the image of God includes centrally significant capacities for moral agency.

This account of the natural law, like that of John Paul II, does not take into account evolutionary views of human nature. This lacuna seems odd if it is the case, as Larry Arnhart maintains, that "much (if not all) of what Aquinas said about the natural inclinations supporting natural law would be confirmed by modern biological research."[55] When extended into the domain of morality, evolutionary theory undercuts its own plausibility to the extent to which it reduces human activity to a quest for reproductive fitness. Christian faith reveals a deeper dimension to life that centers on an infinitely greater human destiny than is imagined by evolutionists. The moral reductionism of evolutionary psychology and sociobiology would make Pinckaers suspicious, but he does not consider any nonreductionistic evolutionary alternatives.

Pinckaers holds that St. Paul's belief that the Gentiles possessed a moral law "written on their hearts" (Rom. 2:15) applies to every human being. Having the same ultimate source, natural law and "divine law" reinforce one another to form a unitary approach to the good life. This theological conviction suggests that human intellectual powers are cognitively insufficient for meeting the task of knowing the human good in a comprehensive sense. The mind is weakened by sin, and nature itself gives no hint of our divinely given supernatural end. As a result, when we think of ourselves as only biological beings, we are unable to understand the full significance of the moral life. Naturalists neither appreciate the truth of some Christian moral teachings nor find it possible to grasp the meaning of the Christian life as a whole.

Natural law is the foundation of Christian moral teaching. God as Legislator, Pinckaers thinks, has ordered human nature in all its dimensions, including its biological inclinations. Natural human inclinations pertain to the physical, sensitive, and rational dimensions of the human good. This perspective might have enabled Pinckaers to incorporate information from the sciences about natural human capacities, desires, and needs, but he fails to do so. Science merely provides information about the material context in which we exist, but human life reaches its most complete perspective in its knowledge of the evangelical law.

---

[55] Larry Arnhart, "Thomistic Natural Law as Darwinian Natural Right," *Social Philosophy and Policy* 18 (winter 2001): 28.

Human intelligence is valued within the larger framework supplied by faith. Reason cooperates with the Spirit to discover ways of acting and living that are most appropriate to the gospel. This process of discovery should lead the Christian to be open to any insights that can be forthcoming from the human and natural sciences. The natural, human dimension of an act is complemented by, and not opposed to, its grace-inspired, spiritual dimension. Just as Thomas Aquinas appropriated Aristotelian science within his account of natural law, so Pinckaers' general Catholic framework should predispose him to presume that moral theology ought to assimilate what is best in evolutionary theory. Since "grace perfects nature," science, as a form of natural human knowledge, needs to be integrated into Christian theology. Christian ethics has to be given neither a foundation nor scientific validation by evolutionary theory, but it can benefit from scientific insight into biological matters. Unfortunately, this possibility is not explored by Pinckaers.

Though in principle Pinckaers recognizes that Christian ethics ought to assimilate what is best from available knowledge, in practice he shows no interest in doing so when it comes to science. This lacuna may be due, in part at least, to having an agenda committed to the retrieval of the vital sources of the tradition for contemporary moral theology. His "supernatural" focus leads him to downplay the natural basis of ethics and, with it, scientific accounts of it.

NATURAL LAW INDIFFERENT IN PRINCIPLE TO EVOLUTION: GERMAINE GRISEZ AND JOHN FINNIS

The "new natural-law theory" pioneered by Grisez, Finnis, and their allies holds that the "first principles of practical reason" give rationality to the process of moral decision-making. Its first principles are known "in themselves" (*per se nota*), comprehended immediately when their meaning and reference are understood, and indemonstrable. Advocates of this theory know that it presumes a supporting metaphysic, for example regarding human freedom, but they eschew any attempt to provide the kind of explicit metaphysical foundations that were developed by the neo-scholastics or Thomas himself. Constructed to speak to a broad pluralistic audience, some of whose members do not accept theistic presuppositions, this ethical theory is closer to the early modern natural-law tradition than to its medieval predecessors.[56]

---

[56] On the differences between classical and modern natural-law theories, see Stephen J. Pope, "Natural Law and Christian Ethics," in Robin Gill, ed., *The Cambridge Companion to Christian*

In one sense the attempt to provide a theory of practical reason without any reference to science might strike some readers as curious, since many philosophers regard science as the most rational and credible mode of public discourse available to us. Finnis and Grisez argue that practical reason, not speculative reason, derives a set of moral implications from a principle that is already normative rather than purely descriptive.[57] This allows them to avoid the "naturalistic fallacy." Rather than draw moral norms from some set of facts about human nature, the new theory derives particular moral norms from more general moral norms. The first principle of practical reason is itself both normative and "underivable." All other principles are derived from it by reason alone (for this among other reasons the "new natural-law theory" is often associated with the ethics of Kant). The independence of the new natural-law theory from evolution is in part generated by a philosophical commitment not to violate the "gap" between "facts" and "values" or that between "is language" and "ought language." Science engages in descriptions and explanations of the natural world, it is assumed, but has little to say about the normative dimensions of human existence and nothing to teach Christian ethics about the theological or moral relevance of its findings. The long-established wall between "is" and "ought" in moral philosophy provides a helpful warning to Christian ethicists to avoid wrongheaded attempts to draw invalid inferences from descriptive to normative statements. Sociobiologists have been guilty of moving illicitly from descriptions of nature to prescriptions about how human beings ought to act.[58]

Grisez and Finnis hold that we have a primary moral responsibility to commit ourselves to "integral human fulfillment" that incorporates "basic goods." Practical reason identifies an array of basic goods: life, knowledge, aesthetic appreciation, play, friendship, practical reasonableness, and religion.[59] Particular items have from time to time been added to or subtracted from this catalogue, so, for example, it later came to include "the marital

---

Ethics (Cambridge: Cambridge University Press, 2001), pp. 77–95, and J. B. Schneewind, *The Invention of Autonomy: A History of Modern Moral Philosophy* (New York: Cambridge University Press, 1998).

[57] See Germaine Grisez, "The First Principle of Practical Reason: A Commentary on *Summa Theologica* Ia IIae, q.94 a. 2," *Natural Law Forum* 10 (1965): 162–201; see also Janice Schultz, "Is–Ought: Prescribing and a Present Controversy," *The Thomist* 49 (January 1985): 1–2.

[58] See, among others, Philip Kitcher, *Vaulting Ambition: Sociobiology and the Quest for Human Nature* (Cambridge, MA and London: MIT Press, 1985), pp. 428–431.

[59] See, among others, John Finnis, *Natural Law and Natural Rights* (New York: Oxford University Press 1980), pp. 86–90.

good."[60] Basic goods are intrinsically valuable and universally recognized as such. The key moral principle holds that it is *always* wrong to choose to act directly against any basic good.[61] Exceptionless rights are correlative of exceptionless duties. Because integrity and community based on truthful communication are both basic goods, for example, we have a duty not to tell lies and a correlative right "not to be positively lied to in any situation ... in which factual communication ... is reasonably expected."[62]

We use other kinds of goods, like pleasure and income, to obtain one or more of the basic goods. The pursuit of integral human fulfillment requires a commitment to acting in accord with second-order practical reasons, also known as "modes of responsibility," and adhering to the "basic methodological requirements" of the natural law. We must strive to live according to a coherent plan of life, eschew arbitrary preferences among the basic goods, apply fairly the standards of natural law to all people, respect the basic goods in every act (i.e., never to violate directly any of the basic goods), promote the common good, and follow our conscience in every act. Finally, third-order or prudential norms guide practical choices about how to act in concrete situations. Policies and laws can be derived from the basic goods and the methodological requirements of the natural law. Since these goods and requirements are self-evident rather than extracted from claims about facts in the world, this morality avoids committing the "naturalistic fallacy."

One should never deliberately do a morally evil deed, Grisez and Finnis argue, and this includes evil acts intended to achieve a good end (the prohibition of which they call the "Pauline Principle")[63] and acts that violate one basic good as a means of promoting another. One basic good is not objectively more valuable than any of the others and is not reducible either to another basic good or to an instrumental good. Basic goods cannot be compared to one another by means of a common criterion of value. Ethics is analogous neither to economics, where all goods are commensurable in monetary terms, nor to sociobiology and game theory in their attempts to analyze human behavior in quantitative terms.

---

[60] See John Finnis, "Is Natural Law Theory Compatible with Limited Government?," in Robert P. George, ed., *Natural Law, Liberalism, and Morality* (New York: Oxford University Press, 1996), p. 5.
[61] See Finnis, *Natural Law and Natural Rights*, pp. 118–123.
[62] Ibid., p. 225.
[63] The Grisez–Finnis school interprets Romans 3:8, what they call the "Pauline principle," to hold that one may never do evil so that a good may come. This should not be confused with the "Pauline principle" that maintains that under certain conditions nonbelievers should be allowed to separate from their Christian spouses (1 Cor. 7:15–16).

Incommensurability rules out "trade offs" between basic goods, for example we cannot violate friendship in order to promote health. This position bears strong anti-consequentialist implications, and it runs directly contrary to positions that want to evaluate all goods in terms of their ability to "maximize" pleasure and "minimize" pain.

The "new natural-law theory" presents a formidable ethical theory in terms that are intelligible to contemporary philosophers. Though this position does not rely on faith in any explicit way and in fact claims to be purely rational, it has been used by many Catholics to provide a contemporary theoretical defense of the moral code taught by the official teaching authority of the Catholic church. It is no coincidence that the content of the "new natural law" happens to agree with almost every item of moral teaching found in the *Catechism of the Catholic Church*.

The "new natural-law theory" has been subject to significant philosophical criticisms. First, lists of basic goods are notoriously ambiguous: are all religions, including cults, instantiations of a basic value? Second, it holds that basic goods are incommensurable and cannot be subjected to "weighing," but it is not clear that one cannot reasonably weigh, say, religion as a more important good than play. Furthermore, the claim that basic goods cannot be "attacked" seems to deny the common human experience of occasions on which one is faced with deep moral conflict between competing goods.

Grisez and Finnis fail to recognize that basic goods can conflict.[64] They maintain, for example, that married couples may under no circumstances ever use artificial birth control, even in cases in which pregnancy or an HIV-positive partner would put the woman's life at risk. There is no moral conflict here for Grisez and Finnis because one may never directly attack the good at stake. The "new natural-law theory" takes as "self-evident" what many critics simply deny: that basic goods ought never to be violated and that they cannot be subordinated to one another with ethical integrity. It is not apparent to many ethicists, for example, that procreation is a good that ought never to be subordinated to the good of responsibly spacing the births within a family. Nor is it apparent to all reasonable people that any and every use of artificial birth control damages the moral integrity of married couples who use it.[65]

---

[64] See David Hollenbach, SJ, *The Common Good and Christian Ethics* (New York: Cambridge University Press, 2002).
[65] See Robert P. George and Gerald V. Bradley, "Marriage and the Liberal Imagination," *Georgetown Law Journal* 84 (1995): 301–320.

A critical weakness of the position is its attempt to interpret natural law without giving any account of human nature itself. Finnis opines, for example, without offering any evidence, that same-sex relations of every kind fail to offer intelligible goods of their own but only "bodily and emotional satisfaction, pleasurable experience, unhinged from basic human reasons for action and posing as its own rationale."[66] This theory reasons a priori from principles to what must in fact be the empirical case, but not in the reverse direction. It presents moral law but no account of nature.[67]

The "new natural-law theory" attends to a certain kind of human experience – an experiential grasp of "self-evident" basic goods – but isolates itself from the data and insights about human goods coming from knowledge of human evolution. It avoids committing the "naturalistic fallacy," but does so in such a way that nothing of normative significance can be learned about human behavior from science. It is thus an abstract theory imposed on human nature rather than an account of the human good informed by current scientific knowledge of human behavior.

CONCLUSION

While Christian ethics ought to be wary of moving in unwarranted ways from descriptive to normative language, it is impossible to develop an ethical perspective without developing some descriptive account of human behavior. The question is not whether Christian ethics will employ descriptions of human nature, but whether those implicit or explicit accounts of human nature strive to be relatively consonant with the best available scientific knowledge.

This book offers an alternative to the positions sketched in this chapter. I would like to suggest that Christian ethics offers a vantage point that attends to the theological and moral significance of the findings of science. I believe we have an obligation to think in the most appropriate ways possible about human nature, including those regarding the conditions and limits of human flourishing. This agenda depends of course on distinguishing well-established findings and theories from the purely

---

[66] John Finnis, *Aquinas: Moral, Political, and Legal Theory* (New York: Oxford University Press, 1998), pp. 153, 151.
[67] On this topic, see Etienne Gilson, *The Christian Philosophy of St. Thomas Aquinas* (New York: Random House, 1956), part III. The major Thomistic critique of the "new" natural-law theory was given by Russell Hittinger, *A Critique of the New Natural Law Theory* (Notre Dame, IN: University of Notre Dame Press, 1989).

speculative hypotheses also made in the name of science. The problems attending a failure to make this distinction will be examined in the next chapter. Pinckaers correctly argues that Christian ethics intends to help us attain genuine happiness. Since this end pertains to human beings, we need to understand as adequately possible what it is that makes us "human" in the complete sense of the term. The virtues are the habits by which this happiness is attained.

CHAPTER 3

Varieties of reductionism

The single most imposing obstacle to understanding the proper relation between Christian ethics and human evolution lies in the inappropriate forms of reductionism presumed by sociobiology and evolutionary psychology. The critical methodological issue for Christian ethics concerning evolution thus concerns "reductionism," a difficult and sometimes confusing term but one that is used so often that an examination of it is unavoidable. There are at least three basic ways of speaking about reductionism: methodological, epistemological, and ontological.[1] This chapter argues that methodological reductionism is entirely legitimate, but that epistemological and ontological reductionism are not; it also argues that Christian ethics can accept the results of the former without developing the latter.

METHODOLOGICAL REDUCTIONISM

According to Francisco Ayala, Darwin's "fundamental discovery" is that "there is a process that is creative though not conscious. And this is the conceptual revolution that Darwin completed: that everything in nature, including the origin of living organisms, can be accounted for as the result of natural processes governed by natural laws."[2] The premise of methodological reductionism, then, is that the natural sciences can explain the workings of

---

[1] See Francisco J. Ayala: "Introduction," in Francisco J. Ayala and Theodosius Dobzhansky, eds., *Studies in the Philosophy of Biology: Reduction and Related Problems* (Berkeley: University of California Press, 1974); "Reduction in Biology: A Recent Challenge," in David Depew and Bruce Weber, eds., *Evolution at the Crossroads* (Cambridge, MA: MIT Press, 1985): 67–78; and "Biological Reductionism: The Problems and Some Answers," in F. E. Yates, ed., *Self-Organizing Systems: The Emergence of Order* (New York: Plenum Press, 1987), pp. 315–324.

[2] Francisco Ayala, "Darwin's Devolution: Design without a Designer," in Robert J. Russell *et al.*, eds., *Evolutionary and Molecular Biology: Scientific Perspectives on Divine Action* (Vatican City State: Vatican Observatory Publications, and Berkeley, CA: Center for Theology and Natural Sciences, 1998), p. 109.

physical, chemical, and biological processes without recourse to nonscientific ways of thinking.

Methodological reductionism functions as a research strategy for examining various wholes in terms of their constituent parts. Biochemists, for example, study cells in terms of their macromolecules. Geneticist Steven Rose points out that we usually find it easier to understand phenomena "if we can hold them relatively isolated from the rest of the world and alter potential variables one at a time."[3] It is simpler to examine the enzymic interactions of one protein when it is isolated from the many other small and larger molecules that would normally influence it when studied in living cells. Methodological reduction helps scientists to establish chains of cause and effect and express these relations in mathematical terms.

This methodology is "reductionistic" because it involves breaking down complex wholes into their component parts, for example the attempt to understand the mechanics of the heart in terms of its pumps and valves. This kind of reductionism is intrinsic to all scientific inquiry. As scientist-theologian Arthur Peacocke points out,

The breaking down of unintelligible, complex wholes into their component units, the determination of the structures of those pieces and what functions they can perform, and then the fitting of them together as best one can, hypothetically at least, in order to see how they function together in a complex whole, are such common ploys in experimental science that most practicing scientists would consider it scarcely worth remarking upon.[4]

Complex systems function within the parameters made possible by their component parts, and these parts have their own constraints, for example the heart is not infinitely durable because muscle tends to break down over time.

Methodological reductionism has inherent limits, however, and it is therefore of limited explanatory value in many domains of investigation. It focuses narrowly on particular aspects of how specific natural phenomena function under certain conditions. It is not always possible comprehensively to explain the actions of a whole by accounting for the traits of its parts. For this reason, paleobiologist Simon Conway Morris argues that sociobiology is "not so much wrong, as seriously incomplete."[5]

---

[3] Steven Rose, *Lifelines: Biology beyond Determinism* (New York: Oxford University Press, 1998), p. 77.
[4] Arthur R. Peacocke, *God and the New Biology* (London: Dent, 1986), p. 6.
[5] Simon Conway Morris, *The Crucible of Creation: The Burgess Shale and the Rise of Animals* (New York: Oxford University Press, 1998), p. 8.

Accounting for genes does not suffice for an explanation of human behavior. Genetics does not make unnecessary an account of specific and concrete developmental pathways within which genes contribute to the growth of living organisms. Conway Morris explains Dawkins' shortcoming in terms of the image of an oil painting:

> In this analogy, Dawkins has explained the nature and range of pigments; how the extraordinary azure colour was obtained, what effect the cobalt has, and so on. But the description is quite unable to account for the picture itself. This view of evolution is incomplete and therefore fails in its side-stepping of how information (the genetic code) gives rise to phenotype, and by what mechanisms. Organisms are more than the sum of their parts ...[6]

The key issue here concerns the part–whole relation. Dawkins and others who think more atomistically can be appreciated for their effort to explore how far one can make progress in understanding nature in terms provided by the mechanistic metaphor. They correct the older assumption that nature strives for what is "best" for the species and that, as Thomas put it, the "part" seeks the "common good" of the species more than its own good. Biologist Ernst Mayr expressed the consensus in evolutionary biology when he wrote: "Nothing is ever selected 'for the good of the species.'"[7] Even though the behavior of "fit" organisms, and even the destiny of "unfit" organisms, tends inadvertently to promote the good of the species, benefits are simply by-products of the ordinary tendency of organisms to pursue their own fitness. To seek "the good of the species" would be to detract from one's own fitness, thus giving a reproductive edge to one's competitors – a policy that would ultimately lead to the extinction of one's own lineage. Even group selectionists hold that when group selection does obtain, organisms do not promote the good of any entity larger than their groups.

This having been said, it is also the case that just as the "whole" cannot be understood in abstraction from its "parts," so the "parts" cannot be understood in isolation from their place in the "whole." "Holism" presupposes a complex and multidirectional interaction between wholes and parts. Living beings are "open systems" influenced by dynamic interactions between and within all levels of their organization. They are organized hierarchically in that their constituent parts (e.g. cells, organs) themselves function as wholes comprising sub-parts. The holistic view offers a powerful alternative to ontological reductionism. It acknowledges the complexities of

---

[6] Ibid., p. 9. [7] Mayr, *One Long Argument*, p. 88.

causal relations that work from the "top down" as well as from the "bottom up."[8]

Living organisms manifest properties that are not reducible to the properties of their physical and chemical components. They manifest emergent, holistic capabilities that can only be grasped in distinctively biological terms. Wholes are subjected to the laws governing their parts, but wholes can possess "emergent properties" that do not exist in their individual components taken as isolated units. Even strict methodological reductionism, as seen, for example, in the geneticists' understanding of evolution as an alteration in the frequency of genes in a given population, has certain limitations. The effects of genes have to be studied in terms of their influence on organisms as wholes and vice versa. Holism recognizes the dynamic nature of reciprocal interaction between organisms influencing their environments as well as environments influencing their organisms. The entire evolutionary process is best interpreted as the gradual emergence of more and more intricate and complex webs of reciprocal influence.

Some of those whom Mayr calls "naturalists" advocate a more holistic approach to evolution as "a process relating to organs, behaviors, and the interactions of individuals and populations."[9] Rose argues that the organism constructs itself through a "lifeline trajectory" in a process he calls "autopoiesis."[10] "Holism" provides a more balanced alternative to reductionism: "Parameters are not fixed; properties are non-linear. And the living world is highly non-uniform ... [We must] remember that what happens in the test-tube may be the same [as], the opposite of, or bear no relationship at all to what happens in the living cell, still less the living organism in its environment. It all depends."[11]

All living systems are sensitive to the role of change and contingency. Organisms are "thermodynamically open systems," and living processes are "homeodynamic"[12] rather than "homeostatic."[13] Organisms, and not genes, are at the center of life. Living processes are "radically indeterminate," and this is especially the case for human beings. Rose argues: "we, like but to an even greater degree than all other living organisms, make our own future, though in circumstances not of our own choosing."[14] Rose thus

---

[8] See also Donald Campbell, "Downward Causation in Hierarchically Organized Biological Systems," in F. J. Ayala and T. Dobzhansky, eds., *Studies in the Philosophy of Biology: Reduction and Related Problems* (Berkeley: University of California Press, 1974), pp. 179–186.
[9] Mayr, *One Long Argument*, pp. 138–139.   [10] Rose, *Lifelines*, p. 18.
[11] Ibid., p. 79.   [12] Ibid., p. x.   [13] Ibid.
[14] Ibid., p. 7. This is not to confuse indeterminacy with freedom. Kenneth Miller argues that quantum indeterminacy explains how free will can exist in a world of law in *Finding Darwin's God*.

acknowledges human freedom – "we make our own future" – without asserting the existence of an uncaused entity that is immune from the influence of our bodies – "though in circumstances not of our own making." Rose criticizes excessively reductionistic evolutionists for taking "homeostasis" as a model of living systems. "Lifelines," he argues, are "homeo*dynamic*." Seeing organisms as "merely homeostasis is to deny them lifelines, to fall into the empty-organism trap that the gene's-eye view of the world demands."[15]

Philosopher Daniel Dennett describes "good reductionism" (what Ayala calls "methodological reductionism") as the presupposition that science need not seek to explain natural processes by appealing to the causal influence or intervention of extra-natural entities. "Good reductionism," Dennett explains, is "the commitment to non-question-begging science without any cheating by embracing mysteries or miracles at the outset."[16] Reductionism turns bad when its "overzealousness leads to falsification of the phenomena."[17]

Methodological reductionism is neither theistic nor atheistic. Science could not have achieved its remarkable successes if it assumed that the universe contained totally inexplicable events that are not subject to causal laws. It strives to understand natural phenomena without referring to the action of God but neither denies nor affirms anything about God's existence or nature.

Methodological reductionism has roots in Christianity as well as in Greek, Roman, and Muslim thought. The doctrine of creation in particular regards the cosmos as a contingent reality freely created by an omnipotent, omniscient, and omnibenevolent God. The orderly and intelligible nature of creation makes it amenable to investigation by human intelligence. Christians in antiquity and the Middle Ages believed that the creation reflects, in a complex and refracted way, something of the nature of the Creator. As Stanley Jaki explained in his Gifford Lectures, "Science found its only viable birth within a cultural matrix permeated by a firm conviction about the mind's ability to find in the realm of things and persons a pointer to their Creator. All great creative advances of science have been made in terms of an epistemology germane to that conviction."[18] The contingent character of this freely created world – the fact that God did not have to make it the way it is – means that it must be investigated empirically. When contrasted with alternative religious

---

[15] Rose, *Lifelines*, p. 157.
[16] Daniel Dennett, *Darwin's Dangerous Idea: Evolution and the Meanings of Life* (New York: Simon and Schuster, 1995), p. 82.
[17] Ibid.
[18] Stanley Jaki, *The Road of Science and the Ways to God* (Chicago: University of Chicago Press, 1978), p. vii.

traditions, the Jewish and Christian Scriptures reveal a creation that is "desacralized" and therefore a legitimate object of scientific examination and technological manipulation. Acknowledging that science has historical roots that are in part religious is not to say, of course, that scientists require theological or historical justification for assuming that the natural world is intelligible and that the human mind is equipped to investigate it. It does suggest that methodological reductionism is compatible with Christian ethics.

EPISTEMOLOGICAL REDUCTIONISM

A second form of reductionism is epistemological. It holds that traits found in higher levels of complexity can be explained entirely in terms of what is discovered on lower levels of complexity: biological events can be fully accounted for in terms of less complex chemical reactions; biochemistry, in terms of chemistry; and chemistry, in terms of physics. This "theory reduction" attempts to explain events studied by biology as special cases of the laws of chemistry, and events studied by chemistry as special cases of the laws of physics. Epistemological reduction proceeds from the essentially unprovable presupposition that all phenomena of life, society, and mind are explicable by a unified set of physical laws.

Sociobiologists typically accept not only methodological but also epistemological reductionism. If the theory of natural selection is correct, some argue, virtually everything about the human mind should be intelligible in evolutionary and biological terms. Wilson insists that, "there is intrinsically only one class of explanation" for all phenomena.[19] The proper class of explanation for human behavior, Wilson holds, is provided by natural science rather than the humanities or the social sciences. The sociobiological attempt to render unnecessary any description of human behavior in terms of the agent's intentions, motives, or self-understanding causes it significant problems in ethics, as we will see in later chapters. Only one class of explanations is true, so interpretations of human action proposed by the humanities are erroneous and misleading. Neurobiology will "cannibalize" sociology and sociobiology will "cannibalize" ethics (whatever that actually means), Wilson writes, but neither neurobiology nor sociobiology is "cannibalized" by physics. He does not think that the biological disciplines are superfluous and he does not want to go as far as Francis Crick did when he declared that "the ultimate aim of the modern movement in biology is to explain all biology in terms of physics and chemistry."[20]

---

[19] E. O. Wilson, *Consilience*, p. 266.
[20] Francis C. H. Crick, *Of Molecules and Man* (Seattle: University of Washington Press, 1966), p. 10.

Some of the most persistent and telling criticisms of sociobiology focus on its insufficient appreciation of the impact of culture on human behavior. Sociobiological counter-responses have proposed a variety of theories of social evolution. Wilson is committed to explaining social and cultural phenomena as contributing to the inclusive fitness of individuals. He and his colleague Charles Lumsden offered a theory of "gene-culture co-evolution," according to which genes influence the generation of culture. Their *Genes, Mind, and Culture* argued that evolved "epigenetic rules" involve biological mechanisms that guide the organization of the human mind.[21] What they called "culturgens," the "array of transmissible behaviors, mentifacts, and artifacts,"[22] emerge in the interaction between the culture and genes that takes place in human minds. However, this attempt to respond to the charge of inappropriate reductionism failed because it continued to interpret culture in an overwhelmingly biological manner.[23] Though Lumsden and Wilson acknowledged that culture has some independence from biology, they also presumed that, in the long run, cultural tendencies are preserved only if they comport with our genetic interests. Culture remains in the service of genes and represents a refraction, extension, and elaboration of the biological propensities of our hunter-gatherer ancestors.[24] This is not to say that there is absolutely no evolutionary basis for some cultural practices, such as incest avoidance or dietary preferences, only that Lumsden and Wilson have not provided a convincing sociobiological argument that long-term social and cultural evolution reflects the strong influence of genetic interests.

The more sophisticated "dual inheritance" theory of Robert Boyd and Peter Richerson acknowledges that cultural pressures can incline people to act in ways that reduce their own genetic fitness.[25] They continue the evolutionist assumptions that culturally acquired behavior is constrained by fitness limits and that some cultural variations are preferred for biological reasons, for example individuals who conform to group practices under many circumstances gain an adaptive advantage in so doing. But they do not attempt to argue that biological evolution can account for important features of long-term social and cultural evolution.

---

[21] Charles J. Lumsden and Edward O. Wilson, *Genes, Mind, and Culture* (Cambridge, MA: Harvard University Press, 1981).
[22] Ibid., p. 7.   [23] See ibid., pp. 24 and 178–179.
[24] See Kitcher, *Vaulting Ambition*, ch. 10, "The Emperor's New Equations."
[25] Robert Boyd and Peter J. Richerson, *Culture and the Evolutionary Process* (Chicago: University of Chicago Press, 1985).

Dawkins' "meme theory" provides another approach to the gene-culture relation. He proposes that "memes,"[26] cultural "units" that are not biologically inherited but rather learned and transmitted culturally, spread in a way analogous to genetic replication. This effort naïvely presumes that ideas function as atomistic units that survive and reproduce in ways analogous to genetic replication. Though culture presents a nonmaterial, ideational kind of information, Dawkins believes that natural selection presents the best model for understanding it. Some "memes" are preferred over others in a particular culture, he argues, because they better suit the fitness needs of people in a given time and place. Thus the "God meme" has survived because of the benefit it has bestowed on believers in the past, but Dawkins argues that this meme has outlived its usefulness and must be eliminated for the good of the species.

The "meme" argument is vulnerable to two fatal objections. One is that there is no scientific evidence that memes replicate in anything like the ways in which genes are copied. Dawkins' position amounts to a philosophical postulation, not an argument justified by compelling evidence. Second, the "meme" postulate is easily turned on itself in that one could argue that sociobiology itself is a meme that serves certain functional purposes but has no bearing on reality. If Dawkins will not accept this argument, he is forced also to allow other cultural positions immunity from his mimetic epistemological reduction. More generally, Dawkins' depiction of "memes" as "viruses of the mind"[27] denigrates the power of human intelligence, memory, and imagination.

All three of these approaches to the relation of genes and culture suffer from a similar epistemological reductionism that denies the autonomy and even credibility of nonscientific academic disciplines. Scientific epistemological reductionism, or "scientism" in its most pure form, regards science as the only path to all knowledge. Sociobiology is but one major exemplification of scientism. Geneticist Richard Lewontin exemplifies it when he asserts that scientists have a responsibility to persuade the public, "to reject irrational and supernatural explanations of the world, the demons that exist only in their imaginations, and to accept a social and intellectual apparatus, *Science*, as the only begetter of truth."[28] Lewontin believes that the general public will always be scientifically illiterate and therefore prone to make

---

[26] See Dawkins, *The Selfish Gene*, ch. 11.
[27] See Dawkins, "Viruses of the Mind," in B. Dahlbom, ed., *Dennett and His Critics*, ed. (New York: Blackwell, 1993), pp. 13–27.
[28] Richard C. Lewontin, "Billions and Billions of Demons" (review of Carl Sagan's *The Demon-Haunted World: Science as a Candle in the Dark*), *New York Review of Books* 44.1 (January 9, 1997), p. 28; emphasis in the original.

unreliable judgments. The only hope for the human race, he seems to think, is for scientists and those who support science to hold political power. Lewontin, of course, cannot provide the slightest scientific argument for his claim that science has an inherent and universal epistemological superiority, let alone exclusive epistemological competence. His naïve and uncritical faith in science as the ultimate "arbiter of truth" is accompanied by the unfortunate implication that what ordinary people suppose they know about the world they live in is not actually knowledge at all. Yet there is no compelling reason to think that the common-sense or "first-order" understanding that allows most people to live their lives is necessarily false or illusory. People do have insights into realities that are not capable of being expressed or justified in scientific terms.

"Scientism" is a philosophical choice, not a necessary presupposition of science *per se*. Wilson and Lewontin are well aware that there is no single method called "science," and that the sciences employ multiple methods. But they do not acknowledge that there are many ways to be "rational" and many shades and degrees of rationality that include but are not exhausted by the forms of rationality exhibited in scientific research. Aspects of human experience that can be analyzed scientifically will not really be understood sufficiently through science. We cannot expect to understand a Vivaldi concerto simply by mapping its sound waves or a Giotto fresco by analyzing its chemical composition. There are even biological reasons for the nonreducibility of all knowledge to science, rooted in the nature of the brain and mind. Neuroscientist Gerald Edelman points out that the historical and symbolic structure of the human mind implies "there can be no fully reducible description of human knowledge."[29] Nobel laureate Arno Penzias concurs when he writes that science is by its very nature "always an incomplete description of the world," and certainly not "the only way to think about the world."[30]

Recent philosophy of science and studies in the history of science have shown that science is not as purely rational and objective in the ways assumed by early modern philosophers.[31] Science is a thoroughly human enterprise that reflects the humanity of the inquirer with all the advantages and the limits of his or her perspective. Because understanding and judging

---

[29] Gerald M. Edelman, *Bright Air, Brilliant Fire: On the Matter of the Mind* (New York: Penguin Press, 1992), p. 177.

[30] Arno Penizias, "The Elegant Universe," in Mark W. Richardson and Gordy Slack, eds., *Faith in Science: Scientists Search for Truth* (London: Routledge, 2001), p. 20.

[31] See, among many others, Stephen Toulmin, *Cosmopolis: The Hidden Agenda of Modernity* (Chicago: University of Chicago Press, 1990).

are acts of human knowers, facts themselves do not stand independently of their interpretation. The recovery of the sense of particularity in recent philosophy of science underscores the importance of noticing the history of the "sciences," which are plural and complex rather than single and monolithic. Scientists employ metaphors and overarching theories to interpret data but they do not give an exact account of where metaphors and theories end and data begin.

Epistemological reductionism fails properly to acknowledge the social context of scientific inquiry and its habit of seeking knowledge of the natural world through what Rose calls "socially organized hypothesis making."[32] Scientists propose hypotheses in particular communities, where they are cumulatively shared, examined, tested, and validated. These communities, in turn, are deeply influenced by cultural, political, economic, and other kinds of forces that shape how their participants view the world.

Sociologists and historians of science have shown ways in which cultural values and expectations that come from outside science are often confused with scientific knowledge itself. Acknowledging the social basis of science need not lead to radical skepticism but it does warrant scrutiny and self-awareness. Rose notes that, "[a]lthough the observations we make about the world are theory- and ideology-laden before we start, and the joints into which we carve nature are provided less by a priori definitions than by operational need, nonetheless they must make some more-or-less good fit with the world or we could not proceed."[33]

The critical-realist alternative to scientism rejects epistemological reductionism without denying the value of methodological reductionism. The flaws of "scientism" do not undermine science itself as the primary way of knowing about the empirical structures of the physical world. Appreciation of the socially located and "theory-laden" character of science is consistent with believing that it provides a relatively reliable set of methods for understanding nature, even if this understanding is provisional and open to correction in light of new discoveries.

Christian ethics can recognize the validity of scientific discoveries, accounts of natural history, and scientific theories of evolution. It must do so within a proper understanding of the epistemological status of claims about evolution. Christian ethics can respond appropriately to evolution if understood in the context of a nonreductionistic epistemology.

Peacocke provides a helpful understanding of the relation between academic disciplines. He relates the sciences to one another in proportion

---

[32] Rose, *Lifelines*, p. 66.   [33] Ibid., pp. 68–69.

to the complexity of what they study, different objects of study having greater or lesser degrees of complexity. Greater degrees of complexity signal the need to acknowledge a kind of hierarchy in which higher sciences reflect the increased complexity of their subject matter. The sciences ought to be related to one another in a way that recognizes both their limitations and the emerging capacities of the kinds of objects they investigate. As Ayala explains:

> A majority of biological problems cannot be as yet approached at the molecular level. Biological research must then continue at the different levels of integration of the living world, according to the laws and theories developed for each order of complexity. The study of the molecular structures of organisms must be accompanied by research at the levels of the cell, the organ, the individual, the population, the species, the community, the eco-system. These levels of integration are not independent of each other.[34]

Lower-level entities impose constraints on characteristics that can exist in the higher-level entities of which they are constituent parts. The functioning of living systems abides by fundamental physical and chemical laws, but at the same time these laws represent necessary but not sufficient conditions for the functioning of higher-level entities. Relatively more complex entities are characterized by capacities that cannot be derived simply from the capacities of their constituent parts – they are thus beneficiaries of "emergent" capacities.

Living systems can only be understood in terms of principles provided by the biological sciences. The laws governing biological interactions obviously cannot violate the laws of physics and chemistry. But, as Eileen Barker of the London School of Economics explains, "over and above the laws of physics and chemistry the biologist can look for further principles which might help us to understand why, from the potential universe of what is permitted by physics and chemistry, certain combinations do occur and others do not."[35] As biological systems become more and more complex, new combinations are made possible and the rapidity of feedback from higher to lower levels of organization increase the probability of change. Higher-level entities are said to be irreducible to their parts when full knowledge of the latter does not explain the former. Barker continues: "When units are arranged together in particular ways it sometimes happens

---

[34] Francisco J. Ayala, "The Autonomy of Biology as a Natural Science," in A. D. Breck and W. Youngrau, eds., *Biology, History, and Natural Philosophy* (New York: Plenum Press, 1972), p. 6.
[35] Eileen Barker, "Apes and Angels: Reductionism, Selection, and Emergence in the Study of Man," *Inquiry* 19 (1976): 378.

that the whole which they form has certain properties which are not present in the units themselves or even in different combinations of those units. Such properties are said to 'emerge' at this level of organization."[36]

"Downward" causality refers to the ways in which new kinds of wholes at higher levels of organization can influence their constituent parts.[37] Wholes are to some extent constrained by the traits of their component parts, but the function of the parts is also to some extent constrained by the traits of the whole. Causation works from "top-down" as well as "bottom-up" directions.

The emergence of new kinds of wholes can be accompanied by new kinds of capacities, processes, and traits.[38] Ian Barbour explains that "emergence" refers to "the claim that in evolutionary history and in the development of the individual organism, there occur forms of order and levels of activity that are genuinely new and qualitatively different. A stronger version of emergence is the thesis that events at higher levels are not determined by events at lower levels and are themselves causally effective."[39]

The notion of emergence suggests neither that nature is inevitably progressive in an absolute sense nor that it is structured in a linear hierarchy, but only that, over long periods of time, evolution tends to increase the complexity of structures and to move from homogeneity to heterogeneity. Evolution is a "branching" process whose pace is slow, irregular, haphazard, and unpredictable, but it does – at least sometimes – move life toward higher and higher levels of organization.

Science recognizes different kinds and degrees of complexity. Peacocke envisions the sciences as forming a hierarchy that contains both vertical and horizontal dimensions. The former is composed of four strata of escalating complexity, beginning with the material world, then moving up to encompass the structures of relatively simple organisms, then to the conduct of more complex organisms, and finally to the very complex interactions made possible by human culture. The horizontal dimension of the hierarchy of the sciences is based on the kinds of part–whole organization of entities. So the science of biology moves from smaller-level entities such as

---

[36] Ibid., 381.   [37] See Campbell, "Downward Causation."
[38] A great deal of work has been done in quantum physics. David Bohm's *Wholeness and the Implicate Order* (Boston: Routledge, 1980), for example, argued that everything that exists is dynamically evolving from a non-material or "implicate" order into a material or "explicate" order.
[39] Ian Barbour, "Neuroscience, Artificial Intelligence, Human Nature," in R. J. Russell *et al.*, eds., *Neuroscience and the Person: Scientific Perspectives on Divine Action* (Vatican City State: Vatican Observatory Publications, 2002), p. 271.

macromolecules and organelles to macro-level objects of investigation such as populations and ecospheres.[40]

Peacocke's placement of Christian theology at the top of the hierarchy indicates both its all-encompassing nature as a discipline and its assimilative and integrative responsibility in relation to the findings and theories of other disciplines. The other disciplines raise questions for theology, for example what does the known age of the earth imply for the doctrine of providence? What does genetic similarity between humans and other species, particularly chimpanzees, imply about the doctrine of the *imago Dei*? The conclusions of theology are in their own way both constrained by the knowledge provided by the other disciplines and at the same time the subject of emergent insights that cannot be generated by the other disciplines.[41]

Peacocke provides a concrete example of this issue from the field of physical biochemistry. The structure of pyrimidine and purine bases gives them certain properties that by themselves are unable to convey hereditary information. Yet when organized within the double helix, and functioning within the matrix of the cytoplasm, these bases do transmit hereditary information. Peacocke emphasizes the fact that this capacity was "absent from the component individual nucleotides."[42] Indeed, the "concept of 'information,' originating in the mathematical theory of communication (C. E. Shannon), had never been part of the organic chemistry of nucleotides, even of polynucleotides."[43]

Peacocke offers strong reasons for rejecting epistemological reductionism and in doing so intentionally makes room for an approach to scientific knowledge that is compatible with Christian faith. Science, by itself and unaided by other disciplines, cannot provide a comprehensive framework for explaining everything relevant to human existence. This approach to science (like Rose's) is consistent with philosopher Mary Midgley's account of knowledge as various kinds of conceptual frameworks or "maps" that help us understand human experience. We attempt to understand features of the world by locating them in various ways on "maps." The sciences have become increasingly detailed and specialized fields of study, so they are less

---

[40] See Arthur R. Peacocke, *Creation and the World of Science* (Oxford: Clarendon Press, 1979), Appendix C, and *Theology for a Scientific Age* (London: SCM Press, 1993), ch. 12, esp. p. 216, fig. 3. See also Peacocke, *God and the New Biology*, ch. 1, esp. p. 16, fig. 2.

[41] Peacocke, *Creation and the World of Science*, Appendix C; Peacocke, *Theology for a Scientific Age*, ch. 12, esp. parts 6–9.

[42] Arthur R. Peacocke, "A Map of Scientific Knowledge: Genetics, Evolution, and Theology," in Ted Peters, ed., *Science and Theology: The New Consonance* (Cambridge, MA: Westview, 1998), p. 191.

[43] Ibid.

and less able to provide anything like the comprehensive breadth of vision involved in the kind of "map" that orients us to life as a whole.[44] Drawing the map on the basis of perspectives made possible by biology alone has its own narrowness, just as do all attempts to draw the maps exclusively from the point of view provided by theology or ethics. What Midgley says about moral philosophy is also true of Christian ethics and evolutionary theory: "Philosophy and biology are not in competition; they are different aspects of one inquiry . . . What we need is to attend more, in moral arguments, to the biological facts. But this does not make philosophy unnecessary; it just makes it harder."[45]

The "maps" by which we live, Midgley argues, inevitably involve us in a kind of faith, though not necessarily in religious belief:

The faith we live by is something that you must have before you can ask whether anything is true or not. It is a basic trust. It is the acceptance of a map, a perspective, a set of standards and assumptions, an enclosing vision within which facts are placed. It is a way of organizing the vast jumble of data. In our age, when the jumble is getting more and more confusing, the need for such principles of organization is not going away. It is increasing.[46]

ONTOLOGICAL REDUCTIONISM

A third form of reductionism pertains to what is "real." In contrast to the limited domains examined by particular natural sciences, metaphysics seeks to understand the broadest and most comprehensive context for understanding all that is real. Ontology, a branch of metaphysics, seeks to account for the different kinds of beings that exist.[47] Questions such as "what are human beings?," "what is free will?," "why do human beings exist?," and "what is the meaning of human progress?" are questions for ontology, not for the natural sciences, and the only way that they can be answered in scientific-sounding language is by writers who smuggle their own ontological presuppositions, moral beliefs, and social visions into their accounts of what constitutes science.

"Ontological reductionism" holds that more complex, higher-level traits or entities are nothing more than a particular way in which simpler traits or

---

[44] See Midgley, *Science as Salvation*, ch. 3.
[45] Mary Midgley, *Beast and Man: The Roots of Human Nature* (Ithaca, NY: Cornell University Press, 1978), p. 174.
[46] Midgley, *Science as Salvation*, p. 57.
[47] See E. J. Lowe, "Ontology," in Ted Honderich, ed., *The Oxford Companion to Philosophy* (New York: Oxford University Press, 1995), p. 634.

entities are organized. It insists that the character of wholes is determined entirely by the traits of their constituent parts. It would seem committed to the claim that only micro-entities are real. "Ontological reductionism" entails "ontological naturalism," the unproven assumption that only entities or processes found in nature are real. It rejects the "vitalistic" claim that entities are influenced by souls or what are taken to be other nonnatural metaphysical forces. It asserts that the natural world is all that exists.

Methodological, epistemological, and ontological reductionisms are often blurred together.[48] Dawkins presents a prime illustration of this tendency. Describing present-day genes as descended through millions of years from the first "replicators," he writes that while these "replicators" once drifted on primeval oceans, now "they swarm in huge colonies, safe inside gigantic lumbering robots, sealed off from the outside world, communicating with it by tortuous indirect routes, manipulating it by remote control. They are in you and in me: they created us, body and mind; and their preservation is the ultimate rationale for our existence ... we are their survival machines."[49] Dawkins' habit of mixing ontological claims with scientific observation displays what Philip Kitcher calls "vaulting ambition."[50] He announces that, "[t]he universe we observe has precisely the properties we should expect if there is, at bottom, no design, no purpose, no evil and no good, nothing but blind, pitiless indifference ... DNA neither knows nor cares. DNA just is. And we dance to its music."[51]

Despite his adamant objection to the epistemologically reductionistic theories of Dawkins and E. O. Wilson, Stephen Jay Gould shared their reductionistic and naturalistic ontology. The problem with the acceptance of evolution, he argued, does not lie in a deep resistance to science but rather in the "philosophical content of Darwin's message – in its challenge to a set of entrenched Western attitudes that we are not yet ready to abandon."[52] This philosophical content can be summarized in the claim that evolution is purposeless, directionless, and meaningless. According to Gould, "Darwin applied a consistent philosophy of materialism to his interpretation of nature. Matter is the ground of all existence; mind, spirit, and God as well, are just words that express the wondrous results of

---

[48] See Karl W. Giberson and Donald Yerxa, *Species of Origins: America's Search for a Creation Story* (Lanham, MD: Rowan and Littlefield, 2002), pp. 119–150.
[49] Dawkins, *The Selfish Gene*, pp. 19–20. A strong case against genetic reductionism is found in Ernst Mayr, *What Evolution Is* (New York: Basic Books, 2001).
[50] Kitcher, *Vaulting Ambition*.   [51] Dawkins, *River out of Eden*, p. 133.
[52] Stephen Jay Gould, *Ever since Darwin: Reflections in Natural History* (New York: W. W. Norton, 1977), pp. 12–13.

neuronal complexity."⁵³ Gould did not derive moral pessimism or fatalism from this materialism; on the contrary, he advocated a kind of existentialist resistance not far in spirit from Camus: "if we cannot find purpose in nature, we will have to define it for ourselves."⁵⁴ At the very least, Gould insisted, we ought to have the honesty to admit that arrogant anthropocentrism finds no support in the evolutionary understanding of our place in the world. Yet one can agree with Gould's denunciation of human arrogance without conceding his materialism.

It is unclear how Gould or Dawkins think that their ontological claims are justified on the basis of science. It is philosophy and theology, not evolutionary biology or paleontology, that can assert that "the ground of all existence; mind, spirit, and God as well, are just words that express the wondrous results of neuronal complexity."⁵⁵ At times materialists concede that their position is based on unproven ontological assumptions, or even that they result from a kind of "faith." Lewontin's exuberant rhetorical advocacy of materialism as the philosophical basis of science almost resonates with the tone of the fideist dictum attributed to Tertullian (c. 160–c. 230), "*Credo quia absurdum est*" – "I believe because it is absurd."⁵⁶ Lewontin writes:

> We take the side of science in spite of the patent absurdity of some of its constructs, in spite of its failure to fulfill many of its extravagant promises of health and life, in spite of the tolerance of the scientific community for unsubstantiated just-so stories, because we have a prior commitment, a commitment to materialism. It is not that the methods and institutions of science somehow compel us to accept a material explanation of the phenomenal world, but, on the contrary, that we are forced by our a priori adherence to material causes to create an apparatus of investigation and a set of concepts that produce material explanations, no matter how counterintuitive, no matter how mystifying to the uninitiated.⁵⁷

Lewontin emphasizes the need to adhere to scientific methods despite seeming anomalies and apparent violations of common sense. One is able to have such commitment, according to Lewontin, only if one has faith in materialism. Materialism allows one to approach the natural world with the proper attitude, a scientific parallel to Anselm's famous "I believe in order to understand."⁵⁸ Lewontin advocates "materialism," but his

---

⁵³ Ibid., p. 13.   ⁵⁴ Ibid.   ⁵⁵ Ibid., pp. 12–13.
⁵⁶ Tertullian actually wrote: "... the Son of God died; it is by all means to be believed, because it is absurd"; *De Carne Christi* 5.
⁵⁷ Lewontin, "Billions and Billions of Demons."   ⁵⁸ *Proslogium*, 1.

approach to scientific method could just as well be supported by a more modest and restricted methodological reductionism. Though he is more frank than some authors, Lewontin is not alone in assuming that evolution carries metaphysical implications. As Don S. Browning notes, "The basic difference between theological anthropology and the anthropologies of other human societies is not that theology is morally and metaphysically freighted and the others are not; the difference is rather that theological anthropology takes responsibility for its moral and metaphysical judgments while many contemporary theologies do not."[59] What Browning says is true of theological anthropology and other sciences also applies to Christian ethics and sociobiology.

The methods of the natural sciences do not require ontological reductionism. Evolutionary ontological reductionists share with the "scientific creationists" a tendency not to distinguish properly scientific from philosophical claims. As we have seen, there is a difference between trying to use science to explain as much about phenomena as possible (methodological reductionism) and assuming that science alone has the ability fully to explain all phenomena (epistemological reductionism). There is also an important difference between employing science to discover whatever can be known about the material world (methodological reductionism) and assuming that the material world alone is real (ontological naturalism or ontological reductionism). Methodological reductionism does not require the further ontological belief that the only things that exist are to be found in the physical world.

My argument is that Christian ethics can accept the methodological reductionism of evolutionary theory while rejecting its epistemological and ontological reductionism. The main counter-argument to the position proposed here holds that methodological naturalism logically implies epistemological and ontological reductionism. Analytic philosopher Alvin Plantinga challenges the extensive epistemological credit given to modern science, and he takes exception to the claim of naturalists who hold that a broadly Darwinian point of view is more accurate than an account of nature provided by a theory of special creation.[60] Plantinga calls into

---

[59] Don S. Browning, "Christian Ethics and the Premoral Good," in *Christian Ethics and Moral Psychologies* (Grand Rapids, MI: Eerdmans, 2007), p. 202.

[60] Alvin Plantinga, "When Faith and Reason Clash: Evolution and the Bible," in David Hull and Michael Ruse, eds., *The Philosophy of Biology* (New York: Oxford University Press, 1998), p. 693. See also Alvin Plantinga, *Warrant and Proper Function* (New York: Oxford University Press, 1993) and James Beilby, ed., *Naturalism Defeated? Essays on Plantinga's Evolutionary Argument against Naturalism* (Ithaca, NY: Cornell University Press, 2002).

question the grounds of the "Grand Evolutionary Hypothesis" based on the following theses: (1) the earth is very old; (2) evolution is progressive; (3) common descent; (4) the evolution of life can be explained in purely natural ways; and (5) life has a natural cause.

Plantinga's primary strategy is to argue that scientific claims to knowledge are not so much based on evidence as on a blind faith in "methodological naturalism." He holds that there is no basis for assuming that the human cognitive capacities exercised in natural science are reliable or that naturalism is accurate. Plantinga argues that it is equally plausible to accept "theistic science" or "Augustinian science" that would combine hypotheses proposed by "methodological naturalism" with other hypotheses provided by "theistic science." If naturalistic science cannot explain some facet of life, he argues, then it is reasonable to appeal to divine agency. The application of these twofold criteria, Plantinga argues, shows the problematic nature of exclusively naturalistic explanations.

Yet Plantinga's attempt to combine evolutionary and theistic modes of explaining traits of organisms (like that of the "intelligent design" position) is self-defeating and self-contradictory. As Bowler explains, "The whole thrust of modern evolutionism has been to eliminate the need for a supernatural purpose in accounting for the present structure of living things."[61] Appealing to divine agency as a response to a scientific question does not yield any kind of scientific account of how x rather than y or z happens to be the case. Appeals to divine action do not explain anything in a scientific sense of the term. "Theistic science" suffers from vicious circularity. Plantinga argues that recognition of the incompleteness of methodological naturalism justifies admission of hypotheses from "theistic science," but there is no reason why it should. Failure to find a cure for a certain kind of cancer does not make it scientifically legitimate to seek the ministrations of a witch doctor. "Methodological naturalism" (or "methodological reductionism") has been fruitful for scientific inquiry into the origin of species, and "theistic science" has not. As the distinguished historian of science Owen Gingerich points out, those opposed to evolution "fail to understand that evolution offers biologists and paleontologists a coherent framework of understanding that links many wide-ranging elements, that it is persuasive, and that any critique of evolution will fall on stony ground unless it provides a more satisfactory explanation than evolution already

---

[61] Peter J. Bowler, *Evolution: The History of an Idea* (rev. edn., Berkeley: University of California Press, 1989), p. 6.

does."[62] Appeals to "theistic science" are not part of what is ordinarily meant by "science."[63]

A second, more popular criticism of evolutionary reductionism comes from anti-evolutionist Phillip Johnson's *Darwin on Trial*. Johnson is critical of any attempt to retain methodological reductionism detached from epistemological and ontological reductionism. He argues that if evolution is true, the natural world cannot have been created by a personal and all-powerful God and, furthermore, that human life cannot have any intrinsic meaning.[64] Since faith teaches that life has intrinsic meaning, Johnson reasons, evolution by natural selection cannot have occurred. He identifies "evolution" as a version of atheistic naturalism that attributes human life to chance. Though willing to accept naturalistic explanations in other domains, for example the passing of fluids through cell membranes, he insists that science must be prepared to acknowledge supernatural intervention in the natural world. Johnson invokes no alternative scientific account to the neo-Darwinian approach to the origin of species. When it comes to the existence of complex bodily structures, and even more, to the origin of life, he wants scientists to leave room for a "whimsical creator"[65] and God's "inscrutable purpose."[66]

Plantinga and Johnson treat the term "evolution" as always necessarily implying ontological reductionism or scientific materialism. Ironically, and for all their other considerable differences, they share this erroneous assumption with E. O. Wilson and Dawkins. According to theologian-physicist Ian Barbour, scientific materialism is characterized by two assertions, the first is epistemological and the second is ontological: "(1) the scientific method is the only reliable path to knowledge; (2) matter (or matter and energy) is the fundamental reality in the universe."[67] Barbour shows that neither of these is necessary for the conduct of scientific investigation or to evolutionary theory.

The next chapter expands on this point by arguing that the term "evolution" has at least four uses: (a) as a fact (i.e., it is the case that species have evolved), (b) as a course of natural history (i.e., how this or that species actually evolved), (c) as a scientific theory (involving common descent, the evolutionary history of particular organisms, and the mechanisms of evolutionary change), and (d) as scientific materialism (also known as

---

[62] Owen Gingerich, "Truth in Science: Proof, Persuasion, and the Galileo Affair," *Perspectives in Science and Christian Faith* 55.2 (June 2003): 87.
[63] See Ernan McMullin, "Evolution and Special Creation," *Zygon* 28.1 (September 1993): 299–335.
[64] P. E. Johnson, *Darwin on Trial*.   [65] Ibid., p. 31.   [66] Ibid., p. 71.
[67] Ian Barbour, *Religion in an Age of Science* (San Francisco: Harper, 1990), p. 4.

"ontological naturalism" or "ontological reductionism"). Some opponents of evolution, including Johnson, use the term in a global and unitary sense. They regard the first three uses as explicitly or at least implicitly entailing the fourth. Yet there is no philosophical or theological justification for a wholesale rejection of evolution; on the contrary, one is bound in intellectual honesty at least to come to terms with the basic discoveries of evolutionary biology and its allied disciplines. We need not repudiate all appeals to evolution in order to avoid ontological reductionism.

CHAPTER 4

Faith, creation, and evolution

This chapter examines three meanings of the term "evolution" and then turns to a more extended discussion of its theological interpretation. It concerns the theological perspective from within which Christian ethics responds to claims about human evolution.

THE MEANINGS OF EVOLUTION

The meanings of "evolution" include (1) the fact of evolution, (2) the scientific theory of evolution, and (3) ontological or other ideological extrapolations from evolution. This book will use the first two major senses of the term "evolution" and reject the third.

The National Academy of Science describes evolution as "the most important concept in modern biology, a concept essential to understanding key aspects of living things."[1] It is no longer possible, the National Academy points out, "to sustain scientifically the view that living beings did not evolve from earlier forms or that the human species was not produced by the same evolutionary mechanisms that apply to the rest of the living world."[2]

The term "evolution" is used in many ways. The most general use of the term is found in its etymological root, the Latin, *evolutio*, which means to unfold or develop from pre-existing structures.[3] In this very broad sense we talk about the evolution of painting, the evolution of democracy, cosmic evolution, and so forth. The pre-Darwinian use of the term usually

---

[1] National Academy of Science, *Teaching about Evolution and the Nature of Science* (Washington, DC: National Academy Press, 1998), p. viii.
[2] Ibid., p. 16.
[3] The treatment of evolution in this chapter, and throughout the book, is indebted to a number of very helpful texts. These include the various writings of Francisco J. Ayala, particularly "Biological Evolution: An Introduction," in James B. Miller, ed., *An Evolving Dialogue: Scientific, Historical, Philosophical, and Theological Perspectives on Evolution* (Washington, DC: American Association for the Advancement of Science, 1998), pp. 9–54.

Evolution as fact

One biological use of the term "evolution" refers to the fact that all forms of life as we know them today have descended from pre-existent life forms. The first multi-celled organisms appeared around 670 million years ago, vertebrates 490 million years ago, reptiles 310 million years ago, mammals 200 million years ago, the earliest primates 60 million years ago, and the first apes 25 million years ago.[4] The first hominids, or human-like bipedal species, the australopithecines, appeared around 4.4 million years ago,[5] and anatomically modern *Homo sapiens* came on the scene around 100,000 years ago. The distinguished evolutionary biologist Francisco Ayala, among many others, describes the evolutionary origin of species as a

> scientific conclusion established with the kind of certainty attributable to such scientific concepts as the roundness of the Earth, the motions of the planets, and the molecular composition of matter. This degree of certainty beyond reasonable doubt is what is implied when biologists say that evolution is a "fact"; the evolutionary origin of organisms is accepted by virtually every biologist.[6]

Evolution as theory

A second biological use of the term "evolution" refers more precisely to the *theory* of the origin of species via "descent with modification" proposed by Darwin and developed by his scientific heirs. According to Ernst Mayr, the "first Darwinian revolution" contained five theoretical components or sub-theories: the fact of evolution as such (i.e. the transformation of species), multiplication of species, common descent, gradualism, and natural selection.[7] Some of these claims can stand independently of the others, but Darwin held all of them together as necessary aspects of the comprehensive scientific explanation of the origin of species. Observation of nature, Darwin thought, shows that "favorable variations would tend to

---

[4] These estimates are in ibid., p. 22.
[5] What follows relies upon Ian Tattersall, "Human Evolution," in Miller, ed., *An Evolving Dialogue*, pp. 99–210.
[6] Ayala, "Biological Evolution," in Miller, ed., *An Evolving Dialogue*, p. 18.
[7] What follows depends on Mayr, *One Long Argument*, ch. 2.

be preserved, and unfavorable ones to be destroyed."[8] He understood, Mayr explains, that evolutionary change is "due to the production of variation in a population and the survival and reproductive success ('selection') of some of these variants." Darwin's key contribution was to claim that the origin of species lies in "natural selection" working on biological variations in the species, not in the direct intervention of a divine Creator.[9] Darwin of course understood "natural selection" to be a metaphor that expresses that nature works something like animal breeders; nature is incapable of consciously "selecting" anything.

As he worked on the theory of natural selection, Darwin gradually changed his mind about other significant matters, some scientific and some religious. First, he replaced typological fixity according to which each member of a species is basically similar to every other member of the same species, with what is now called "population thinking,"[10] which attends to the differences between individuals within a given population and sees variation as an essential mark of any given population. Second, Darwin became convinced that nature is not appropriately conceived as a static or even stable order. Balances in the natural world are always potentially changeable. Third, Darwin moved from believing that inheritance was influenced by the environment of the organism shaping inherited traits (Lamarck's "inheritance of acquired characteristics") to the thesis of "hard inheritance," the belief that inherited traits are not directly influenced by use or nonuse (e.g., the blacksmith's son does not inherit bigger arms because of his father's exercise).[11] Finally, Darwin moved from conventional Christianity to religious agnosticism. Though he long felt that adherence to doctrinal orthodoxy impedes science, his gradual embrace of

---

[8] *The Autobiography of Charles Darwin*, ed. Nora Barlow (London: Collins, 1958), p. 120; cited in Mayr, *One Long Argument*, p. 70.

[9] Mayr points out that the term "Darwinism" has been used in multiple and often inconsistent ways. He identifies nine major uses of the term, most of which are misleading or inaccurate. See *One Long Argument*, ch. 7. Mayr proposes that the most meaningful nineteenth-century use of term refers to "explaining the living world by natural processes" and the most meaningful recent use concerns "adaptive evolutionary change under the influence of natural selection, and variational instead of transformational evolution"; ibid., p. 106. By "variational" evolution he means the "orderly change in a lineage over time, directed toward the goal of perfect adaptation" (ibid., p. 93).

[10] Bowler argues that *laissez-faire* individualism is "almost certainly reflected in Darwin's decision to treat biological species as a population of diverse organisms rather than copies of an ideal type." Conway Zirkle shows that the notion of a complete fixity of species began in the eighteenth century and became prevalent in the nineteenth. See Conway Zirkle, "Species before Darwin," *Proceedings of the American Philosophical Society* 103.5 (October 15, 1959): 636–644.

[11] Mayr, *One Long Argument*, p. 110. Historians, however, think that Darwin tended to be ambiguous on this issue and never completely rejected the Lamarckian thesis that acquired characteristics can be inherited.

natural selection, in tandem with events in his personal life, was accompanied by a growing rejection of Christian theism, especially belief in a provident Creator.[12] His *Autobiography* expressed doubt that a natural order so pervaded by waste and pain could have been the creation of a benevolent and all-powerful God.[13]

Today, the theory of evolution is the scene of a variety of significant disputes. Niles Eldredge and the late Stephen Jay Gould's "punctuated equilibrium" offers an account of the irregular nature of evolutionary change against the dominant neo-Darwinian gradualism;[14] the theory of "self-organization" proposed by Stuart Kaufman and his colleagues regards as inadequate the "orthodox" neo-Darwinian focus on random mutation and natural selection;[15] the holism of Steven Rose and his colleagues corrects the gene-selection emphasis that predominates in sociobiology;[16] the revived "group selection" thinking of Elliot Sober and David Sloan Wilson expands evolutionary theory to embrace not only the gene and individual organism but also the group;[17] finally, paleobiologist Simon Conway Morris advances a theory of convergent evolution according to which the emergence of human intelligence is nearly inevitable.[18] The diversity of these positions constitutes movements within evolutionary theory, and not a wholesale repudiation of the role of natural selection and variations within the origin of species.

A sharp distinction between "fact" and "theory" has been invoked by religious objectors to evolution for some time, and with decreasing levels of plausibility. Fundamentalist and fiercely anti-Darwinian crusader William Bell Riley (1861–1947) held that it is an "unquestioned fact that evolution is not a science; it is hypothesis only, a speculation."[19] William Jennings Bryan made the same argument in the last century: "Evolution is not truth; it is merely a hypothesis – it is millions of guesses strung together."[20]

---

[12] See Gillespie, *Charles Darwin and the Problem of Creation*. See also E. Janet Browne's magisterial two-volume work: *Charles Darwin: Voyaging* (Princeton: Princeton University Press, 1996) and *Charles Darwin: The Power of Place* (New York: Knopf, 2002); and Adrian Desmond and James Moore, *Darwin: The Life of a Tormented Evolutionist* (New York: W. W. Norton, 1994).
[13] See *Autobiography of Charles Darwin*, ed. Barlow.
[14] See Stephen J. Gould and Niles Eldredge, "Punctuated Equilibria: The Tempo and Mode of Evolution Reconsidered," *Paleobiology* 3 (1977): 115–151.
[15] See Stuart Kaufman, *At Home in the Universe* (New York: Oxford University Press, 1995).
[16] See Rose, *Lifelines*.
[17] See Elliot Sober and David Sloan Wilson, *Unto Others: The Evolution and Psychology of Unselfish Behavior* (Cambridge, MA and London: Harvard University Press, 1998).
[18] Conway Morris, *Life's Solution*.
[19] Cited in Ronald R. Numbers, *The Creationists: The Evolution of Scientific Creationism* (Berkeley: University of California Press, 1993), p. 50.
[20] Cited in Edward J. Larson, *Summer for the Gods: The Scopes Trial and America's Continuing Debate over Science and Religion* (San Francisco: Basic Books, 1997), p. 7.

The dismissal of evolution as a mere "theory" is based on a great oversimplification of the fact–theory relation as well as on a misunderstanding of how science functions. The common-sense use of the word "theory" means something like an unsubstantiated opinion, guesswork, or conjectural speculation unsupported by empirical evidence. If evolution is a theory in this sense, the argument runs, then scientists have not proven it to be true.[21]

This criticism rests on an inadequate notion of how scientists understand their own task. According to physicist George Ellis, scientific theories are usually evaluated in light of four standards: simplicity, beauty, accuracy in prediction and verifiability, and explanatory power – a capacity for giving the most adequate account of problematic data.[22] Along similar lines, philosopher of science Ernan McMullin offers six criteria as "epistemic values" in the analysis of the scientific status of a given theory: predictive accuracy, internal coherence, external consistency, unifying power, fertility, and simplicity.[23] Slightly modifying McMullin, Van Till argues for a similar set of criteria: cognitive relevance, predictive accuracy that a given physical system should exhibit, internal coherence, explanatory scope, unifying power, and fertility.[24]

Theory assessment varies from one science to another. As a "historical science," evolutionary biology is not suited to testing predictions, but it meets the other criteria. As Ellis explains: "The upshot is that in the case of historical sciences, an element of interpretation is implied in any theory ... There is no way to avoid this. We minimize this as far as possible by demanding consistency with present day scientific theory and understanding, together with the requirement of broad explanatory power."[25] The best scientific theory is not one that provides irrefutable "proof," but rather one

---

[21] Richard John Neuhaus, "A Continuing Survey of Religion and Public Life," *First Things* 69 (1997): 56–70, which suggests that evolutionists themselves claim that evolution is a fact, not a theory, and, further, that John Paul II considered "intelligent design" a theory of evolution.

[22] See Ellis, "The Thinking Underlying the New 'Scientific' World-Views," pp. 251–280. Ayala offers four similar criteria: internal consistency, explanatory value, consistency with commonly accepted scientific knowledge, and coherence between predictions and observations. See Ayala, "Darwin's Devolution," pp. 113–114, n. 9. Finally, Barbour proposes that four criteria for successful science include agreement with data, scope, coherence, and fertility. See Ian Barbour, *Religion and Science: Historical and Contemporary Issues* (San Francisco: HarperCollins, 1998), p. 113.

[23] See Ernan McMullin, "Values in Science," in Peter D. Asquith and Thomas Nikles, eds., *PSA 1982: Proceedings of the 1982 Biennial Meeting of the Philosophy of Science Association*, vol. II (East Lansing, MI: Philosophy of Science Association, 1982), pp. 1–25.

[24] Howard J. Van Till, "The Character of Contemporary Natural Science," in Howard J. Van Till et al., *Portraits of Creation: Biblical and Scientific Perspectives on the World's Creation* (Grand Rapids, MI: Eerdmans, 1990), ch. 5, pp. 141–145.

[25] George F. R. Ellis, *Before the Beginning: Cosmology Explained* (London and New York: Boyars/Bowerdean, 1993), p. 17.

that gives the most plausible account of all the relevant data. "Evolutionary theory" is thus not opposed to "fact," but rather refers to a broad and well-established scientific explanation for data received from various sources such as the fossil record, genetics, comparative anatomy, and physiology. In these terms the "theory of evolution" has been demonstrated to have broad explanatory power. Evolution is the unifying theory of the biological sciences and the indispensable key to understanding the proliferation of life. As the great geneticist Theodosius Dobzhansky put it, "Nothing in biology makes sense except in the light of evolution."[26]

Pope John Paul II pointed out that faith in creation and belief in evolution are not mutually contradictory as long as each is properly understood: "Evolution presupposes creation: creation in the light of evolution is a temporally expansive event – *creatio continua* – in which God is visible to the eyes of faith as the 'creator of the heavens and the earth.'"[27] The first Genesis account of creation, including the sixth-day text, is a religious vision presented in a form that is accessible to us. According to the pope, "In no way do these statements contradict a general world evolution, if evolution is limited to naturally provable findings . . ."

At his famous speech before the Papal Academy (October 22, 1996) on "The Origin of Life and Evolution," John Paul II moved from Pope Pius XII's claim that evolution could be "taken seriously as a hypothesis" to the less circumspect if still modest affirmation that "it is to be seen as more than a hypothesis."[28] He acknowledged that the theory of evolution is widely accepted by scientists from a variety of different fields. He also argued that we ought to speak of "theories of evolution" rather than of *the* "theory of evolution," since the variety of evolutionary "worldviews" often involves more philosophy, and sometimes ideology, than science. The pope's teaching is that science can be used to explain the evolution of the body as long as the direct and independent creation of the soul by God is acknowledged. He explained that "It is by virtue of his spiritual soul that the whole person possesses such a dignity even in his body."[29] This position

---

[26] *The American Biology Teacher* 35 (March 1973): 125–129.
[27] April 26, 1985, cited in Rainer Koltermann, "Evolution, Creation, and Church Documents," *Theology Digest* 48.2 (2001): 127.
[28] See Pope John Paul II, "Message to the Pontifical Academy of Sciences on Evolution": 349–352. See more recently the July 2004 Vatican Statement on Creation and Evolution, "Communion and Stewardship: Human Persons Created in the Image of God," *Origins* 34.15 (September 23, 2004): 233–248.
[29] John Paul II, "Message to the Pontifical Academy of Sciences on Evolution": 352, no. 5. Wolfhart Pannenberg challenges this Platonizing interpretation that suggests an independently existing soul is inconsistent with the Hebrew concept of "*nephesh*," the spirit that is seen in the constant hunger

fits the pope's exegesis of Genesis 2:7, where God breathes life into the man, the basis of the patristic theory that human beings propagate the bodies of their children and God directly inserts souls into the bodies of the unborn. This anthropology provides a high doctrine of free will and human dignity, but proposes a supernatural ontogenetic alternative to the Darwinian naturalistic account of the origin of our moral capacities.

Evolution as ontology

The pope's caveat indicates that a third use of the term can be distinguished from evolution as fact and as theory. The infamous ideological use of evolution was the social Darwinian construction of a moral and social theory based on "survival of the fittest." This term was coined by Herbert Spencer, not Darwin; Darwin himself was not an advocate of "social Darwinism."[30] Drawing an analogy between the natural order and the social order, social Darwinians in the nineteenth century argued that for the good of the society, government should not interfere with the workings of natural social processes that tend to eliminate the "unfit" and reward the most "fit." Progress occurs as a result of natural conflict. Individuals struggle to exist in nature; they became dominant by the intelligent use of technology, and especially weapons; modern society allows for the use of capital to expand this power; and the best social arrangement allows maximum liberty to social action so that the most "fit" individuals will rise above their less "fit" competition. This ideological use of evolution obviously represents the antithesis of Christian ethics and has drawn sharp and sustained criticism from Christian moralists (unfortunately it has also occasionally been adopted by other Christian moralists with a less acute sense of justice).

Reductionistic ontologies are often grouped under the label of "naturalism," a term that has multiple meanings and that is used in philosophy of science, epistemology, metaphysics, and ethics. In the broadest possible sense of the term, advocates of "naturalism" hold that the natural world is the whole of reality and that reliable knowledge is about it rather than

---

for life in each person. For Pannenberg, Gen. 2:7 is better interpreted as indicating that God animates both soul and body. See "Human Life: Creation versus Evolution?," in Ted Peters, ed., *Science and Theology: The New Consonance* (Cambridge, MA: Westview, 1998), p. 143.

[30] Interpretations of Darwin's political views have run a broad spectrum from right to left. See Michael Ruse, "Darwinian Evolutionary Ethics: between Patriotism and Sympathy," in Philip Clayton and Jeffrey Schloss, eds., *Evolutionary Ethics: Human Morality in Biological and Religious Perspectives* (Grand Rapids, MI: Eerdmans, 2004), p. 57, and Stephen R. L. Clark, *Biology and Christian Ethics* (Cambridge: Cambridge University Press, 2000), ch. 2.

about spiritual entities or supernatural powers. The philosopher Peter F. Strawson distinguishes "soft" or "non-reductive" from "hard" or "reductive" naturalism.[31] The former concerns our common-sense experience of things in the world, for example the warm breeze, the hard table top, the rising sun. Naturalism in this sense does not appeal to nonnatural factors as explanatory causes of the phenomena that we experience in the course of living our lives: no magical witchdoctor healings, no miraculous cures by patron saints, no demonic possessions, just the world as we (moderns) experience it.

"Hard" or "reductive" naturalism approaches events in nature in a detached and objective perspective. The term "naturalism" in this sense is often used interchangeably with "scientific materialism," which claims that matter (or matter and energy) is the "fundamental reality" in the universe and that the methods of the natural sciences provide the only reliable way to know about matter.[32] "Soft" and "hard" naturalistic approaches to reality are compatible as long as they are assigned competence over different aspects of reality: one, the world as encountered by a perceiver; the other, natural events studied by scientific methods. The "reductive" or "hard" naturalist, however, maintains not only that the latter explains the way that common sense perceives the world the way it does and corrects the former in cases when they conflict about the character of natural events, but also that it presents a more reliable and accurate view of reality as such (though of course, always with a "fallibilist" proviso that leaves it open to scientific correction and revision).

In an evolutionary context, hard reductive naturalism tries to show that we perceive the world the way we do in terms of the adaptive advantages of our cognitive, social, and perceptual capacities. Evolutionists argue that some opinions which common sense takes to be obviously true are illusions, but that others can be retained as functionally appropriate within the intuitive framework of everyday life. They might say, for example, that it is an illusion to think of a strong and certain judgment of one's conscience as the "voice of God" or a "divine command" (possibilities ruled out by their ontological naturalism), but that we should still take seriously the moral authority of conscience on other grounds (because it signals, for example, the requirements of honor, or of promises we ought to keep).

---

[31] P. F. Strawson, *Skepticism and Naturalism: Some Varieties* (New York: Columbia University Press, 1985), pp. 38ff. Strawson himself rejects reductive naturalism.
[32] See Barbour, *Religion in an Age of Science*, p. 4.

It is important to recognize that ontological naturalism goes beyond the findings of the natural sciences in its attempt to provide a general account of reality as a whole and of our place in it. Unlike major philosophical systems, however, evolutionary ontological naturalism typically avoids extended chains of philosophical reasoning and instead tends to take for granted its major metaphysical commitments (presuming them to be obviously true) and/or claim that certain metaphysical judgments (about the soul, "free will," etc.) are fairly clearly implied by the findings of evolutionary biology or its allied disciplines. Dennett's depiction of Darwinism as a "universal acid" exemplifies the reductive naturalist agenda to eliminate any claims from religious or moral traditions that do not comport with their view of what is real. Evolutionary ontological naturalism strives to trim down our metaphysical affirmations to what can be justified by the natural sciences. Theological naturalism, as we will see in Gustafson and Peacocke, takes the same approach to warranted theological claim.

The alternative epistemological position relied upon in this book holds that there will always be some phenomena in the universe that cannot be explained exhaustively in scientific terms. This is especially the case for attempts to provide a complete scientific explanation of human beings. The anti-naturalist ontological position holds that physical reality (matter and energy) is the most elementary aspect of reality but is neither the only existing nor the most significant reality. Christian ethics is committed to the claim that while human experience as we know it is always dependent upon underlying material states, the most meaningful dimensions of reality are irreducible to claims about matter.

Physicist Howard Van Till distinguishes evolution itself from ontologies and ethics that fall into the broad category of "evolution*ism*."[33] This book follows Van Till's use of the term "evolutionism" as the ideology of "those who indenture the scientific concept of evolution in the service of their naturalistic creed."[34] Christians reject this naturalism, he writes, "because its materialist creed puts the material world in place of God, because it asserts that the cosmos is self-existent and self-governing ... because it asserts that cosmic history has no purpose, that the idea of purpose is only an illusion."[35]

As a physicist, Van Till is concerned with the evolution of the universe, but most of what he writes about cosmic evolution can also be applied directly to speculation based on biological evolution. It might be helpful at this point to note Van Till's objections to "evolutionism" and his own

---

[33] Van Till et al., *Portraits of Creation*, p. 121.   [34] Ibid.   [35] Ibid.

alternative position. He argues that scientific accounts of biological evolution are by definition silent on matters of morality and ontology. Scientific cosmology, similarly, can furnish an explanation of how stars are formed but not of why the cosmos exists in the first place. Van Till is particularly critical of the unsubstantiated evolutionist assumption that we must consider scientific accounts of evolutionary processes as competing with or replacing the theological notion of divine governance. For Van Till, divine governance works in and through natural processes, not outside them. He insists, furthermore, that evolution cannot fairly be depicted as the product of sheer chance alone but rather proceeds through the interaction of contingencies and law-like regularities that can be interpreted theologically as allowing for the fulfillment of divine purposes.[36]

Evolutionism advances an ontological as well as a scientific agenda. The passionate faith with which some naturalists cling to their positions, along with the thoroughness with which they employ naturalism as a "world picture" and as an explanatory principle for all areas of human life, has led Mary Midgley to criticize them for promoting "evolution as religion."[37] Yet despite these distortions and misuses of evolutionary conceptualities, we need to keep in mind the old scholastic axiom, "an abuse of a thing does not prohibit its proper use" (*abusus non tollit usus*). Outrage at social Darwinism, or latter-day evolutionary ideologies, should not incite a complete repudiation of all claims about evolution. The other meanings of evolution, particularly as fact and as theory, do not represent such an immediate and obvious offense to Christian ethics.

EVOLUTION AND CHRISTIAN FAITH

All systematic approaches to Christian ethics have to deal with three central issues: the nature of the good, the nature of the moral agent, and the criteria for determining the rightness or wrongfulness of acts.[38] Ethicists who strive for a reasonable degree of comprehensiveness are forced to address all three issues and can be distinguished with regard to how they understand which of these issues ought to be given the greatest weight. Christian ethicists must also address what Gustafson calls the "theological base," that is, "the understanding and interpretation of God, the ultimate power and value,

---

[36] Ibid., pp. 121–125.
[37] See Mary Midgley, *Evolution as a Religion: Strange Hopes and Stranger Fears* (London and New York: Methuen, 1985).
[38] See James M. Gustafson, *Christian Ethics and the Community* (New York: Pilgrim Press, 1979 [1971]), esp. ch. 2, and *Theology and Christian Ethics* (Philadelphia: Pilgrim Press, 1974).

his relations to the world, and particularly to humans, and his purposes, both 'moral' and 'nonmoral.'"[39] The central "base point" for Christian ethics is theological and envisions God as both calling human beings to goodness and promoting their good. This is not to say that the human good is *the* central focus of divine activity – a claim that would be idolatrous because it suggests that the world exists ultimately for the sake of humanity rather than for God.[40] At the same time, there is no reason to posit an absolute dichotomy between the glory of God and human flourishing, properly understood. Indeed, according to St. Irenaeus the latter contributes to the former: "the glory of God is a living man; and the life of man consists in beholding God."[41]

I give the highest priority among the three ethical "base points" to the "teleological" concerns and therefore to our conception of the human good. This approach allows the argument to maintain a theological center of gravity while being open to insights from perspectives not directly found in revelation or typically considered by Christian ethicists.

Christian faith

In Christian ethics the normative response to the offer of divine grace lies in the threefold "theological virtues" of faith, hope, and charity. "Faith" here will be taken to be primarily a matter of trustfully believing God and, secondarily, believing in the truth about God and humanity given in revelation.[42] "Hope" is the theological virtue that moves toward a goal that is difficult but not impossible to attain, namely, eternal union with God. "Charity" is the grace-infused love of God and love of neighbor "in God."

Three important facets of Christian faith need to be mentioned. First, Christian faith involves a deep personal response to a personal God. The primary function of religious language lies in speaking *to* God rather than *about* God. Religious faith affirms the existence of God as an intellectual conviction, but it also, and more profoundly, responds to God in worship,

---

[39] James M. Gustafson, *Protestant and Roman Catholic Ethics: Prospects for Rapprochement* (Chicago: University of Chicago Press, 1978), p. 140.
[40] Gustafson concludes that "[m]y argument radically qualifies the traditional Christian claim that the ultimate power seeks the human good as its central focus of activity"; *Ethics from a Theocentric Perspective*, vol. I, p. 271.
[41] "Irenaeus against Heresies," in Alexander Roberts and James Donaldson, eds., *The Ante-Nicene Fathers* (Edinburgh: T. & T. Clark, and Grand Rapids, MI: Eerdmans, 1989), p. 490. See also Jn 10:10; 6:48; 7:8; Acts 5:20, etc.
[42] See *ST* II-II, 1,2 ad 2.

prayer, and devotion. The deeper reference of faith involves a willingness to respond with acceptance to God's offer of fellowship. The Christian moral life flows from and leads to a communion with the divine, a point missed by evolutionists who think that reflection on God is concerned primarily with explaining the origin of species, the structure of nature, or the binding power of moral norms.

Second, Christian faith has important social dimensions in that it is nurtured, expressed, and challenged to develop within the faith community. Christian faith is not reducible to private feelings or individual religious convictions. Faith is essentially and profoundly a life in community, not only an individual quest or solitary enterprise. The faith of the individual Christian is always shared with all other Christians and with the whole church, understood as past (the "apostolic tradition"), present (the tangible "Body of Christ" in the world), and future (the coming "Kingdom of God"). It involves the living church with all its flaws, weaknesses, and sins, and also the church that dwells with God in eternal life (the "communion of saints").

Christianity, as much as Hinduism, Islam, and other major "world religions," is a large-scale cultural reality with deep relevance to the cultures within which it thrives. The attempt to limit religion to the private sphere is a modern endeavor that has its roots in the political and moral response to the Thirty Years War.[43] The desire to establish peace on rational foundations led to the devaluation of the particular narratives that bind together concrete communities and traditions and to replacing them with religion based on reason alone or to reducing religion and religious morality to a public moral code that would be acceptable to all "rational citizens." As we saw in the first chapter, this project reaches one expression in E. O. Wilson's attempt to construct an evolutionary ethic, based on the "epic of evolution," that is meant to give a scientific alternative to the fractious particularity of what he considers the parochial, intellectually exhausted, and morally impoverished monotheistic religions of the west.[44]

Third, Christian faith makes cognitive claims grounded in "divine revelation." The term "revelation" refers primarily to the process of God's self-manifestation and self-communication to human beings. The tradition distinguishes between "natural" revelation given in the order of nature and the voice of the individual conscience and the "historical" revelation to individuals and communities, transmitted through Scripture and tradition, and interpreted by the church (this sense of revelation is used in Jude 3; 1 Tim. 6:20; 2

---

[43] Lash, *The Beginning and the End of "Religion."* [44] See E. O. Wilson, *Consilience*.

Tim. 1:12, 14). The self-communication of God takes place within history and in a way that contains what Rahner calls a "dynamism toward its own objectification."[45] Thus the term "revelation" refers not only to the process of God's self-disclosure to human beings but also to the wisdom that is gained by the believing community that results from this process: the articulation of this wisdom in the form of "public, official, particular, and ecclesially constituted revelation and its history."[46]

Trust in God leads to trust in, and acceptance of, God's revelation in the Scriptures and Christian tradition, the center of which is the self-disclosure of God's purpose to bring human beings into fellowship with God and even to "become sharers in the divine nature."[47] Faith assents to certain beliefs grounded in Scripture and formulated and interpreted by the tradition. The classic Christian creeds of the patristic era articulate the essential commitments shared by members of the Christian community. Christian theology relies on a body of authoritative teaching that does not need to be given prior validation by reason, philosophy, or natural science. Yet this reliance need neither be naïvely uncritical about the challenges of interpretation nor underestimate the relevance of other disciplines to Christian ethics. Revelation takes place in historical circumstances and is communicated in particular languages and through the symbolic resources made available by particular cultures. Theology provides a vantage point that other disciples do not have, but it can be enhanced, informed, and developed by critically appropriating insights and information made available by non-theological disciplines.

Faith recognizes a kind of "natural" or "general" revelation as well as the historically particular revelation given in the Christian Scriptures and tradition. "Natural" revelation is seen in the ordering of the natural world (e.g., as described in Paul's speech at Lystra in Acts 14:15–17) and in the inner prompting of the inspired human conscience (usually seen as indicated in Paul's recognition of Gentiles who "do instinctively what the law requires," Rom. 2:14).[48] The "special" or "historical" revelation given first to Israel and then also to the New Testament communities offers a much more clear avenue to Christian faith. Divine purposes are displayed with special poignancy and clarity in the history of God's interactions with Israel, and

---

[45] Rahner, *Foundations of the Christian Faith*, p. 173. [46] Ibid., p. 174.
[47] *Dei Verbum* no. 1, in Austin Flannery, OP, ed., *Vatican Council II* (rev. trans., Northport, NY and Dublin: Costello Publishing Company, 1996), p. 98.
[48] "God, who creates and conserves all things by his Word (see Jn 1:3), provides constant evidence of himself in created realities (see Rom. 1:19–20)"; *Dei Verbum* no. 3, in Flannery, *Vatican Council II*, p. 98.

Faith, creation, and evolution

also, and most powerfully, in the life, death, and resurrection of Jesus and in the continued work of the Holy Spirit in the church and throughout the world.

Gustafson properly warns us that theology becomes distorted, as it often did in the twentieth century, when it focuses exclusively on the presence of God in history and ignores the role of God in the natural world. God is both the Lord of history – "Did I not bring Israel up from the land of Egypt, and the Philistines from Caphtor and the Arameans from Kir?" (Amos 9:7) – and the source of the universal order of creation – "The one who made the Pleiades and Orion, and turns deep darkness into the morning, and darkens the day into night; who calls for the waters of the sea, and pours them out on the surface of the earth; the Lord is his name, who makes destruction flash out against the strong, so that destruction comes upon the fortress" (Amos 5:8–9). God's immanence is operative in nature as well as in history, and God's transcendence can be interpreted in light of history as well as nature. The affirmation of one contributes to a deeper appreciation of the latter. Knowledge of human evolution, then, can be understood as allowing us a deeper grasp of both God's presence in the world and God's sovereign transcendence of it.

Knowledge of nature can enhance rather than undermine our sense of both immanent and transcendent dimensions of God's relation to the world. The Christian tradition has long recognized awareness of nature as relevant to theology and morality. Direct experiences of nature can inspire in us a sense of dependence, awe, and gratitude in relation to the Creator, the "Maker of heaven and earth." Gustafson speaks eloquently of the sense of dependence evoked by nature:

> We did not create the fundamental conditions for animal life and finally human life to develop; we do not create ourselves (even in-vitro fertilization is a technology which manipulates ova and sperm to enable an embryo to come into being); we have not determined our own fundamental genetic endowments; we are mortal and bound to die, though our dying can be prolonged by medical technology; we cannot fully control our individual destinies but are subject to constraints and possibilities objective to ourselves. All this evokes profound religious, affective sensibilities.[49]

Knowledge of evolution enhances our awareness of dependence on nature. It can also deepen our awe at the majesty of creation, the power of the Creator, the finitude and frailty of life, our connection to other

---

[49] Gustafson, *A Sense of the Divine: The Natural Environment from a Theocentric Perspective* (Cleveland: Pilgrim Press, 1994), p. 101.

creatures, and so forth. It can also highlight the special place that humanity has in the care of the Creator – "When I look at your heavens, the work of your fingers, the moon and stars that you have established; what are human beings that you are mindful of them, mortals that you care for them?" (Ps. 8:3–4). Knowledge of nature can also help us to understand more deeply the action of the Son of God, as the creed puts it, "through whom all things were made," and the Holy Spirit, "the Lord and Giver of Life." Scientific knowledge can enhance our understanding of some aspects of what is maintained in Christian doctrine and affirmed in religious faith. As Jesuit cosmologist William B. Stoeger, SJ, points out, "material reality is on every level more vast, more intricate in its structure and development, more amazing in its evolution, in its variety flowing from fundamental levels of unity, and in its balance of functions, than we could ever have imagined without the contributions of the sciences."[50]

Christian faith and human evolution

Christian faith affirms God as Triune, as Father, Son, and Spirit existing eternally in relations of knowing and loving. It also affirms God as Creator and Redeemer of the world. As Creator, God is the cause of the existence of the world, and the cause of its basic structures, patterns, and law-like regularities.

The Christian creed professes the Father as "Maker of heaven and earth," but all three "persons" are involved in the work of creation. Christian faith professes belief in the creation of all things by the Creator, but, as Thomas Aquinas observed, the details concerning the "manner and order" of this creation are not essential to faith.[51] The Son, "through whom all things were made," constitutes the highest meaning of creation as well as the definitive expression of divine wisdom. The Spirit, "the Lord and Giver of life," pervades the created order, offers divine grace to every life, and inspires the radical transformation of every person and community. As Redeemer, God brings the world to the goodness of which it is capable.

God is both immanent – pervading the entire created order from the micro- to the macro-level – and transcendent – existing eternally,

---

[50] William R. Stoeger, SJ, "Contemporary Cosmology and Its Implications for the Science–Religion Dialogue," in Robert John Russell, William R. Stoeger, SJ, and George V. Coyne, SJ, eds., *Physics, Philosophy, and Theology: A Common Quest for Understanding* (Notre Dame, IN: University of Notre Dame Press, and Vatican City State: Vatican Observatory Publications, 1988), p. 240.
[51] For the general distinction between what of Scripture is essential to faith and what is not, see *ST* II-II, 1,6, ad 1.

continually sustaining the world in being and ordering it through the inherent structures of nature. The doctrine of creation professes God as the cause of all that exists, as continually sustaining the whole world in existence. The doctrine of providence refers to God's governance, ordering, and caring for creation.

God orders the world through the evolutionary process. Ayala notes that "[t]here is no more reason to consider anti-Christian Darwin's theory of evolution and explanation of design than to consider anti-Christian Newton's laws of motion. Divine action in the universe must be sought in ways other than those that postulate it as the means to account for gaps in the scientific account of the workings of the universe."[52]

Seen in this context, Christian ethics engages in disciplined analysis and reflection on the moral life of individuals and communities in light of the Christian message. Christians, like all human beings, are called by God to become what they are created to be: flourishing human beings. This flourishing constitutes a fulfillment of our created natures, and it is promoted through a forgiveness of sin and a healing of our sinfulness. The most powerful human flourishing is achieved when God's grace works to transform the heart and mind of the human person and the human community. In classical scholastic language, human beings are not only forgiven and healed but also "elevated" beyond their purely natural capacities.[53] The Christian moral life, then, is one based on a transformation in grace that brings out what is best and even most God-like in the human person. And as Rahner explains, even the highest degree of transformation

does not leave behind or reject the corporeality of the one being of man as if it were something non-essential ... The whole man (and hence also his spirit) will be changed. The whole man (and hence also his corporeality) will be saved. That we cannot imagine the preserved salvation of the bodily man is not surprising: the whole glorified man is withdrawn from us in the absolute mystery of God.[54]

As Thomas Aquinas put it, grace "perfects" rather than destroys or replaces human nature.[55] The authentically Christian moral life is one that is more deeply human than a life marked by infidelity, deception, and injustice. Grace moves the person toward the highest flourishing possible for a human being, and since our end is eternal it relativizes the

---

[52] Ayala, "Darwin's Devolution," p. 113.   [53] *ST* I-II, 109,2.
[54] Karl Rahner, "The Secret of Life," trans. Karl-H. Kruger and Boniface Kruger in *Theological Investigations*, vol. VI: *Concerning Vatican Council II* (New York: Seabury Press and London: Darton, Longman and Todd, 1974), p. 152.
[55] See *ST* I, 62,5.

merely biological good. At the same time, the movement toward the greatest flourishing in this life involves not only grace but also a human nature that has been shaped by millions of years of evolutionary history.

The Christian moral life is structured by the human good, the ultimate end of life in which is found the highest form of flourishing. As William Spohn explained, "The gospel is not the only norm, as shown in Paul's frequent appeals to human values, practical logic, common secular wisdom, and the like."[56] Paul and other New Testament texts (and also Hebrew Bible texts) presume that "general human standards" are normative. At the same time, Spohn emphasized, "The life of Jesus as related in the Gospels ... is the fundamental norm for Christians."[57] This means that the human good in its deepest meaning is pursued through following Christ and striving to be Christ-like. Christian identity is shaped by the virtues, taught by communities, displayed in the acts of moral exemplars, and communicated in paradigmatic narratives upheld by the Christian tradition. The Christian moral life involves the ongoing development, formation, conversion, and habituation of the person in ways that allow for the incorporation of the virtues in his or her life; in an analogous way, the same is true of vital and authentic Christian communities.

Christian life is also shaped by the "moral law" – by norms, rules, principles, or standards that inform decision-making, the formation of intentions, and moral judgment regarding what to do in concrete situations. Moral law and virtues thus ideally exist in spiral relation to one another.[58] Moral standards reinforce virtues, and virtues extend our understanding of moral standards.

The moral law itself comprises the principles and rules of action expressed in the gospel, the Christian tradition, and the contemporary church. It includes not only distinctively Christian teachings, but also moral insights found in reason, philosophy, and human experience broadly construed. This array of teachings includes what the tradition calls the "natural law." The natural law is ordered with regard to human flourishing, and its standards are intended to shape human action in ways that allow the agent to live the "good life" as it is understood in Christian terms. The natural-law tradition, in other words, understands moral claims to be warranted by their contribution to human flourishing. Since it concerns the flourishing of human

---

[56] William C. Spohn, *Go and Do Likewise: Jesus and Ethics* (New York: Continuum, 1999), p. 11. The spiral image is preferred by James Keenan, *Moral Wisdom: Lessons and Texts from the Catholic Tradition* (New York: Sheed and Ward, 2003), and Thomas R. Kopfensteiner, "The Metaphorical Structure of Normativity," *Theological Studies* 58 (1997): 331–346.
[57] Spohn, *Go and Do Likewise*, p. 11.   [58] See Keenan, *Moral Wisdom*.

beings, natural law is rooted in an account of human nature. The reason that Christian ethics ought to consult scientific knowledge of evolution, then, is precisely because it provides an important source of information for understanding human nature, human flourishing, the natural law, and the virtues.

The doctrine of creation

The focus of the faith of the people of Israel was not on God as Creator but on God as acting to rescue Israel from oppression in Egypt, giving the covenant, and founding a monarchy. The central Christian teaching is not that God is the Creator of the world, but rather that God is Triune, that there is one God who exists eternally as Father, Son, and Spirit. The Christian creeds profess belief in God as Creator, Redeemer, and Sanctifier of the world, in Jesus Christ as the Incarnation of God, and in the Spirit as the "Lord and Giver of life."

As a noun, the word "creation" refers to whatever exists that is not God, the entire finite world.[59] The essential affirmation of the doctrine of creation is that every created thing depends fundamentally on the Creator for its existence.[60] As a verb, the act of creating is not simply the act of changing one thing into another but instead the act of giving being to something.[61] An ordinary act of making a new substance involves changing pre-existing matter into a new form, in the way in which, say, a carpenter makes a bench out of raw wood or a potter makes a pot by shaping clay; the carpenter educes the form of the table that already pre-exists in the potency of matter that is the wood. Creation is not the production of one mode of

---

[59] This "scientific creationism" takes on a variety of different forms but all suffer from well-documented and devastating theological and scientific problems. For Biblical studies, see Richard J. Clifford, "The Hebrew Scriptures and the Theology of Creation," *Theological Studies* 46 (1985) and "Creation in the Hebrew Bible," in Robert J. Russell *et al.*, eds., *Physics, Philosophy, and Theology: A Common Quest for Understanding* (Vatican City State: Vatican Observatory Publications, 1988), pp. 160–164; Claus Westerman, *Genesis 1–11: A Commentary*, trans. John J. Scullion (second edn., Minneapolis: Augsburg Publishing House, 1984); James Barr, *Biblical Faith and Natural Theology* (Oxford: Clarendon Press, 1993). The philosophical case against literalist creationism, and related anti-evolutionary movements, has been made by, among others, Philip Kitcher, *Abusing Science: The Case against Creationism* (Cambridge, MA: MIT Press, 1982), Ashley Montagu, *Science and Creationism* (New York: Oxford University Press, 1983), Michael Ruse, *Darwinism Defended: A Guide to the Evolution Controversies* (Reading, MA: Addison-Wesley, 1982), Michael Ruse, ed., *But Is It Science?* (Amhurst, NY: Prometheus Press, 1996). Steven Goldberg points out in *Seduced by Science* (New York: New York University Press, 1999) that creationists and others abandoned religion's proper tasks for quasi-science, as if it provides the only kind of truth that matters.
[60] See Augustine, *Confessions* 12.7.   [61] See *ST* I, 46,1 ad 6; I, 45,2.

being from another but the beginning of a relation of absolute dependence.[62] This is the core meaning of creation *ex nihilo*.[63] In Thomas Aquinas' language, God is pure act without any potency and God's essence is God's existence;[64] creatures, in contrast, are marked by potency and they participate in being. Divine creation involves no "passive potency" in matter from which to make any created being. God is the source of everything – the "universal cause of all being"[65] – and thus is the cause of the existence of all finite beings and the world in which they exist. This is why Rahner could say not only that God has created everything that exists but also that God "continues to create them out of nothing."[66] The creation, he explained, "must be radically dependent on God, without making him dependent on it as a master is dependent on his servant."[67] This radical ongoing dependence, however, is at the same time marked by an empowerment to proportionate and proper creaturely "autonomy."[68]

God as Creator, the primary cause of everything that exists, communicates the power of existence to every creature and in so doing endows every creature with its own active power. As Herbert McCabe puts it, "God is not a separate and rival agent in the universe. The creative causal power of God does not operate on me from outside, as an alternative to me; it is the creative causal power of God that makes *me*."[69] The divine activity pervades creation and no aspect of it exists outside God's action – "For from him and through him and to him are all things" (Rom. 11:36).

Creation is not the work of the Father alone. All three divine "persons" are involved in the creation, just as all three are involved in the work of redemption. The doctrine that God creates the world through the Word – "through him all things were made," as the Nicene Creed puts it – implies that God is not required by some kind of necessity to create.[70] God could have remained eternally satisfied with the divine activity produced in the self-communication eternally generated in the Trinitarian processions. In contrast to the Neoplatonic doctrine of emanation, which conceived of creation as following by necessity from the divine nature, the Christian teaching that the Son eternally "proceeds" from the Father implies that the creation of the temporal world is the result of God's free decision. The

---

[62] See *Summa contra Gentiles* 2, 16 (hereafter *SCG*).
[63] 2 Macc. 2:28 is often identified as the biblical warrant for the doctrine of creation *ex nihilo*, but this passage, showing the influence of Hellenism, should probably not be used as an interpretative key for the understanding the "authorial intent" or the meaning of the earlier Genesis accounts of creation.
[64] *ST* I, 3,4.   [65] See *ST* I, 45,1.   [66] Rahner, *Foundations of the Christian Faith*, p. 76.
[67] Ibid., p. 78.   [68] Ibid., p. 79.
[69] Herbert McCabe, *God Matters* (London: Cassell Publishers Limited, 1987), p. 13.
[70] See also Augustine, *De Civ. Dei* XI,24.

Creator does not create out of need, in order to gain some good not already possessed, but does so freely and in order to communicate the divine goodness;[71] God creates solely for the purpose of sharing divine goodness with creatures.

Divine creation offers no theory about the initial conditions of energy that led to the formation of the natural order and it should not be confused with some account of the temporal beginning of the universe. Thomas took it as a revealed truth that creation has a "starting point," and that prior to that point there was no such thing as time or place.[72] Today, the Christian doctrine of creation does not understand the "Big Bang" as *the* creative act of the Creator. As intelligent a thinker as Stephen Hawking misses this point when he writes: "So long as the universe had a beginning, we could suppose it had a creator. But if the universe is really completely self-contained, having no boundary or edge, it would have neither a beginning nor an end: it would simply be. What place, then, for a creator?"[73] Yet, as Polkinghorne points out,

> Theology is concerned with ontological origins and not with temporal beginning. The idea of creation has no special stake in a datable start to the universe. If Hawking is right, and quantum effects mean that the cosmos as we know it is like a kind of fuzzy spacetime egg, without a singular point at which it all began, that is scientifically very interesting, but theologically insignificant. When he poses that question, "But if the universe is really completely self-contained, having no boundary, or edge, it would have neither beginning nor end: it would simply be. What place, then, for a creator?," it would be theologically naïve to give any answer other than: "Every place – as the sustainer of the self-contained spacetime egg and as the ordainer of its quantum laws. God is not a God of the edges, with a vested interest in boundaries. Creation is not something he did fifteen billion years ago, but it is something that he is doing now."[74]

Divine providence

Theologians distinguish the doctrine of creation, according to which God brings something into existence *ex nihilo*, from the doctrine of providence, the affirmation that God continually governs, sustains, and cares for

---

[71] See *ST* I, 44,4. For further reflections on *ex nihilo*, see David B. Burrell, CSC, *Freedom and Creation in Three Traditions* (Notre Dame, IN: University of Notre Dame Press, 1993).
[72] See *ST* I, 46,2. It should be mentioned that the notion of creation *ex nihilo* as interpreted in classical theism is not unproblematic in contemporary Christian theology, but this is not a dispute that can be resolved here.
[73] Stephen Hawking, *A Brief History of Time* (New York: Bantam Books, 1988), pp. 140–141.
[74] John Polkinghorne, *Science and Christian Belief* (London: SPCK, 1994), p. 73.

creatures. Christians have an understandable tendency to assume that the proper point of contact between Christian theology and evolution lies in the doctrine of creation, but for Christian theology the doctrine of providence is also critically important. The central theological issue in the Darwinian controversy was not the interpretation of Genesis, though that was important, but rather the existence and power of a providential God in the face of nature as Darwinians described it. The mechanistic world explained by natural causation alone made it difficult to see why God was needed, and, as we saw in chapter one, the "red in tooth and claw" characterization of the evolutionary process made it difficult to know if God cares.

Faith in divine providence lies at the core of Christianity.[75] Thomas held that all the particular articles of faith are implicitly contained in the twofold belief that God exists and that God is providential.[76] Theologians — and indeed Scripture itself — often employ the language of creation in tandem with the language of divine providence (e.g. Ps. 104:30; 148:7; Isa. 45:7; Heb. 1:2–3) and they are not always distinguished. Scripture speaks of God as engaged in the continuous creation of all things (Exod. 34:10; Num. 16:30; Isa. 42:5; Job 10:8–11), especially through their preservation and governance. It also testifies to special providence. God is said to have knowledge of each person (Ps. 139), to be watchful and caring (Ps. 121:5–8), to rescue from evil (Ps. 107), to answer personal prayer (Ps. 120:1), and so forth. The special providence of God is revealed most poignantly and significantly in salvation history, especially in the election of Israel (Exod. 19:4–6), its deliverance from slavery (Exod. 12:31–36), survival in the desert (Exod. 14:10–31), and the covenant at Sinai (Exod. 19–20). But God also extends providential care to the whole world and shapes the destiny of the "nations" to fulfill the divine plan (Amos 1–2; Isa. 7:17–19; Jer. 25:9–14). Providence brings good out of evil and ultimately orders everything to the glory of God (Rom. 8:9–22; Eph. 1:3–14).

Theologian John H. Wright, SJ, understands divine providence as an expression of God's wisdom, power, and love: "wisdom as the perception of all possibilities is first; then love of the divine goodness as a gift to be given, expressed in the choice to create, is next; and finally divine power, in the actual execution of the choice is last."[77] Against the primacy of divine

---

[75] *ST* II-II, 1,7. [76] *ST* II-II, 2,5.
[77] John H. Wright, SJ, "God," in Joseph A. Komonchak *et al.*, eds., *The New Dictionary of Theology* (Wilmington, DE: Michael Glazier, Inc., 1987), p. 432. See also John H. Wright, SJ, "The Eternal Plan of Divine Providence," *Theological Studies* 27 (1966): 27–57 and "Divine Knowledge and Human Freedom: The God Who Dialogues," *Theological Studies* 38 (1977): 450–477.

sovereignty, Wright points out, Christian faith understands divine power in the service of love informed by divine wisdom.

Creatures are called, in ways appropriate to the kinds of creatures they are, to participate in God's ordering of creation. In contemporary terms, creatures are produced as a result of the gradual eliciting of the inherent potentialities of matter through the working of the evolutionary process. Every creature is internally ordered appropriately to share in the divine goodness; the more a creature flourishes, the more it shares in divine goodness and, thereby, gives glory to God. Divine providence intends the flourishing of all creation, including that of human creatures.

The divine goodness is communicated in varied ways according to the nature of the creature with whom it is shared. The evolutionary emergence of higher kinds of creatures, including human beings, brings with it an enhanced capacity to cooperate with the divine will and to share in divine goodness. As "rational animals" made in the image of God, we are called not only to participate in but also, more distinctively, to cooperate freely and intelligently with God's plan for the creation.[78] Whereas evolutionists speak about human behavior as purely physical events, Christian doctrine speaks also of the interior aspects of human agents when we form intentions, make judgments, and intentionally move toward our chosen ends. Theologically, God's providential action works in and through these human acts in a way that respects our intelligence and freedom.

The notion of the divine "plan" sounds as if it were at odds with both the contingency of nature and human powers of free choice, but in this context the word "plan" simply means an intelligible order of things to the divine end.[79] This order does not have to be a detailed itinerary, because divine providence works in and through the contingencies of nature and the free choices of human beings. In the evolutionary world, divine providence determines neither the exact shape of particular events in nature nor the precise details of particular human acts, but rather always responds to the free decisions of human beings as well as to the genuine contingencies that condition events in nature.

Wright helpfully conceives of the pattern of divine–human interaction in terms of a "three-step pattern." First, in any given event God takes the initiative to communicate divine goodness. Secondly, the creature responds to grace with some degree of acceptance, or refusal, or some mixture of these possibilities. Every truly and fully human act, in some way or another, knowingly or not, involves the dimension of accepting or rejecting divine

---

[78] *ST* I-II, 91,2.   [79] See J. Wright, "God," p. 434 and "Eternal Plan": 28.

grace. Finally, in the third "step," God responds in judgment to the human person's choice of response to the preceding initiative of grace. Each step usually involves a continual interaction with, and interpenetration of, another.

This same pattern is also found in divine providence. In the first step, the divine initiative is expressed in the act of creation and in God's offer of redemption. In describing this step, Wright draws from a distinction between antecedent divine will and consequent divine will developed earlier by John Damascene and Thomas Aquinas. What Wright calls the "antecedent divine plan" constitutes God's invitation to each person to enter more deeply into divine goodness. We experience this movement of grace in any internal inclination to what is just, good, true. God provides every possible avenue to advance the salvation of every human being.

Secondly, the human agent responds freely to the providential divine initiative, either moving in cooperation with it or moving away from it. Providence here does not replace human responsibility, but rather works in and through it. The human person who accepts the prompting of grace functions as an instrument of God's goodness and thereby concretely assists the work of divine providence in the world. Human freedom in its most profound meaning is rooted in our assent to grace. This component of the divine plan gives wide scope to human action, and therefore allows for the possibility that we will misuse our capacity for choice in doing what is destructive to ourselves and others, intentionally thwart the internal human impetus for goodness, and reject divine love.

God allows for moral evil, and actual sin, without being its direct cause.[80] The Creator directly wills only good, but allows evil to exist as part of the overall order of the universe.[81] "Evil" in the broadest sense refers to any kind of harm and destruction. It thus includes but extends far beyond the more narrow sense of moral evil brought about through deliberate human wrongdoing and sinful acts. As finite creatures, we are inherently subject to decay and other forms of harm and at some point are overcome by destruction. We are subjected to the possibility of moral evil as well as physical evil. This vulnerability to moral evil is an unavoidable effect of the fact that God wills that there be finite creatures with free choice. God wills that, as a possibility, we be able not only to choose to do good but also to choose to do evil. God does not directly will failure, injury,

---

[80] See *ST* I, 19,8; 22,2 ad 2.
[81] *ST* I, 49,2. On God's tolerance or "permission" of evil, see Thomas Aquinas, *De Veritate* 5.

or accidents, but God does will a world in which events with these qualities exist as part of the "order of the universe."

Ultimately the suffering caused by various sorts of harm, and experienced with special intensity by the most intelligent and sensitive beings, cannot be "explained" by a moral calculus that makes it all "worthwhile" for pedagogical and/or eschatological purposes. The "problem of evil" cannot be taken away, or made non-problematic, by rational solutions. From a Christian standpoint, though, the most we can hope is that suffering is informed by the passion, death, and resurrection of Christ, an event in which God embraces human suffering and brings life out of death. Christian hope trusts that God's provident and redeeming love is ultimately completed in our own bodily resurrection and eternal joy in the vision of God and communion of saints.

Finally the third providential "step" in the work of providence takes place when God responds to the human person's response to God's original initiative. God opens possibilities for goodness in each event in history, whether positive or negative in character. Where human beings are simply dependent on God's act of creation, God's providence is conditioned by the choices of human beings to whom God responds. God is not ontologically changed by our human choices, but God is affected intentionally, and changed intentionally, by them. In the moment of "divine judgment," Wright explains, "God achieves his gracious purpose to the extent and in the way that created wills are willing."[82] God, who is more intimate to each of us than we are to ourselves, continues to offer grace in the mode appropriate to our prior decisions, such as in forgiving, healing, and encouraging. This aspect of providence, Wright explains, is the "effective expression" of the "consequent plan of God," namely "that portion of the antecedent plan actually put into execution through the free acceptance of the creature."[83] This is what each of us actually witnesses, in all kinds of ways, in concrete daily experience.

Divine conservation and government of human life is realized when divine goodness is actually achieved through the free acts of rational creatures. The accomplishment of this purpose is called "divine sovereignty." Particular individuals cooperate with God's grace (and mercy is accomplished) or defect from it (and justice is accomplished). God's ultimate purpose is accomplished without fail. There is a sense in which God is affected by our actions because divine actions are tailored to provide opportunities for goodness that are appropriate to concrete human

---

[82] J. Wright, "God," p. 434.   [83] Ibid.

situations. Thus, as noted already, instead of being in absolute control of every event, and thereby bearing complete responsibility for it, God interacts with creatures who possess their own kind of quasi-independence and who can exercise limited but real responsibility for their own acts. This is why Wright speaks of the "God who dialogues."

This raises the question of God's relation to the world. As pure act and complete perfection, God can be neither enhanced nor diminished by events in the world. God knows, loves, and cares for the world and its creatures, but God does not do so because of divine inadequacy or to acquire some enrichment that God does not already possess. God creates simply because God wills to communicate divine goodness to what is not God.

Christian faith also affirms that God knows and is present to the creaturely experience of pain, suffering, and loss. Classical Christian theology affirms that God suffers in the humanity of Jesus Christ. "We could say," wrote Gerald Vann, "that God became man precisely in order to suffer." Because in the Incarnation God fully embraces the human condition in all its dimensions, writes Vann, "We know that God is involved in the suffering of creation."[84] Since the humanity of Christ is, in Trinitarian theology, also the humanity of God, we can speak in some sense of "divine suffering." The proviso "in some sense" marks the implication that God does not possess a body and so has no "emotions" or "passions" as we experience them. Theologians and philosophers who write about divine suffering – most famously, Whitehead's the "fellow-sufferer who understands"[85] – often do so without giving much of an indication of how this phrase applies to God and so can lend itself to an anthropomorphic sentimentality that fails to do justice to God. The phrase "divine love" refers to God's will for the good of all creation and its creatures, not to divine "feelings";[86] God responds to our plight by working providentially for our healing.

---

[84] Gerald Vann, OP, *The Pain of Christ and the Sorrow of God* (Oxford: Blackfriars, 1947), pp. 64, 65. See also Michael J. Dodds, OP, *The Unchanging God of Love: A Study of the Teaching of St. Thomas Aquinas on Divine Immutability in View of Certain Contemporary Criticisms of this Doctrine* (Fribourg, Switzerland: Editions Universitaires, 1986) and "Thomas Aquinas, Human Suffering, and the Unchanging God of Love," *Theological Studies* 52 (1991): 330–344. The theme of divine suffering is treated in profound ways by a number of theologians who would offer an alternative to Vann's and Dodd's Thomistic approaches. These include Arthur McGill, *Suffering: A Test of Theological Method* (Philadelphia: Westminster Press, 1982); William C. Placher, *Narratives of a Vulnerable God: Christ, Theology, and Scripture* (Louisville, KY: Westminster John Knox, 1994).

[85] Alfred Whitehead, *Process and Reality* (New York: The Free Press edn., 1969), p. 413.

[86] See *ST* I, 20,2.

This doctrine of providence shows the great distance of Christian doctrine from deistic and pantheistic alternatives. The Christian affirmation of God as Creator is often confused with a deistic view of a remote, unaffected, and indifferent God who begins the process of evolution and then lets it run on its own without any divine conern. Identifying creation with the temporal beginning of the universe can encourage this assumption. A permutation of this theology regards God as normally uninvolved in creation but able periodically to intervene for various purposes. In an evolutionary context, this interventionist deity works to modify the evolutionary process at certain key points in natural history (e.g. at the origin of cellular life) to guide the evolutionary process to the planned result. While very different in some ways, these deistic and interventionist theologies share a view of God as existing "outside" nature; they differ over whether God is a personal agent with the capacity to intervene at a particular time and place. The doctrine of providence, in contrast, regards natural history neither as proceeding outside the influence of God, nor as a self-sustained, completely independent system. The doctrine of creation regards everything that exists as radically dependent on God every moment of its existence; the doctrine of providence holds that God interacts with the world in a way that moves it to fulfill the divine purposes.

Christian doctrine also avoids the opposite, pantheistic view of God as one who acts within the space and time limitations that condition all creatures, who creates in order to increase the divine goodness or to acquire benefits that God would not otherwise enjoy, and who develops and becomes more divine as the evolutionary process proceeds through time. Christian faith, however, takes the evolutionary process to be neither identical with nor lying outside the scope of divine providence. God's action pervades the creation and the lives of creatures.

The full scope of divine providence must be related to the divine eternity. The word "eternal" etymologically indicates the absence of time, not simply an everlasting existence. The term taken by itself presents a negation of our natural human way of thinking in temporal terms, but it does not suggest any positive way of imagining God's mode of being. God's eternity is rooted in the complete fullness of the divine being, as pure act.[87] God creates the being of each creature and pervades the depths of the temporal world and so cannot be said to be "removed" or "abstracted" from

---

[87] Based on *ST* I, 3,3, which argues that God's essence is God's very act of existing, Thomas Aquinas argues that God is immutable (q. 9) and eternal (q. 10).

the creation. God's eternity in fact means that God is immediately and simultaneously present to each creature.

Ernan McMullin endorses Augustine's view of this matter and expresses it concisely: "The act of creation is a single one, in which what is past, present, or future from the perspective of the creature issues as a single whole from the Creator."[88] Describing God as "eternal" does not mean that God has no beginning and no end, but rather that God is not subjected to the condition of temporality, that time is not in God. This bears very important implications for our understanding of creation, among the most important of which is that God does not learn about evolution or human history in the order that it takes place. "God knows the world in the act of creating it," McMullin explains, "and thus knows the cosmic past, present, and future in a single unmediated grasp."[89]

This single unmediated comprehension implies a fundamental disanalogy between human plans and divine plans and between human knowledge and divine knowledge. We learn in a temporal sequence, where, for example, understanding one thing makes us infer that another might be the case, but God does not reason, deduce, or otherwise engage in chains of reasoning that suggest a movement from a state of ignorance into knowledge. God knows everything at once in a single act of eternal divine understanding. We make plans, then carry them out, or fail to do so, and then think about what to do next. The divine plan is accomplished without any "gap" between the divine will and the divine end. The "divine plan" thus refers to a given outcome, a divinely willed end, and not to any process in which God works through specific stages to steer courses of events in the desired direction. Evolution does not have to be pre-programmed to be described as reflecting the divine plan. As McMullin points out, "It makes no difference ... whether the appearance of *Homo sapiens* is the inevitable result of a steady process of complexification stretching over billions of years, or whether on the contrary it comes about through a series of coincidences that would have made it entirely unpredictable from the (causal) human standpoint."[90]

---

[88] Ernan McMullin, "Evolutionary Contingency and Cosmic Purpose," in Michael Himes and Stephen J. Pope, eds., *Finding God in All Things: Essays in Honor of Michael J. Buckley, S.J.* (New York: Crossroad, 1996), p. 155.

[89] Ibid., p. 156. McMullin here accords with Thomas, *ST* I, 14,13, who argues that God knows future contingent things (though obviously without determining their outcomes). God knows the outcome of contingent events that are future to us but not to God; God knows them "in their presentness as their Creator"; ibid., p. 161.

[90] Ibid., p. 157.

McMullin's point is that the divine plan is accomplished in the achievement of the divine end, not in the execution of a particular strategy. He admits: "There is nothing about the evolutionary process in itself that would lead one to recognize in it the deliberate action of a Planner. It does not look like the kind of process human designers would use to accomplish their ends."[91] Evolutionists deny evolutionary purpose and cosmic finality because the world lacks the design features they associate with a human plan. Ayala does not believe that evolution entails purposelessness but agrees that evolution shows no sign of being a pre-planned rational and orderly process: "The fossil record shows that life has evolved in a haphazard fashion. The radiations, expansions, relays of one form or another, occasional but irregular trends, and the ever present extinctions, are best explained by natural selection of organisms subject to the vagaries of genetic mutation and environmental challenge."[92]

Evolutionists infer that since there is no obvious preordained plan, evolution is purposeless. These critics, however, identify "plan" and specific "strategy," the means to the divine end, rather than understanding it as the achievement of a "goal," the divine end itself. If the divine end includes not only the proliferation of energy, matter, and life, but also and especially the emergence of intelligent and loving life, then the divine end for evolution has been accomplished. If this end is accomplished, then the existence of chance, and even of Gould's radical contingency, does not defeat claims of evolutionary purposiveness. As McMullin puts it, "if one maintains the age-old doctrine of God's eternality, the contingency of the evolutionary process leading to the appearance of *Homo sapiens* makes no difference to the Christian belief in a special destiny for humankind."[93]

Primary and secondary causation

The notion of creation concerns not only the dependence of all things on God but also ongoing creation, or *creatio continua*, via intermediary means or "secondary causes." The notion of "secondary causes" was developed by Thomas Aquinas but also transmitted by way of many intermediaries to Darwin.[94] Darwin wrote that the notion "accords better with what we know of the laws impressed on matter by the Creator, that the production

---

[91] Ibid. [92] Ayala, "Darwin's Devolution," p. 108.
[93] McMullin, "Evolutionary Contingency," p. 157.
[94] See *ST* I, 19,5,8; I, 22,4, and 4, ad 2; I, 49,2; *SCG* 3, 64–77, and John Calvin, *Institutes*, I, xvi, 5. On Thomas, see Etienne Gilson, "The Corporeal World and the Efficacy of Secondary Causes," in Owen Thomas, ed., *God's Activity in the World: The Contemporary Problem* (Chico: Scholars Press,

and extinction of the past and present inhabitants of the world should have been due to secondary causes, like those determining the birth and death of the individual."[95] Darwin believed that the evolutionary process is more "ennobling" for creatures than special creation would have been.[96]

Darwin's Christian followers also adopted the language of secondary causality to speak about how God produces creatures through intermediaries. Botanist Asa Gray, a devout Christian and Darwin's most staunch American ally, insisted that the Creator's power is in no way compromised by the fact that God chooses to use secondary causes in the ordering of life. In fact, he judged, "The record of the fiat – 'Let the earth bring forth the living creature after his kind' – seems even to imply them."[97]

The notion of secondary causation is not found in Scripture, which often describes God as acting directly within the biological order, for example: "When you hide your face, they are dismayed; when you take away their breath, they die and return to the dust. When you send forth your spirit, they are created; and you renew the face of the ground" (Ps. 104:29–30). The notion of secondary causation was brought into Christian theology by Albert the Great and his student Thomas Aquinas, who had adopted the notion from what was probably a ninth-century Neoplatonic Islamic text, *The Book of Causes*.[98] Because the primary cause, God, gives being to whatever exists, the significance of the primary cause on its effect is much greater than that of secondary causes. God's radical transcendence is seen in the fact that in God alone is being itself (*esse*), God's essence is God's existence. God's immanence is reflected in the fact that all creatures have being through participation in God's being. Yet while God is in all things insofar as they have being, God is not their essence. They have their own distinct powers and modes of behavior that are their own and not God's. Thus God as primary cause is the immediate cause of all being; the secondary causes of the created order have their own real causal efficacy. Medieval universities had strong theological reasons for sponsoring the natural sciences as ways of discovering real causes in nature.

1983), pp. 213–230. For a historical treatment, see Armand Maurer, "Darwin, Thomists, and Secondary Causality," *The Review of Metaphysics* 57 (March 2004): 491–514. For a contemporary exposition of this tradition, see Elizabeth Johnson, CSJ, "Does God Play Dice? Divine Providence and Chance," originally published in *Theological Studies* 56 (1996): 3–18, and republished in Miller, ed., *An Evolving Dialogue*, pp. 355–373.

[95] Darwin, *The Origin of Species* (New York: Penguin, 1986), p. 458.   [96] Ibid.
[97] Cited in St. George Mivart, *On the Genesis of Species*, second edn. (London: Macmillan, 1871), pp. 291–292.
[98] For historical background and significance, see Etienne Gilson, *History of Christian Philosophy in the Middle Ages* (New York: Random House, 1955), pp. 235–237.

Secondary causes are genuine causes but their power is radically derivative; when they are involved in the production of being, they are so because of the power of the primary cause.[99] The causal role of secondary agents is to give concrete specificity, particularity, or definiteness, to the effect of the primary cause. The teeth of the dog, the wings of the bird, and the intelligence of a person are all the proper effects of secondary causes.[100] It is appropriate before meals to thank God, the primary cause who lies behind the secondary causes (farmers, climate, soil, commerce, food preparers, etc.) that contribute to the production of food about to be eaten. Particular secondary causes contribute the products of their own specific powers, none of which would exist were it not for the primary cause.

Notice also that one does not speak of the primary cause acting first, and then the secondary causes acting later, or as a consequence, of the initial operation of the primary cause. The term "primary" refers to a metaphysical relation, not to a temporal sequence. God acts whenever secondary agents act and the effect of the secondary agents is also a divine effect – though in very different ways.[101] The language of primary and secondary causation functions analogously rather than univocally; God and creatures do not function as "causes" in the same way. Most importantly, creatures modify matter in activities that involve making things but God alone can cause creatures to *be* in the first place.

The doctrines of creation and providence provide the theological context for the natural-law moral tradition. Thomas held that divine power is demonstrated in God's ability to communicate power effectively to creatures as causes. God endows creatures with their own capacities, inclinations, and needs. Out of the divine goodness, God empowers creatures with the "dignity of causality."[102] "If God were to govern alone," Thomas Aquinas wrote, "the capacity to be causes would be missing from creatures."[103] Causality has a special dignity because it bears a certain analogous resemblance to God in that creatures "would not only exist but be the cause of others."[104] As human agents we are the genuine causes of our own acts who nevertheless act only in virtue of the power given to us by the Creator. We are able to come to some knowledge of the world by using our own intelligence, so we are able to attain some level of moral goodness by

---

[99] See *SCG* 3,66.   [100] See *SCG* 3,66.
[101] See *SCG* 3,70. In *SCG* 3,69 Thomas argues against the Muslim Ash'arite teaching that God alone is the cause of all things. See Gilson, *History of Christian Philosophy in the Middle Ages*, pp. 184–185.
[102] *ST* I, 22,3.   [103] *ST* I, 103,6.   [104] *SCG* 3,70,7.

making good choices, doing good deeds, and shaping our lives to be more virtuous.[105]

This understanding of the real but limited power of human agency is related to the distinction between creaturely procreation and the divine act of creation. We do not "create" when what biologists call our "reproductive acts" bring about the "generation" of offspring. The act of creation, for Thomas, not only "makes" something out of what existed prior to it, for example in the gametes of the couple, but more importantly, causes it to be. The human "procreation" that takes place through the exercise of the natural reproductive capacities of a man and a woman is the means by which God creates a human person. God and the human couple are both properly said to be "causes," but in different senses of the term.

The distinction between primary and secondary causality does not imply that God somehow renounces divine power. John Polkinghorne's suggestion that God willingly "withdraws" from the world assumes that divine power is essentially similar to creaturely power and that the exercise of the latter could compete with the former. Divine "withdrawal" in Thomas' ontological sense would lead to the end of being. For him, God acts in and through secondary agencies so is in no way diminished by their causal efficacy. The analogical interpretation allows us to see that secondary causation refers to a kind of cause that is able to exert some kind of power in the world while at the same time standing in ontological dependence on the primary cause. The procreative couple is ontologically dependent on God; they exist in virtue of God; God is not intrinsically dependent on them and does not exist in virtue of their power. Yet they do exert powers of agency to procreate and are responsible for their act and its consequences.

Because of this fundamental ontological difference, Thomas pointed out, "the same effect is not attributed to a natural cause and to divine power in such a way that it is partly done by God, and partly by the natural agent; rather, it is wholly done by both, according to a different way, just as the same effect is wholly attributed to the instrument and also wholly to the principal agent."[106] Had he been aware of this distinction, E. O. Wilson could have avoided his forced dichotomy between nature and God, as when he writes: "If human kind evolved by Darwinian natural selection, genetic chance and environmental necessity, not God, made the species."[107] This distinction is critically important for own understanding of

---

[105] See Thomas, *De Veritate* 11,1; *De Potentia* 3,7.
[106] *SCG* 3,70,8.   [107] E. O. Wilson, *On Human Nature*, p. 1.

the theological status of the evolutionary process. Thomas held that God created each distinct species as such, either all of them at once "in the beginning" or in such a way that they would appear over the course of time.[108] He assumed that the Creator intentionally created the essences of creatures with the right powers for their role in the order of creation. Species are given the ability to "be fruitful and multiply" and in so doing they contribute, as secondary causes, to the conservation of creation.

In light of our knowledge of evolution, on the other hand, we say that the human species, like all other species, has been produced by the vast and intricate set of secondary causal events that compose natural history. Secondary causes not only reproduce pre-existing forms of species but also lead to the evolution of new substantial forms in the process of evolutionary "speciation." God continues to exert divine influence on the world through the creative powers inherent in nature itself. As the primary cause, the Creator does not "intrude" into natural processes for their proper operation, but works from within the inherent capacities, structures, and tendencies of creatures.

Divine action in the world

A number of contemporary Christian thinkers have called into question the value of secondary causality as a way of thinking about the relation of God and creatures. The two major objections are focused, on the one hand, on accusations of "double agency"[109] that eliminates creaturely causality, and, on the other hand, on its excessive vagueness or modesty. The first objection finds it implausible to think that God acts through creatures without abolishing the creature's own relative autonomy. It would seem that creatures themselves are not genuine causes but only conduits for divine causality. This criticism, though, makes the mistake of presuming a univocality of divine causality and creaturely causality. The criticism of "double agency" holds that the creature's efficient causality is overpowered by divine efficient causality. Yet, as Thomas Aquinas pointed out, creatures act as true causes by exercising the powers of their natures.[110] The efficient causality of creatures moves things in the way that their natural capacities allow them to move them, for example by moving their fins or arms or legs

---

[108] See *SCG* 3,70,8.
[109] Austin Farrer is the direct target of these criticisms. See his *A Science of God?* (London: Geoffrey Bles, 1966) and *Faith and Speculation* (New York: New York University Press, 1967).
[110] See *ST* I, 105,5.

to effect a motion in something. God is said also to be the source or cause of this creaturely movement because God gives the creature its nature, the matter of which it is composed, and the ends that it naturally seeks – that is, God, the source of its act of existence and its formal, material, and final cause, is the source of its behavior. This meaning of "cause" or "source" is radically different from its use when we speak of the secondary causality attributed to the instrumentality of nonhuman animals or the agency of human beings.

The second objection is that secondary causality fails to explain how God acts in the world. Defenders of this position give no explanation of the "causal joint" that connects primary and secondary causes, so they take an excessively "apophatic" stance on God's action in the world.[111] Critics themselves, however, have not provided any "explanation" of how God acts in the world, and they have not shown that it is even possible to provide such an account. They do offer some images intended to give some intelligibility to what faith affirms about God's action in the world, but these are imaginative conjectures rather than explanations.

Polkinghorne, for example, believes that divine faithfulness to the order of creation respects the laws of nature, but that the structure of the world allows for "divine manoeuvre" within the physical order.[112] God, in this view, can best be thought of as shaping the events of nature and history by working from the "bottom-up," through the unpredictabilities of indeterminate systems both on the level of micro-events and on the level of large-scale chaotic systems, in such a way that the ordering of nature is fully respected. He suggests that God might give direction to biological evolution by shaping genetic variations through influencing underlying quantum mechanical processes. God could direct evolution by somehow promoting types of genetic mutations that would be liable to provide later organisms with adaptive advantages. Polkinghorne does not pretend that he can explain how God shapes evolution; he only intends to propose a fruitful image of how it might be intelligible to say that God does so.

Arthur Peacocke describes divine action in terms of a "general (top-down) interaction with the world."[113] Peacocke believes that Polkinghorne's

---

[111] See John Polkinghorne, "The Metaphysics of Divine Action," in Robert Russell *et al.*, eds., *Chaos and Complexity: Scientific Perspectives on Divine Action*, vol. II (Vatican City State: Vatican Observatory Publications, 1996), p. 150, citing Peacocke's complaint of "apophatic" resignation.

[112] John Polkinghorne, *Science and Providence* (London: SPCK, 1989), p. 31.

[113] Peacocke, *Theology for a Scientific Age*, p. 179; see also pp. 182–183. Consider Thomas Aquinas' observation that "the whole human soul is in the whole body, and again, in every part, as God is in regard to the whole world" (*ST* I, 93,3).

location of divine action in micro-events tends unhelpfully to suggest that God acts to influence the natural order by inserting a divine "finger" into nature on the micro-level. This vision can slide into another version of the "God of the gaps."[114] Peacocke thinks God is better envisioned as the "unifying, unitive source and centred influence on the world's activity,"[115] in a way similar to how the mind exerts a continual influence on brain states. God acts upon the whole of creation as its ultimate "boundary"[116] condition and, as such, can shape the occurrence of particular "patterns of events."[117] God's continuous influence on the world is modeled along the lines of how our mental activity continually influences our bodily states.[118] Creatures within the natural world are subjected to God's will in something like the way our arm is moved by our mind when we choose to reach for a glass of water. The whole affects the disposition of its parts in ways that are real even though we cannot identify the "causal joint"[119] between thought and action.

There are significant limits on the value of attempting to understand how God acts in the world on the analogy of how our minds influence our bodies. One can wonder about the usefulness of attempting to use one not very well understood relation, that between the mind and the body, to shed light on another, even less well understood relation, that between God and the world. In none of these cases do we have a clear account of what we are trying to understand, let alone a definitive explanation of these phenomena. Polkinghorne and Peacocke are both aware of the fact that we cannot comprehend in any clear way the "precise causal modality" of God's interaction with the world. Indeed, if God is essentially and fundamentally incomprehensible to us, so also, it would seem, would be God's ways of influencing the world. Christian faith affirms that God influences the world so that it serves the divine purpose. It does not offer an explanation of how this influence works or provide an account of the "mechanics" of providence. Such a topic is entirely beyond the range of our intelligence or the scope of our experience.

The tentative and exploratory nature of this suggestion makes it reasonable to ask what theological and philosophical work it can really be

---

[114] See Peacocke, *Theology for a Scientific Age*, pp. 154–155.   [115] Ibid., p. 161.   [116] Ibid., p. 163.
[117] Ibid., p. 386. Peacocke would provide a negative answer to Gustafson's question, "Is the idea of God as person a projection from the human so that God is conceived in the image and likeness of the human?"; Gustafson, *An Examined Faith*, p. 91.
[118] See Peacocke, *Theology for a Scientific Age*, p. 161.
[119] See ibid., pp. 139–183, pp. 148f. on the "causal joint," and pp. 156–157 for why this is not a helpful way of putting the matter.

expected to accomplish, and whether it offers any advantage to the theological modesty of analogical ways of talking about God's causal efficacy on the world. None of the images offered by these contemporary theologians have helped us to understand either general or particular divine providential action. They differ from the traditional notion of secondary causality by believing that God has to act in the lacunae resident in the natural causal order and so seem to be vulnerable to the "God of the gaps" problem, especially to the extent that the "gaps" close.

In either case, Christians believe that God not only creates the world but also guides and directs it toward its proper end. We do not comprehend God's purposes beyond the most general sense. The Book of Isaiah is relevant to this topic: "For my thoughts are not your thoughts, nor are your ways my ways, says the Lord. For as the heavens are higher than the earth, so are my ways higher than your ways and my thoughts than your thoughts" (Isa. 55: 8–9; also Job 38:1–42:6). We cannot pretend to understand exactly the divine purposes that are achieved in the course of time; as Paul says, we see now "dimly" (1 Cor. 13:12). Christian faith affirms that God sustains the world in being, governs through created natural-law-like regularities and contingencies, acts indirectly through creatures that behave according to their natures, and acts to bring about events in the world (e.g. the resurrection of Jesus) that would otherwise not have occurred because they fall outside what unfolds according to the law-like regularities of the natural order. Though it is not easy to determine what constitutes the natural range of events outside of which one can speak of the "miraculous," it seems theologically presumptuous to maintain that God cannot in principle influence events in the natural order except through the possibilities inherent in that order. It is reasonable to affirm that while not restricted to this mode of causal efficacy, divine governance typically acts within the prevailing patterns and possibilities of the natural order. Christian faith approached in this way can thus understand God as working in and through the evolutionary process to create, sustain, and guide all of creation, including human creatures.

CHAPTER 5

Chance and purpose in evolution

The major counter-position to the previous chapter's theological interpretation of evolution comes from evolutionists who argue that the evolutionary process is so pervaded by chance and contingency that it cannot possibly be described as a "purposeful" divine creation. If true, neo-Darwinism would undercut the religious foundations of Christian ethics. If evolution is not part of a purposeful world, Christian convictions are false and Christian ethics is an exercise in futility and self-deception.

The neo-Darwinian approach to purpose in nature is best understood in light of the broad shift of modern philosophy and theology away from ancient and medieval habits of discerning teleology, or final causality, or intrinsic ends, in nature. The modern project proceeded on the assumption that nature is best explained mechanistically, that is, in terms of efficient and material rather than formal and final causes. Christian ethics, on the other hand, has traditionally understood the natural world to constitute an overarching finality reflecting the purposive wisdom of the Creator.

This chapter addresses the question of whether knowledge of evolutionary contingency does indeed provide evidence for the supposition that nature is purposeless. I will first examine the argument for this position advanced by S. J. Gould and then the replies to it by biologist Simon Conway Morris and by astronomer William Stoeger, SJ. The latter two thinkers provide reasonable support for thinking that evolutionary directionality is consistent with a religiously based belief in the divine purpose of evolution.

The universe has been "fine tuned" to give rise to at least one planet with physical conditions that allow for the emergence and maintenance of life. The earth provided conditions that were hospitable first to beings marked by some elemental forms of information processing, then to beings capable of consciousness, and finally to beings who possess self-consciousness. The earliest forms of life gave rise to organisms with increased capacities for movement, sensitivity, awareness, and responsiveness. Organisms moved only by chemical reactions gave rise to organisms moved by apprehensions, drives, and emotions. Spontaneity was complemented by restraints imposed by the social ordering of animals living in groups. Increased environmental demands called forth expanding behavioral repertoires, increasingly complex emotional responses, and more and more sophisticated mechanisms of information processing.

The constructive heart of this chapter begins with the fact that in the course of evolution lower levels of order give rise to higher levels of emergent complexity. The presence of increasing complexity in nature, including cognitive complexity, allows for capacities and inclinations that can serve in increasingly clear ways the purpose of the Creator. Nature as a whole is organized in such a way as to produce more and more complexity and higher and higher capacities for responsiveness, intelligence, and consciousness. One of the most impressive products of this directionality is *Homo sapiens*. A Christian theological perspective can interpret the inherent directionality of evolution as structured to serve the divine purpose for creation.

EVOLUTION AS DIRECTIONLESS AND PURPOSELESS: GOULD

Gould developed a thorough, consistent, and incisive discussion of the implications of evolutionary contingency for the question of purpose. Many traits of organisms happen to be what they are because of the "luck of the draw," he noted.[1] The pentadactyl limb commonly possessed by mammals "just happens to be," and there is nothing particularly adaptive in the fact that we have five rather than four or six digits as the result of unpredictable and unrepeatable events of history. Evolution, he thought, is usually generated by fairly rapid speciation, the separation of one lineage from a parental origin, in a geographically isolated region. In his view, evolution is best envisioned not as a steadily constructed ladder moving from lower to higher, but as an oddly shaped "bush."[2] And of the three once coexisting limbs of the "human bush," we alone continue to exist.

Gould's most famous treatment of the impact of chance on natural history concerned the Cambrian fauna discovered in the Burgess Shale sediment in

---

[1] See Stephen Jay Gould, *Eight Little Piggies* (New York: Penguin, 1993), p. 77.
[2] See Stephen Jay Gould, "Bushes and Ladders," in *Ever since Darwin: Reflections in Natural History* (New York: W. W. Norton, 1977), pp. 56–62.

the Canadian Rockies.[3] His *Wonderful Life* discusses the initial production of most of the modern animal phyla within what is, in geological terms, a fairly short time span of several million years, the onset of which probably took place around 570 million years ago. Only four of the anatomical frameworks that were discovered in the shale have modern representatives, but any one of the disappeared structures could have become the basis for a distinct phylum. The four surviving plans became the founders of the modern animal phyla.

Gould emphasized the fact that the survival of one lineage rather than another was a matter of contingent circumstances. Other circumstances might have eliminated these founders and given rise to others, and everything would have been different for animal species as we know them today; for example, the earth under different conditions might have been host to insects alone. Gould held that the unlikelihood of the multitude of contingent events that lined up to make our appearance possible is made even more poignant when we keep in mind the thousands of unrealized possibilities that would have precluded our evolution: "Replay the tape a million times from a Burgess beginning, and I doubt that anything like Homo sapiens would ever evolve again."[4] The appearance of the human species within the evolutionary process is extremely fortuitous – the supreme stroke of luck, at least for our species.

Gould's agenda was long concerned with countering the claim that evolution is "progressive" or marked by an in-built trend toward greater and greater complexity that would eventually give rise to human beings. Cultural evolution can make progress because the rapid communication and diffusion of ideas, traditions, and other kinds of information can contribute to cumulative improvement, but biological evolution cannot be described accurately as "progressive."

The lengthy course of evolution has generated more complex and intelligent species, Gould conceded, but, measured in terms of simple biomass, simple organisms have been much more dominant on our planet than highly intelligent organisms. Gould accounted for the appearance of complexity in terms of what he called the "left wall" phenomenon.[5] He asked readers to imagine a Saturday night drunk stumbling along a street. The drunk is not walking with any purpose in mind, but every time he stumbles into the wall of the buildings to his left he happens to bounce to

---

[3] Gould, *Wonderful Life: The Burgess Shale and the Nature of History* (New York and London: W. W. Norton, 1989).
[4] Ibid., p. 289.
[5] Stephen Jay Gould, *Life's Grandeur: The Spread of Excellence from Plato to Darwin* (New York: Random House, 1997), pp. 167–171.

his right because has nowhere else to go. Applied to evolution, Gould's point was that evolution generates more complex organisms because all the niches for the simple organisms – the "left wall" of the continuum – are already filled. Nature's fecundity has to go "somewhere" and an increase of complexity is the only avenue left open to the diffusion of nature. This generative capacity, Gould concluded, is no more purposive than the drunk's meandering into the street after he collides with the buildings on his left. Thus, the "vaunted progress of life is really a random motion away from simple beginnings, not directed impetus toward any inherently advantageous complexity."[6]

Gould's "left wall" argument, together with his general focus on contingency and chance, was intended partly to deflate the arrogant human tendency to view the entire evolutionary process as deliberately leading to our species. Evolution does not "aim" at us; we are here by "the luck of the draw, not the inevitability of life's direction or evolution's mechanism."[7] Considered in strictly evolutionary terms, Gould wrote, our species is no more than a "tiny twig on an improbable branch of a contingent limb on a fortunate tree."[8] "If life had always been hankering to reach a pinnacle of expression as the animal kingdom," he argued, "then organic history seemed to be in no hurry to initiate this ultimate phase. About five sixths of life's history had passed before animals made their first appearance in the fossil record some 600 million years ago."[9] Gould's sense of the small place of humans in natural history constituted a counterweight to excessive anthropocentrism. If the 4.5 billion-year history of organic life on the planet is imagined to be a single twenty-four hour day, human beings appear only in the last thirty-eight seconds.[10] Like Gustafson, Gould believed strongly that the value of all other life forms is not dependent on its instrumental value to humanity.

Gould's argument, however, proceeds by something of a sleight of hand. He identifies all forms of "purpose" with the modern doctrine of progress and then collapses the meaning of "progress" into biological adapation. He does not acknowledge the (modest and context-relative) sense of progress implied in the notion of "adaptation." Dawkins argues against Gould that evolution as a whole can be described as progressive in the sense that natural selection over time produces organisms better adapted to their

---

[6] Ibid., p. 173. [7] Ibid., p. 175. [8] Gould, *Wonderful Life*, p. 291.
[9] Gould, "On Embryos and Ancestors," *Natural History* (July/August 1998): 58.
[10] Miller, *Finding Darwin's God*, p. 244.

particular environments. Maladaptive organisms are eliminated, adaptive organisms "replicate." In the long run, the most adapted organisms will become successful in the "evolutionary arms race." Gould's point, though, is that survival or destruction (as in the case of the massive dinosaur extinction in the Cretacious period) can result from fortuitous events and should not always be simplistically credited to the adaptive powers of the organisms in question. Here Gould corrects Dawkins' gratuitous assumption that organisms at one stage are better adapted than organisms had been at an earlier stage. It is possible that, as environments shift over time, a given organism that had been well adapted to an earlier habitat becomes less adapted to the new environment. Time and shifting environments can lead to the decline and even the elimination of a species.

Dawkins' emphasis on the progressive character of natural selection is as overstated as was Gould's emphasis on contingency. Gould was right to argue that when considering the notion of progress we have to take into account the "full house" of variation within the whole system being evaluated and not just its most unusual parts (or most distinctive species). Yet Dawkins' attentiveness to natural selection properly accounts for why some features are preferred over others and why organisms that have these traits are relatively more successful than those that do not. Evolution is a tinkering process that sometimes generates good "solutions" to evolutionary problems.

It seems to me that natural selection operating on random mutations to produce relatively advantaged organisms can be considered a legitimate candidate for the mechanism of directionality. This helps to account for why the emergence of new kinds of complexities is not just the product of increased variation or random diffusion, but rather an intelligible process that generates organisms that are more successful than their competitors because they are better adapted to their surroundings. As Ayala puts it, evolution is the outcome of the interaction of random and *non-random* processes. There is a "'selecting' process," he writes, "which picks up adaptive combinations because these reproduce more effectively and thus become established in populations. These adaptive combinations constitute, in turn, new levels of organization upon which the mutation (random) plus selection (non-random or directional) process again operates."[11]

---

[11] Ayala, "Darwin's Devolution," p. 106.

PROGRESS RECONSIDERED

The evolutionary process as a whole generates increased levels of complexity over time. It also produces relative homeostasis in some contexts and regressions in others. Complexity *per se* is not always adaptive. Some strategies for specialization amount to degeneration if they make it difficult for organisms later to adapt to new challenges. Since what constitutes an "adaptation" is relative to particular contexts, complex traits that happen to be adaptive in some contexts might not be adaptive in others.

Nature is inherently structured to give rise to emergent complexity. Evolution led to the most complex of all species, *Homo sapiens*, and our distinctive kinds of social, emotional, cognitive, moral, and religious capacities. While methodological reductionism ought to be employed to understand as much about our behavior as possible, we need to acknowledge that these emergent capacities allow us to engage in activities and to strive for goals that cannot be explained in exclusively biological terms.

Human beings may not be the goal of the evolutionary process as a *whole*, but the emergence of humanity is a central part of the evolutionary story. The Darwinian mechanism of natural selection is "blind," but what it has produced is not. This is not to say that human nature itself has been steadily improving for the last 150,000 years. The kinds of improvements reflected in human society and in human conduct, such as it is, are much more a reflection of cultural evolution than ongoing biological evolution. Whatever moral progress has been accomplished in the course of human history is not the result of a significant shift in gene frequencies, but of developments in moral culture.

EVOLUTION AS DIRECTIONAL: CONWAY MORRIS

Simon Conway Morris provides an account of another important aspect of nature that can be considered in discussions of evolutionary directionality. Conway Morris responds to Gould with his theory of "evolutionary convergence," by which he means "the recurrent tendency of biological organization to arrive at the same 'solution' to a particular 'need.'"[12] Conway Morris does not posit some kind of general "drive" to complexity in species, but he does find that the structural limits to evolutionary possibilities lead to the recurrent evolution of certain biological properties. The metaphor of convergence expresses the fact that evolution

---

[12] Conway Morris, *Life's Solution*, p. xii. See also Conway Morris, *Crucible of Creation*.

"navigates the combinatorial immensities of biological 'hyperspace.'"[13] Examples of evolutionary convergence are abundant: the independent evolution of sensory systems such as the camera-like eye (which evolved separately in at least six different cases), olfaction, hearing, and insect "gyroscopes," to name a few examples. It is also exemplified in complex behavioral traits such as hunting patterns in many species, matriarchal social structures in sperm whales and elephants, and "farming" in ants and humans.[14]

Conway Morris is particularly interested in the convergent evolution of intelligence in many species, and most notably the "toothed whales" (and in this group especially dolphins) and the great apes, especially humans. He does not of course detect a simple linear development from slime mold to *Homo sapiens*:

> even though you and I are, by definition, lineal descendants of a bunch of African apes, hominid phylogeny can no longer be construed as an evolutionary railway: from the heavily browed and small-brained tree-climber to smooth-headed bicyclist. Despite the limitations of the fossil material and the taxonomic squabbles, it is clear that we are sole survivors of quite a substantial "bush" of hominid diversification.[15]

At the same time, he does not regard intelligence as a "by-product" of evolution, a lucky "bounce" off the left wall of organic life. Intelligence often provides adaptive advantages, so it is not surprising that it has evolved in many independent species. The larger-brained primates are more behaviorally flexible than others, able to use tools, and marked with greater capacities for social learning.[16] The gradual increase of the size of the neo-cortex among many primates indicates a "clear trend, parallel and independent, that is consistent with selective pressure."[17] The dramatic increase of pre-human brain size two million years ago gave rise to a long and complex chain of events that eventually allowed our Paleolithic predecessors to develop unprecedented levels of complexity in social organization and cognition, including the language-based symbolic capacities that provide the wherewithal for abstract communication and thought today.

A growing number of scientists recognize the transmission of learning and the rudiments of culture in nonhuman animals. In contrast to Thomas Aquinas' clear line separating "rational" from "irrational" animals, Conway Morris argues that our knowledge of the complex social networks and

---

[13] Conway Morris, *Life's Solution*, p. 127; see also ch. 1. [14] Ibid., p. xii.
[15] Ibid., p. 270.   [16] Ibid., p. 246.   [17] Ibid., p. 168.

developing cognitive powers in many species indicate that cultural capacities lie along a continuum.[18]

Conway Morris' argument centers on four claims: first, complexities are usually a way of organizing traits that are "inherent" in simpler systems; second, evolution cannot move in an infinite number of directions – the "number of evolutionary end-points is limited";[19] third, what is evolutionarily possible is "usually" arrived at many times, which, Conway Morris suggests, means that the emergence of various biological traits is "effectively inevitable"; and fourth, evolution takes a great deal of time and over the course of time what were once slim possibilities increasingly become probable and even, as time goes on, inevitable. Evolution often works in a "step-like arrangement: once one stage is achieved, other things then become so much more likely."[20] Contrary to Gould, Conway Morris argues that the evolution of sentient animals and then mammals led to a situation in which the emergence of human beings became a "near-inevitability."[21] We are not a "cosmic accident," he insists. Indeed, he argues against Gould, "Rerun the tape of life as often as you like, and the end result will be much the same."[22] He does not of course mean that evolution would inevitably produce a five-fingered, two-legged featherless *Homo sapiens*, only that something like human intelligence would again evolve. This conclusion also leads Conway Morris to speculate that other hospitable planets constrained by the same "landscape" of physical possibilities could give rise to roughly similar kinds of intelligent life. In any case, he properly notes, caution is in order: "it is far more prudent to assume that we are unique, and to act accordingly."[23]

Conway Morris provides reasons for thinking that scientific knowledge of biological evolution is congruent with the Christian belief that the world is God's creation. Evolutionary pathways are constrained and shaped by the way that nature is organized. "The organic world is a plenitude and a marvel, but it still has a rational structure."[24] The suggestion that evolution has an inherent directionality need be taken neither to minimize the significance of contingency nor to identify the evolution of the human race as the climax of the entire evolutionary process. "Yet, when within the animals we see the emergence of larger and more complex brains, sophisticated vocalizations, echolocation, electrical perception, advanced social systems including eusociality, viviparity, warm-bloodedness, and

---

[18] See ibid., p. 259. "Culture" includes tool use and other technological developments.
[19] Ibid., p. xii.   [20] Ibid., p. 18.   [21] Ibid., p. xii.
[22] Ibid., p. 282.   [23] Ibid., p. 328.   [24] Ibid., p. 303.

agriculture – all of which are convergent – then to me that sounds like progress."[25] And of course this notion of directionality has nothing to do with the "progress" read into nature by the social Darwinians or sought by eugenicists and utopian social engineers.

One would be hard pressed to find a stronger scientifically based alternative to Gould's estimate of evolutionary probabilities than that found in Conway Morris, who disputes every step of Gould's argument. Both biologists examine the same data, but they come to contradictory conclusions about its implications for the likelihood of human evolution and the presence of a general directionality in evolution.

EVOLUTION AS PURPOSIVE: STOEGER

I now turn to accounts of directionality proposed by explicitly Christian thinkers, and, more strongly, to construals of evolution as purposive. Astronomer William Stoeger, SJ, argues that the order of nature in its entirety, its laws and regularities, and its tendency to generate complexity all give us reason to think that it is structured according to an "immanent directionality." He tends to avoid the term "teleology" because of its ambiguity and philosophical connotations, but he believes that the notion of "immanent directionality" is reasonable, well-supported from evidence in a variety of scientific disciplines, and consistent with legitimate methodological reductionism.

The immanent directionality of evolution can be discerned in the regularities and processes studied by science. Understanding the structure of the natural world, and the course of evolution, does not require us to posit some supernatural power or "hidden forces"[26] that can channel the course of nature so that it achieves its end. "Each time a concerted effort to discern a vital force, a whole–part or top-down controlling factor, or a nomogenetic influence within evolutionary history and process is mounted, the results are negative."[27] To be sure, there is a lot about the structure of nature that escapes human knowing at this point, but on the level of the "big picture," at least, we can be confident that the natural sciences provide a sufficient approach for understanding the way in which the mechanisms,

---

[25] Ibid., p. 307.
[26] William R. Stoeger, SJ, "The Immanent Directionality of the Evolutionary Process, and Its Relationship to Teleology," in R.J. Russell *et al.*, eds., *Evolutionary and Molecular Biology: Scientific Perspectives on Divine Action* (Vatican City State: Vatican Observatory Publications, 1998), p. 166.
[27] Ibid., p. 163.

regularities, and contingencies of nature are expressed in natural history. Stoeger accepts fully the uncertainties and contingencies of natural history and the fact that evolutionary adaptations are the product of neither foresight nor planning.

The world is best understood, Stoeger argues, neither as a "rigidly deterministic" closed physical system nor as "completely open."[28] The natural sciences give abundant evidence of the crucially important fact that random events take place within the dynamic regularities, law-like processes, and patterns of nature. Uncertainty takes place within a structured evolutionary process. Stoeger does not believe that this structure can be identified only by those whose interpretation of nature flows from their Christian convictions, but he does think that Christians ought to be attentive to this directionality. Stoeger believes that scientific knowledge of the structures of the physical and organic world tells us something about God's general providence – "God's creative action," he writes, "is immanent in the processes revealed by the sciences."[29]

Stoeger's account of this directionality begins with cosmology and the Big Bang, the formation and destruction of the stars and the elements, the formation of planets, asteroids, comets, and so forth, and moves to the beginning of life on our planet – thus encompassing cosmological, prebiotic, and biotic stages that generate increased diversification and complexification. Each phase in this development created conditions that made latter stages more likely. In particular, "once we are outside the quantum-dominated real and outside the inflationary epoch, the conditions and regularities obtaining at any stage are, from what we well know, directed towards – are set for – the formation of the structures in the universe we now have."[30] Science studies this order in both the basic forces and structures of nature and in the operation of "nonlinear, nonequilibrium systems, and chaos," which also constitute important kinds of self-organization. Directionality does not depend on strict determinism at any given moment, "but rather a manifold of possibilities together with the capability of gradually actualizing some of those possibilities, which are more and more complex."[31] Stoeger believes that the natural sciences, and the methodological reductionism by which they are disciplined, provide a sufficient basis for accounting for the inherent directionality of cosmic evolution.

Stoeger extends the same principles to the emergence of life on earth and the subsequent course of biological evolution. Scientists cannot yet explain how life first came to appear on earth, how a planet rich in amino acids and

[28] Ibid. [29] Ibid., p. 166. [30] Ibid., p. 169. [31] Ibid., p. 173.

proteins became one pervaded by living organisms.[32] They know that the evolution of life involved an initial stage of chemical evolution, then moved to a stage which allowed for molecular self-organization, and finally to one marked by biological evolution proper.[33] Stoeger provides a helpful and detailed summary of current theory on these phases and the transitions between them. Biological evolution is structured by natural selection constrained by the environment, can be understood as the "steady generation of information" recorded in the genes of organisms, relative to whole organisms at each stage of their lives, marked by feedback loops and processes of self-organization, and stimulated by complex and unpredictable fluctuations of matter and energy. The influence of indeterminate processes is highly significant. Yet, though the actual course of evolution is indeterminate, Stoeger argues, "its general course towards complexity, self-organization, and even the emergence of self-replicating molecules and systems, given the hierarchies of global and local conditions which are given, can be interpreted as inevitable in the universe in which we live."[34]

Stoeger maintains, then, that evolution in all its phases and taken as a whole manifests an inherent long-range direction. He does not mean that evolution contains conscious purpose, nor that it is guided supernaturally to go where it would not have otherwise gone, nor that it has a specific predetermined target or "rigidly defined terminus."[35] Evolutionary events do not proceed according to a scripted plan controlled by a "long-range supervisory feedback mechanism."[36] Stoeger finds what Mayr and others call an "open teleonomic system,"[37] according to which the evolutionary process under various conditions gradually approaches a focused range of outcomes. Over time the evolutionary process develops what Stoeger calls a "nested set of directionalities which gradually emerge with ever greater specificity in certain locales within the overall evolutionary manifold."[38] As clusters of possibilities are actualized on different levels, what was once a remote possibility becomes more and more likely. This of course is exactly what Conway Morris examines in extensive detail in *Life's Solution*. Nature has a tendency over time to actualize more and more possibilities. This tendency involves higher levels of organization, greater complexity, the specification of structures in new forms of life, and

---

[32] Even here, though, Stoeger notes that "There is no indication whatsoever that something more than is studied or presupposed by physics and chemistry is needed in accounting for life"; ibid., p. 175.
[33] See ibid., p. 175.   [34] Ibid., p. 180.   [35] Ibid., p. 182.   [36] Ibid., p. 183.
[37] Ibid., p. 182, citing, among others, E. Mayr, *Towards a New Philosophy of Biology* (Cambridge, MA: Harvard University Press, 1988), pp. 38–66.
[38] Stoeger, "Immanent Directionality," p. 183.

the exploration of more and more potentialities embedded in nature – evolution is thus an open, dynamic process.

The evidence of evolutionary directionality does not of course demonstrate the belief that the evolutionary process serves divine purposes. Reflection on the meaning of the divine plan ought not to limit the pursuit of divine purposes to anthropomorphic images of human planning, goal-seeking, and strategizing. We formulate plans and execute them through controlling variables, and we do not typically make plans involving such things as long-range unpredictability, accidental events, random locations. As Stoeger points out, "Signs of divine purposiveness within creation may be much different and much more subtle than those we associate with human intentionality, and are probably not susceptible to detection by the sciences. Thus there *may be* such a divine intention or purpose driving evolution."[39] This point can be made more strongly: the very notion of "divine plan" falls outside the purview of science; science can neither affirm nor deny the presence of divine purpose in the evolutionary process.

Science, then, can be used neither for nor against claims of divine purposiveness. Stoeger's recognition of the limits of science and his sense of the productiveness of the evolutionary process give him room to affirm, as a matter of Christian faith, that the course of evolution reflects divine purposes. On the level of Christian faith, Stoeger writes, "we become aware that God is somehow working within the immanent dynamisms and interlocking directionalities of the evolutionary process – despite and even through its autonomy, contingency, inner freedom, and apparent blindness. This conscious divine purposiveness is only unambiguously manifest in God's revelation of God's action and intention to us."[40]

The debate over purpose in evolution is centrally related to the question of whether evolution as a whole is an "end-resulting" or "end-directed" process. Stoeger believes that the scientific data suggest that evolution is neither merely a minimal non-directional "end-resulting" process nor, at the other end of the spectrum, an "end-seeking" process (i.e., it is not like the functional organization of a bodily organ). He regards cosmic and biological evolutionary processes, taken in the broadest sense, as "end-directed." Evolution involves "end-directed" processes as part of the regular ordering of nature, and these occur because natural structures orient events to produce a general range of ends. Reference to law-like regularities in nature is just another way of speaking about the place of end-directed processes in nature.

---

[39] Ibid., p. 185; emphasis in original.   [40] Ibid., 186.

Gould and those who stress the role of chance events in natural history adamantly insist that evolution is only an "end-resulting" process. Conway Morris emphasizes that contingencies operate within the constraints, probability structures, and ordering frameworks of nature. Evolutionary directionality is given in the general law-like regularities of physics, chemistry, and biology. The evolutionary process is not directed to result in specific ends (say, the production of manatees) but it is constituted by regularities and patterns of order that are oriented to "general classes of end products."[41]

In Stoeger's judgment, Christian faith reveals most adequately the purpose of evolutionary directionality. Religious convictions lead him to regard the course of evolution as influenced by "overarching holistic principles" that shape long-range end-directedness toward the production of sensate life, conscious life, and finally self-conscious life. Science alone, then, cannot supply sufficient justification for the claim that evolution has an ultimate end or that we can know what it is. The cosmos as a whole hosts millions upon millions of "experiments" in basic physics and chemistry regarding the potentials of matter and energy that might be actualized under the pressure of end-directed process. The divine intentions disclosed in revelation are "not evident in what the natural sciences tell us about evolutionary history and development;"[42] at the same time, Stoeger immediately adds: "Nor are they contradicted by it."[43]

CHANCE IN AN EVOLUTIONARY WORLD: A THEOLOGICAL INTERPRETATION

This discussion of the role of contingency in evolutionary directionality is related to the role of chance in the evolutionary process. The question here focuses on how divine purpose can be reflected in a world marked by chance. The previous chapter argued that divine providence works through the influence of chance on the evolutionary process. At least three meanings of "chance" are relevant to this discussion. "Chance" can refer to an event that we could not have predicted because of the incompleteness of our own knowledge of all the relevant causal factors. The fact that an avalanche kills hikers on a mountain trail is, to its observers, a chance event, but presumably we would know why this event occurred if we had a complete account of all the relevant factors.

---

[41] Ibid., p. 188.   [42] Ibid., 190.   [43] Ibid.

A second meaning of the term "chance" pertains to events that result from the intersection of two previously unrelated causal chains. Their unexpected convergence leads to unforeseen consequences, either for good or for ill. A person who is killed accidentally dies by "chance" because no agent acted with the intent of producing this effect. This sense of chance is also seen in the cosmic evolution that led to the appearance of our solar system and planet and to genetic mutations that constitute the variations on which natural selection operates.

The third meaning of "chance" pertains to the results of micro-events studied in the field of quantum mechanics. Events at the quantum level, and also in some cases of large-scale chaotic systems,[44] can only be understood in terms of probabilities rather than in terms of determinative cause–effect relations. The first two meanings of chance refer to events that either we do not know enough about to have been able to predict from preceding conditions or are not produced from an intention of a rational agent. The third meaning suggests that nature is in some fundamental ways open and indeterminate.

None of these three meanings imply that chance events lie outside God's power. Thomas Aquinas, who tends to speak about chance in the first two senses, argued that God exercises his governing power through both law-like regularities and chance events – the latter work probabilistically according to God's will and the former work by necessity according to God's will. Secondary causes operate within the contingencies of nature, where something might or might not happen. God thus allows for the interplay of contingent events within the overarching framework provided by the law-like regularities of nature. God does not predetermine levels of rainfall and soil nutrients to sustain forests, so the well-being of forests waxes and wanes according to various natural and human contingencies.

The same interrelationship between chance and law, of course, obtains in human society. A pedestrian strolls down a street and a strong wind begins to blow over a dead tree, the tree falls toward the sidewalk on which the man is walking and it crushes him. This chance event is not divinely willed *per se*: it is said to take place "by accident." Yet because the order of creation includes the possibility of this kind of interaction between contingency and law, we say that this event accords with the universal order of things. It can be described as happening according to God's will, in that God wills that there be a world in which creatures act within the context of a vast network of contingencies and law-like regularities.

---

[44] See Polkinghorne, *Science and Providence*, pp. 28–30.

All of creation, then, and every creature in its own ways manifests some aspects of the intelligence and love of God. "All of creation" includes its contingencies as well as law-like regularities, its destructive as well as reproductive activities, its evolutionary "dead ends" as well as "success stories." God wills for the good of the universe that some things occur in nature necessarily and that others take place through the chance events that are part of the natural order. Creatures, subject to defect and failure, behave neither uniformly nor automatically in ways suitable for themselves.[45] God instead wills the communication of goodness to creatures through natural contingencies, including those that pervade the animal kingdom.[46]

Applied to the world of evolution, we can say that God is the primary cause of the entire world. When we speak of God as Creator we refer to the fact that he perpetually sustains the world in being; when we speak of the world as the creation, we refer to its radical ontological dependence on God. The doctrine of providence understands the evolutionary process as including the natural history of the unfolding of divine care for the world. Animals, plants, and inanimate objects constitute secondary causes for one another within the evolving world.

Evolutionary biologists obviously come to a much more adequate scientific understanding of the natural intelligibilities of these secondary causes than did their classical, medieval, and modern predecessors. Evolutionary biology does not need to refer to primary and secondary causality, indeed it cannot do so, because these are ontological terms rather than concepts belonging to the natural sciences. Evolution as the object of science can be understood entirely in scientific terms, but its theological and ontological significance must be described in terms of secondary causality. An adequate theological and philosophical understanding of the relation between God and nature needs to employ this distinction in order to avoid both conflating divine and natural kinds of causality and misinterpreting the Creator's relation to the world as basically similar (though with unimaginably greater power and knowledge) to human operations in the world.

A number of theologically sensitive contemporary writers describe the natural world in ways that comport well with this view of creation and providence. Stoeger understands divine activity as working in and through secondary causes to draw all creatures, and the universe itself, considered as a created whole, to their proper activity. He holds that "God is always acting through the deterministic and indeterministic interrelationships and

[45] *ST* I, 19,8. [46] See *SCG* 3,72.

regularities of physical reality which our models and laws imperfectly describe."[47] God is not bound by necessity to use intermediary conditions to produce certain effects – but as a matter of fact God does. This view of creation does not think of God as "intervening" into the natural course of events to bring about a desired effect, because God is not "outside" nature at all. God's immanence allows Stoeger to speak of divine interaction with creatures. Theologically it can be said that God works in continuing creation through the natural processes and structures that are studied in the scientific disciplines of cosmology, astronomy, geology, and so forth. These natural processes and structures reflect the interplay of chance events and law-like regularities. Since they can be understood entirely in strictly scientific terms, he insists, "There is no need for 'intervention' in these evolutionary processes."[48]

Physicist Howard Van Till takes much the same position as Stoeger. He describes creatures as endowed with capacities to realize a variety of forms conceived for them by the Creator. The structure of creation is "pregnant with possibilities," given a plenitude of potentialities in a "gapless economy."[49] Van Till follows Augustine's appreciation of God's endowment of the creation with a pervasive "functional integrity": "The universe was brought into being in a less than fully formed state but endowed with the capacities to transform itself, in conformity with God's will, from unformed matter into a marvelous array of structures and life forms."[50]

The vastly expanded time and space framework that has been discovered by science in the last century encourages Van Till and Stoeger to understand the structures of creation as unfolded through lengthy temporal process. Like Stoeger, Van Till understands the created order as given a relative autonomy by the Creator, and, for this reason, he understands science to have its own integrity as an intellectual enterprise. Scientific inquiry does not compete with theology, because the primary causality of divine creation constitutes a different order of explanation than that

---

[47] William R. Stoeger, SJ, "Contemporary Physics and the Ontological Status of the Laws of Nature," in Robert John Russell, William R. Stoeger, SJ, and George V. Coyne, SJ, eds., *Quantum Cosmology and the Laws of Nature – Scientific Perspectives on Divine Action* (Vatican City State: Vatican Observatory Publications, and Berkeley, CA: Center for Theology and the Natural Sciences, 1993), pp. 209–234.
[48] William R. Stoeger, SJ, "Faith Reflects on the Evolving Universe," in Michael Himes and Stephen J. Pope, eds., *Finding God in All Things: Essays in Honor of Michael J. Buckley, S. J.* (New York, Crossroad, 1996), p. 171.
[49] See Van Till *et al.*, *Portraits of Creation*. See also *ST* I, 73,1 ad 3 on the production of new species.
[50] Howard Van Till, "God and Evolution: An Exchange," *First Things* (June/July 1993), pp. 32–41, at p. 40.

examined in the natural sciences. God works in probabilistic rather than deterministic ways, and the Creator endows the created order with potentialities that are realized in the constant interplay between "chance and necessity" or "contingency and law."

CONCLUSION

Theologian Keith Ward expresses a characteristically Christian conviction when he writes that God created the cosmos in a way that would lead to the "generation of communities of free, self-aware, self-directing sentient beings."[51] Biologist Kenneth R. Miller applies the notion to the organic world:

> Given evolution's ability to adapt, to innovate, to test, and to experiment, sooner or later it would have given the Creator exactly what He was looking for – a creature who, like us, could know Him and love Him, could perceive the heavens and dream of the stars, a creature who would eventually discover the extraordinary process of evolution that filled His earth with so much life.[52]

Christians accept as a matter of faith the claim that creation is the expression of the divine purposes, but their faith is consonant with what scientists tell us about the nature of cosmic and biological evolution. One does not, of course, have to be Christian to believe that the natural dynamisms of emergent complexity were created to give rise to beings capable of knowing and loving. Christian faith can acknowledge an important truth that can be recognized by other religious believers as well. The antecedent plan of God precedes all action and God's consequent plan draws the acts of creatures into harmony with this purpose. The divine plan aims at the end of sharing God's goodness, but it does not do so by imposing a predetermined order of events on the evolutionary process.

This chapter's answer to Gould is that, for Christian faith, the Creator works in and through the contingencies, indeterminacies, and chaotic systems of nature. The Creator has made a world in which physical events take on a degree of independence from divine control, at least in the sense that God does not direct the course of every event in the physical world. As Barbour points out, "Natural laws and chance may equally be instruments of God's intentions. There can be purpose without an exact predetermined plan."[53] This relative independence is extended in significant ways in the

---

[51] Keith Ward, *God, Chance, and Necessity* (Oxford: Oneworld Publications, 1996), p. 191.
[52] Miller, *Finding Darwin's God*, pp. 238–239.   [53] Barbour, *Religion and Science*, p. 216.

evolution of our species, the concern of the remainder of this book. It is reflected in scriptural narratives where God does not "force" human beings to do what is right. God provides inspiration, edification, and strength, but God does not override the capacity of human beings to make free choices, even when these choices are destructive. The purpose of the cosmos is displayed in a special way in the evolution of human beings, whose distinctive traits make it possible for us to love God and one another.

CHAPTER 6

Human nature and human flourishing

This chapter examines both evolutionary and Christian theological accounts of "human nature," including its innate, biologically based needs and inclinations. I argue that evolutionary insights into our cognitive, emotional, and social capacities can inform Christian ethical understanding of human flourishing.

Evolutionary psychologists think of human nature as the collection of species-typical goal-directed mechanisms designed by natural selection. These mechanisms constitute a "human nature" that is virtually the same in all "modern" human beings, namely those living in the past 100,000 years or so. Biologists have argued that our "sweet tooth" and our natural craving for fatty foods, for example, reflects our predecessors' evolutionary adaptation to ecological conditions of periodic scarcity that rewarded maximum consumption and storage of calories in body fat.[1] These dietary proclivities were adaptive for Paleolithic hunter-gatherer bands but they cause well-known health problems for people living in modern developed sedentary societies.[2]

Central among our species-typical evolved inclinations is our need to form personal attachments and to participate in communal life. Evolutionists hold that the human psyche manifests basic conflicts between and among the teachings of culture and the demands of biologically based proclivities. They stress the inner tensions that reside in our emotional constitution as human beings and not only those due to the particularities of our cultures or personal life histories. Konrad Lorenz used the helpful phrase "Parliament of Instincts"[3] to communicate the dynamic interaction between many different and competing motivations in the human psyche.

---

[1] For a recent treatment, see Sharman Apt Russell, *Hunger: An Unnatural History* (New York: Basic Books, 2005).
[2] See P. G. Kopelman, "Obesity as a Medical Problem," *Nature* 404 (2000): 635–643.
[3] *On Aggression*, trans. Marjorie Kerr Wilson (New York: Harcourt, Brace and World, 1966).

His own research focused on the co-evolution of aggressive and conciliatory instincts, but other conflicts occur as well.

Any reference to "human nature" today is of course highly controversial, especially to those who regard our identity as the exclusive product of learning, socialization, and culture. Those who deny that there is such a thing as an innate set of biologically based traits shaped by Pleistocene pressures regard the human race as simply a vast collection of individuals who share nothing more in common than the most rudimentary genetically encoded physical traits and a wide biological potential for a very extensive range of different behaviors and different cultures.[4] These critics rightly complain that sociobiologists place too much weight on various conjectures about the selection pressures operative in an environment of evolutionary adaptedness about which we actually know very little, and they accuse the evolutionary psychologists of "using one tentative scientific construction (lifestyle and culture of Pleistocene hunter-gatherers) to develop another (fitness-enhancing mental modules that influence behavior)."[5]

As seen in chapter two, many Christian ethicists often assume that our evolutionary past has minimal influence on human nature today. They do not deny that we, like all other species, are the product of the evolutionary process, but they hold that we are no longer significantly influenced by our species' evolutionary past. This perspective avoids the "homogenization" of individuals and honors the distinctive character of particular communities, but it suffers from a kind of disembodied dualism that fails to acknowledge that we are all members of one species and share a long evolutionary history.

Acknowledging the existence of species-wide regularities (a better term than "universals") needs to be accompanied by a healthy sense of the variability and range of differences between cultures and from individual to individual within any given culture. The evolutionary view of human nature, in contrast to older positions, regards every species as marked by variation rather than uniformity. Steven R. L. Clark puts it, "Once we realize that human variety is not an error, that there is no one sort of human being that is 'what a human being should be,' and that we must expect our species always to be variegated, we can begin to think again about

---

[4] See for example Stephen J. Gould, "Biological Potential vs. Biological Determinism," in Arthur L. Caplan, ed., *The Sociobiology Debate* (New York: Harper and Row, 1978), pp. 343–351 (originally printed in *Natural History Magazine*, May 1976).

[5] Gregory Peterson, "Falling Up: Evolution and Original Sin," in Philip Clayton and Jeffrey Schloss, eds., *Evolution and Ethics: Human Morality in Biological and Religious Perspective* (Grand Rapids, MI: Eerdmans, 2004), p. 279.

constructing social orders that will provide a place for all."⁶ This evolutionary view of human nature has important ethical implications, says Clark: "We need a society that will have places – I do not mean asylums – for the aphasic, 'deformed', 'disturbed' and 'eccentric.'"⁷

With the aim of properly acknowledging the diversity within human nature, it might be helpful to distinguish carefully between human *desires* and biologically based *inclinations*. We experience desires in the immediacy of life. Desires are concrete, emotionally experienced attractions to this or that perceived good, for example to this hamburger or beer. Attempting to satisfy desires does not necessarily contribute to human flourishing for either self or others. The term "inclinations," on the other hand, can be used to refer to species-wide orientations to certain important classes of goods, for example food and drink. Inclinations are given, part of human nature; desires are not. These inclinations have been shaped by evolution and are a part of the common humanity that marks our species in general. When they appear in our individual lives it is because we are members of the species. Recognition of our animal heritage is consistent with the fact that people sometimes have quite different desires that have been shaped by an enormous range of factors. Our "material constitution," Rowan Williams points out, does ground certain basic dispositions, needs, and limits,

> but no one element in this exists without cultural mediation. We learn what we are in language and culture – even what we physically are. What I feel is structured by how I have learned to talk; what I want is what I picture to myself in the images I have learned to form from the observation of others, images that are not innocent representations of objects and goals but complex, differentiated constructions existing in potentially tense relation with the world of other subjects.⁸

Human nature is best conceived philosophically in terms of "emergent complexity," a notion whose importance was noted above. Speaking of the human person, Ian Barbour helpfully describes the meaning of emergence: "A living organism is a many-leveled hierarchy of systems and subsystems: particle, atom, molecule, macromolecule, organelle, cell, organ, organism, and ecosystem. The brain is hierarchically organized: molecule, neuron, neural network, and brain, which is in turn part of the body and its wider

---

⁶ Stephen R. L. Clark, *The Political Animal: Biology, Ethics and Politics* (London and New York: Routledge, 1999), p. 50.
⁷ Ibid.
⁸ Rowan Williams, *Lost Icons: Reflections on Cultural Bereavement* (Edinburgh: T. & T. Clark, 2000), p. 141.

environment."⁹ Human nature comprises a set of emergent capacities that make us distinctively human.

EMERGENT COGNITIVE CAPACITIES

The most remarkable factor in hominid evolution concerns the incredible expansion of the size and complexity of the human brain. The australopiths had a brain size of 350–500 ml, *Homo habilis* had a brain size of 800 ml, and *Homo sapiens* 1700 ml. The brain to body ratio found in modern *Homo sapiens* is three times that of our hominid precursors, the australopithecines, a mere (in evolutionary terms) two million years ago.[10]

It is not, of course, the sheer size of the human brain that matters, but the fact that its internal architecture gives us the ability to communicate through the use of symbolic representation and especially through language. Unlike their precursors, Cro-Magnons who migrated to Europe from Africa some forty thousand years ago were "drenched in symbol."[11] Their symbols included the use of decorative sculptures, cave paintings, music, graves, body ornaments, ceramics, utensils, and, most importantly, spoken language (made possible, of course, by anatomical changes in the structure of the throat).

Human cognitive capacities go well beyond those of chimpanzees, but they did not appear out of the blue. Emerging cognitive capacities made it possible for human beings to engage their imaginations, to speculate about various courses of action, and to interpret the mental and emotional states of their interlocutors with subtlety and nuance. "Members of other species often display high levels of intuitive reasoning, reacting to stimuli from the environment in complex ways," Tattersall observes, "but only human beings are able arbitrarily to combine and recombine mental symbols and to ask themselves questions such as 'What if?'" The evolution of cognitive symbolic capacities led to the existence of personal beings possessing unified, self-conscious centers. This "What if?" is extended not only to behavior involving hunting and gathering, but also to modes of personal and group living. Integral to personal and group living is the social

---

[9] Barbour, "Neuroscience, Artificial Intelligence, Human Nature," p. 269.
[10] Recent works include Ian Tattersall, *Becoming Human: Evolution and Human Uniqueness* (San Diego, New York and London: Harcourt Brace and Co., 1998), and *The Monkey in the Mirror: Essays on the Science of What Makes Us Human* (Orlando, FL: Harvest Books, 2002); F. B. M. de Waal, *Tree of Origin: What Primate Behavior Can Tell Us about Human Social Evolution* (Cambridge, MA: Harvard University Press, 2001); and Craig Stanford, *Significant Others: The Ape–Human Continuum and the Quest for Human Nature* (New York: Basic Books, 2001).
[11] Tattersall, *Monkey in the Mirror*, p. 120.

institution known as "morality" and its categorization of acts as right or wrong.[12]

Scientists have proposed the hypothesis that human intelligence underwent this remarkable surge of growth partially because of selection pressures exerted by the increased complexity of social interactions. The larger the social group, the more cognitive skills were needed to understand, monitor, and respond appropriately to social interactions. Large-group living rewards individuals with some ability to engage in the cognitive skill of intuitive "mind-reading" (the "theory of mind")[13] by which an individual comes to understand how his or her acts are viewed from perspectives other than that of the agent. This capacity might play a role in sympathy and pro-social behavior but it also lies behind deception and manipulation (hence its description as "Machiavellian intelligence").[14]

Our central nervous system makes possible unprecedented degrees of behavioral flexibility that allow us to be less restricted by the limits of our local environments than many other species. The development of consciousness and self-consciousness makes possible these new degrees of behavioral plasticity. The term "consciousness" refers to the capacity to experience the world through the senses and to respond cognitively to that experience. Antonio Damasio distinguished "core consciousness" that other primates experience in their immediate interaction with their environments from the "extended consciousness" that human beings can experience because of their capacity to remember the past and anticipate the future.[15] "Extended consciousness" seems to be a condition of "self-consciousness," the second-order capacity to be aware of one's own experience, internal responses, and distinct individuality. Use of the "mirror test" indicates that chimpanzees, and maybe orangutans and dolphins, can be

---

[12] Goodall points out: "When one human begs forgiveness from or gives forgiveness to another there are, however, moral issues involved; it is when we consider these that we get into difficulties in trying to draw parallels between chimpanzee and human behavior. In chimpanzee society the principle involved when a subordinate seeks reassurance from a superior, or when a high-ranking individual calms another, is in no way concerned with the right or wrong of the aggressive act" (*In the Shadow of Man* [New York: Random House, 1974], p. 244).

[13] Simon Baron-Cohen, "The Evolution of a Theory of Mind," in Michael C. Corballis and Stephen E. G. Lea, eds., *The Descent of Mind: Psychological Perspectives on Hominid Evolution* (New York: Oxford University Press, 1999), pp. 261–277.

[14] Richard W. Byrne and Andrew Whiten, eds., *Machiavellian Intelligence: Social Expertise and the Evolution of Intellect in Monkeys, Apes, and Humans* (Oxford: Clarendon Press, 1988).

[15] See Antonio Damasio, *The Feeling of What Happens: Body and Emotion in the Making of Consciousness* (New York: Harcourt Brace and Co., 1999).

aware that they are distinct and different from other conspecifics or group members; most animals are not.[16]

A conscious animal has its own felt experience of what it is like to be itself. A conscious person has both this felt experience and a cognitive capacity to reflect on what it is like to have that experience. Self-consciousness brings new capacities for self-reflective awareness, self-possession, and "centeredness" for the whole human organism. It allows the person to direct his or her energies and capacities to goals on behalf of the whole organism. Most importantly for this book, human consciousness depends on the body but manifests levels of intelligence and love that cannot be explained simply in terms of the needs and goals of the body.

Evolutionists are often tempted to draw superficial analogies between animal and human behavior. David Barash provides a graphic example of misplaced evolutionary analogizing. Immediately after acknowledging that human behavior is not as simple as that of other animals, Barash proceeds to write:

mallard rape and bluebird adultery may have a degree of relevance to human behavior. Perhaps human rapists, in their own criminally misguided way, are doing the best they can to maximize their fitness. If so, they are not that different from the sexually excluded bachelor mallards. Another point: Whether they like to admit it or not, many human males are stimulated by the idea of rape. This does not make them rapists, but it does give them something else in common with mallards.[17]

This bizarre analogy fails to acknowledge that human conduct is based on animal capacities but can only be understood adequately in light of human purposes, motives, and intentions. Attempts to understand human action as nothing more than physical events miss the fact that, as Charles Taylor puts it, "things matter" to us because we are "self-interpreting animals."[18] An adequate scientific theory must therefore be sensitive to the central place of intentions in human actions. It must "save the phenomena" of our

---

[16] See G. Gallup, J. Anderson, and D. Shillito, "The Mirror Test," in M. Berkoff, C. Allen, and G. Burghardt, eds., *The Cognitive Animal: Empirical and Theoretical Perspectives on Animal Cognition* (Cambridge, MA: MIT Press, 2002), pp. 325–333; D. Reiss and L. Marino, "Mirror Self-Recognition in Bottlenose Dolphin: A Case of Cognitive Convergence," in *Proceedings of the National Academy of Sciences of the United States of America*, 98.10 (May 2, 2001): 5937–5942. The mirror test of course shows that some animals can be aware that their bodies are distinct from other animals but not necessarily that they have a distinct sense of "self."
[17] David Barash, *The Whisperings Within: Evolution and the Origin of Human Nature* (New York: Harper and Row, 1979), p. 55.
[18] Charles Taylor, *Human Agency and Language: Philosophical Papers*, vol. 1 (Cambridge: Cambridge University Press, 1985), chs. 1 and 4.

ordinary views of ourselves as responsible agents, and it ought to take into account all of the properties that are characteristic of an "evaluative animal" with its relatively significant powers of self-determination.

We are different from other animals in our possession of higher cognitive capacities, including those that make possible self-consciousness and self-transcendence. Relative to other animals, we possess unique degrees of intelligence and capacities for self-determination. Our designation as *sapiens* was made possible by the significant enlargement of the brain and by its uniquely complex organization. Darwin exaggerated when he claimed that "there is no fundamental difference" of kind in the mental powers of humans and "higher mammals."[19] Our enormous capacity for information processing and for forming symbolic representations of things in the world empower us for kinds of agency that are impossible for other animals. Language and other symbols make possible our unique ability to represent the world and to pursue goals in a conscious and deliberate way.

We are able to present to ourselves reasons for acting the way we do and to form "second-order desires,"[20] or evaluations and choices about desires in light of their moral ranking in relation to moral considerations, most generally their place in our conception of human flourishing. Language is the symbolic medium which makes possible the articulation of this kind of evaluation.[21] This capacity also allows individuals to misrepresent and conceal their own motivations from others. It gives us, evolutionary psychologists argue, an evolved tendency to overestimate our own virtue and to deceive ourselves about the most influential motivations lying behind our acts. Indeed, the most effective interpersonal strategy flows from sincere self-deceived feelings of benevolence, not from crass selfishness or cynical detachment from the well-being of others. Biblical narratives display a significant awareness of human psychological complexity and our capacity for self-deception (the stories of Cain and Abel, Gen. 4:1–16; David and Bathsheba, 2 Sam. 10–12; etc.).

Culture played the decisive role in the development of the potential that resides in the modern human brain. The evolution of cognitive capacities for symbolism, rooted in the higher centers of the brain, brought with it abilities to engage in chains of reasoning, to express thought verbally, to reflect on lessons from the past, to make detailed plans for the distant future, and to communicate from a long distance with others (e.g. by

---

[19] In *Darwin: A Norton Critical Edition*, second edn., ed. Philip Appleman (New York: W. W. Norton, 1979), p. 177.
[20] Taylor, *Human Agency*, p. 28.   [21] See ibid., p. 73.

sending verbal messages through third parties). Brain evolution brought with it vastly greater possibilities for destructive as well as constructive behavior. This cognitive superiority might even have enabled these anatomically modern humans to eliminate Neanderthals.[22] The ancient brain structures that lie beneath the modern neo-cortex put us in a position to experience enormous internal and external conflict, confusion, and ambiguity.

Our ability to use symbolic communication, and especially language, gives us a unique intellectual capacity that is not even remotely approximated in the lives of other animals.[23] Yet the recognition that human beings have special degrees of intelligence need not lead to an underestimation of the different forms of animal intelligence – their capacities for nonverbal communication, information processing, monitoring of the behavior of other animals, deliberate deception, food gathering, the use of tools, and various skills in problem-solving. Jane Goodall has shown that chimpanzees have the ability to solve complex problems, use and make tools for various purposes, communicate in elaborate detail with one another, and show the beginnings of self-awareness.[24] Animals display countless signs of intelligence, even if these do not include "reasoning."

Practical rationality is made possible by language. A hungry dog knows to eat meat and not wood. The dog does not, however, have the ability to advert to the fact that meat is food and wood is not, and he cannot articulate to himself the status of food as a means for sustaining life. It is thus impossible for the dog to make a deliberate choice – to choose after having gone through a process of deliberation over the relative value of various potential means to a given end and to make an informed judgment about the most appropriate means to that end. As Larry Arnhart explains, "Human beings use their unique capacity for rational deliberation to formulate ethical standards as plans of life for the harmonious satisfaction of their natural desires over a complete life."[25]

Brain evolution has made us highly verbal creatures. "Out of its brain," Dennett writes, "it spins a web of words and deeds, and, like the other creatures, it doesn't have to know what it's doing; it just does it."[26] This "spinning" is primarily in the form of story-telling: "Our fundamental

---

[22] Ibid., p. 148.
[23] See Robert W. Sussman and Audrey R. Chapman, eds., *The Origins and Nature of Sociality* (Hawthorne, NY: Aldine de Gruyter, 2004).
[24] Goodall, *In the Shadow of Man*, esp. ch. 19.
[25] Larry Arnhart, *Darwinian Natural Right* (Albany, NY: SUNY, 1998), p. 2.
[26] Daniel Dennett, *Consciousness Explained* (Boston: Back Bay Books, 1992), p. 416.

tactic of self-protection, self-control, and self-definition is not spinning webs or building dams, but telling stories ... about who we are ... Our tales are spun, but for the most part we don't spin them; they spin us. Our human consciousness, and our narrative selfhood, is their product, not their source."[27] For Dennett, the self is the "Center of Narrative Gravity."[28]

Recent work on the "theory of mind"[29] also sheds light on human cognitive capacities. Simon Baron-Cohen holds that there are eight categories of behavior that require "theory of mind": (1) intentional communication with others, (2) the ability to repair problematic communication with others, (3) the ability to teach others, (4) intentional persuasion of others, (5) intentional deception, (6) planning and pursuing goals, (7) intentionally holding a focus or object of attention with others, and (8) pretending something to be the case when it is not.[30]

Cognitive capacities allow for emergent behavior that transcends fitness. Instead of the one-directional sociobiological reductionism that assumes that brain states give rise to mental states, a holistic approach to mind and brain affirms both that brain chemistry can influence conscious states of mind and that conscious states of mind can in various ways shape brain chemistry. Mind and brain interact in "top-down" and "bottom-up" feedback loops.[31] This accords with the non-reductionist account of emergence advanced by Roger Sperry: "As a brain scientist, I now believe in the causal reality of conscious mental powers as emergent properties of brain activity and consider subjective belief to be a potent cognitive force which, above any other, shapes the course of human affairs and events in the world."[32]

According to philosopher Anthony O'Hear, our higher-level intellectual capabilities were "selected" by nature but not directly "selected for,"[33] that is, they were apparently the indirect result of the operation of capabilities that were adaptive in the past. They allow us to prescind from what is immediately helpful in the practical order, not to mention what is fitness enhancing. As O'Hear observes, "the development of our reasoning powers

---

[27] Ibid., p. 418.
[28] Ibid., p. 427. Dennett holds that such "centers of narrative gravity" are "magnificent fictions."
[29] This phrase was first coined in David Premack and Guy Woodruff, "Does the Chimpanzee Have a Theory of Mind?" *Behavioral and Brain Sciences* 4 (1978): 515–526.
[30] Baron-Cohen, "The Evolution of a Theory of Mind," pp. 262–265.
[31] See Fraser Watts, "Are We Really Nothing More than Neurones?" *Journal of Consciousness Studies* 1 (1994): 275–279 – a response to Francis Crick's famous statement in *The Astonishing Hypothesis* (1994) that we are "nothing but a pack of neurones."
[32] Roger Sperry, "The New Mentalist Paradigm and Ultimate Concern [part 1]," *Perspectives in Biology and Medicine* 29.3 (spring 1986): 413, cited in Terence L. Nichols, *Sacred Cosmos: Christian Faith and the Challenge of Naturalism* (Grand Rapids, MI: Brazos, 2003), p. 146.
[33] O'Hear, *Beyond Evolution*, p. 142.

which has been made possible through self-consciousness has given us cognitive goals which have nothing to do with the acquisition of adaptive beliefs or skills."[34]

Among the most remarkable of all human activities are those generated by the desire to understand for its own sake. The human mind has a desire for truth that transcends instrumental values. The desire can of course be derailed, manipulated, biased, and disguised, but it exists nonetheless. The actual performance of evolutionary biologists themselves exemplifies a commitment to discovering what is the case, regardless of its practical application in the future. As Mayr puts it, "Virtually all biologists are religious in the deepest sense of the word . . . The unknown, and maybe the unknowable, instills in us a sense of humility and awe."[35] Wilson and Dawkins claim that their positions are reasonable, cogent, and empirically supported; they do not recommend that we believe them because doing so will maximize our fitness. This is even true of those, such as Ruse, who claim that we ought to have the intellectual courage to face the fact that believing that we can engage in the search for truth is an illusion perpetrated on us by our genes.[36]

The kinds of questions asked by thoughtful people far outstrip the mundane use of the "computational equipment" that evolutionary psychologists say evolved for the purposes of solving problems of survival and reproduction. How is it that a primate with a vast array of tools tailored to promote survival and reproduction on the African savannah can produce a Bach piano concerto, quantum physics, Hegelian metaphysics, or Hamlet? How is it, psychiatrist Peter Kramer asks, that "A traumatized ape is uneasy when it feels far from the familiar," but that "Only a person can feel lost in the cosmos"?[37]

Some evolutionists regard the capacity for these kinds of cognitive activities as merely an accidental and trivial result of the evolutionary process. Gould speculates that a range of human cognitive capacities, including the ability to use language, are not the result of specific adaptations for abstract thinking and speaking, but are by-products of capacities that were adaptive in the past. The human brain possesses some

---

[34] Ibid., p. 69.
[35] Ernst Mayr, *The Growth of Biological Thought* (Cambridge, MA: Harvard University Press, 1982), p. 81.
[36] See Michael Ruse and Edward O. Wilson "The Evolution of Ethics," *New Scientist* 108 (October 17, 1985): 52.
[37] Peter Kramer, *Listening to Prozac* (New York: Penguin, 1997), p. 284. Kramer refers to Walker Percy's *Lost in the Cosmos: The Last Self-Help Book* (New York: Macmillan, 2000).

functionally specialized adaptations that evolved because they conferred reproductive advantages. Yet the mind also shows a wide range of abilities to engage in activities that have nothing to do with solving ancestral problems. Gould likens these capacities to the "spandrels" found in non-functional niches of cathedrals:

> Yes, the brain got big by natural selection ... [but] the brain did not get big so that we could read or write or do arithmetic or chart the seasons – yet human culture, as we know it, depends on skills of this kind ... [T]he universals of language are so different from anything else in nature, and so quirky in their structure, that origin as a side consequence of the brain's enhanced capacity, rather than as a simple advance in continuity from ancestral grunts and gestures, seems indicated.[38]

Abilities to write novels or construct metaphysical systems have nothing to do with fitness and reflect powers of mind that go far beyond fitness advantages. Some intellectual capacities are the non-adaptive side effects of features of the mind that are adaptive. These powers have been extended and developed for other purposes. Gould's "spandrels" thesis is more plausible than the sociobiological hypothesis that poetry and metaphysics confer fitness advantages on their authors through increasing their social status. The modern human mind has taken on a life of its own, one connected only in the smallest way to "fitness interests."

The human brain and central nervous system make possible forms of cultural evolution that allow for kinds of conduct that go beyond the biological needs of organisms. As Elliott Sober puts it, the "brain is able to liberate us from the control of biological evolution because it has given rise to the opposing process of cultural evolution."[39] Cultural evolution can be more powerful than biological evolution because "thoughts spread faster than human beings reproduce."[40] Culture provides a kind of moral leverage against some of the less desirable aspects of our evolutionary heritage.

Human beings came to meet their material needs more effectively in communities, and intellectual abilities initially employed to meet the demands of survival continued to transcend what is instrumentally necessary for fitness. The larger brain no doubt provides more than enough cognitive ability to meet the needs of basic physical survival in many environments, but the increased complexity of changing social

---

[38] S. J. Gould, "Tires to Sandals," *Natural History* (April 1989): 4. The classic article was S. J. Gould and S. Lewontin, "The Spandrels of San Marcos and the Panglossian Paradigm: A Critique of the Adaptationist Programme," *Proceedings of the Royal Society*, series B, 205 (1979): 581–598.
[39] Elliot Sober, "When Natural Selection and Culture Conflict," in Holmes Rolston III, ed., *Biology, Ethics and the Origins of Life* (Boston: Jones and Bartlett, 1995), p. 151.
[40] Ibid., p. 156.

environments came to place new and greater demands on human intelligence. This is true of higher primates in general, but especially true of human beings because of the intensely interactive nature of human social life.

EMERGENT EMOTIONAL CAPACITIES

Human emotional capacities and needs, like human intelligence, are the product of a long evolutionary process marked by emergent complexity. Psychiatrist Antonio Damasio helpfully distinguishes "emotions," "feelings," and "choices."[41] The term "emotion" refers to a collection of chemical and neural responses that begin in some regions of the brain and then influence other regions of the brain and various parts of the body. Emotions are caused by transient changes in the body that are triggered by internal and/or external factors. The emotion of fear prompts the brain to release adrenaline into the bloodstream and to send signals down neural pathways to make muscles prepare for action. This kind of emotional reaction is facilitated by what Damasio calls the "emotion induction sites," including the "ventromedial prefrontal cortices, the cingulated cortex, and the amygdala," which enable the brain to notice an emotionally relevant situation. The brain's "emotion effectors," the brain mechanisms at the level of the "brain stem, the hypothalamus, and the basal forebrain/ventral striatum,"[42] receive messages from the "induction sites" and then in turn send further messages to other regions of the brain and other parts of the body. Natural selection has made sure that such reactions are triggered automatically by the appropriate stimulus and are not dependent on any rational choice of the agent. Attentive hikers do not have to make a deliberate decision to be afraid of the rattlesnake encountered on the path or to avoid walking too close to the edge of a deep canyon. Quick emotional responses to danger are more effective than deliberation.

Emotions are always prompted by some particular situation. This situation is emotionally provocative either because of "innate" or "learned inducers." Human beings are primed by evolution to respond emotionally to certain situations. "Innate inducers" include the sudden appearance of a steep cliff, the smell of delicious food, the smile of a baby, and the like.

---

[41] Damasio does not offer an uncontroversial theory of emotion that is accepted by all philosophers and scientists, but there does not seem to be any consensus on the nature of emotion among experts.

[42] Antonio Damasio, "A Note on the Neurobiology of Emotions," in Stephen G. Post *et al.*, *Altruism and Altruistic Love: Science, Philosophy, and Religion in Dialogue* (New York: Oxford University Press, 2002), p. 266.

Complex emotions can be influenced by a variety of combinations of "inducers." Emotions lead to specific behaviors, such as facial gestures, visceral reactions. Some of these are externally detectable by third party observers, but others are not.

Damasio distinguishes among a number of different kinds of emotions, all of which reflect the operation of a biological system subject to a long process of evolutionary "fine-tuning." Primary emotions recognized across cultures include fear, anger, happiness, sadness, disgust, and surprise. "Secondary" or "social" emotions – emotions generated in relation to other people – include embarrassment, pride, jealousy, and guilt. "Background emotions" include calm and tension, well-being, and malaise. Emotions constitute states of pleasure or pain and take place within the context of a set of underlying behavioral proclivities that are known as "drives" and "motivations."

Emotions evolved as ways to help our evolutionary ancestors cope with their environments. They play two key regulative roles in the biological life of an organism. First, they allow the organism to react to an object that causes an emotion, for example to flee danger, to recognize the presence of food or drink. Second, emotions shape the internal condition of the animal so that it is able to react in the right way when called upon to do so. The surge of adrenaline alters the internal state of an animal so that it can fight to protect itself if confronted with an enemy. Emotions are "action-programs" that can be either innate (e.g. fear of a large animal looming in the shadows at night) or learned from concrete experience (e.g. fear of an electric outlet after having been shocked by it).

Animals have emotions, Damasio holds, but human beings also have "feelings" of emotions, or "mental mappings" of emotional changes.[43] It is one thing to undergo something, and another to feel that one is undergoing it. Emotions are publically observable, but feelings are private, undetectable by third parties. These mental images are based on neural patterns in the various brain structures that "map, in sensory form, events that occur internally in the organism."[44] Mental images are preserved in the mind and associated with the vast array of other mental images with the self's memory bank. Emotions have a pervasive influence on the brain's neural circuits and, more generally, the body, the "theatre" of the emotions.[45]

"Feelings" are defined as the "mental representation of all the changes that occur in the body and in the brain during an emotional state."[46] They help us to recognize a problem that the body is trying to solve or an

---

[43] Ibid., p. 268.   [44] Ibid.   [45] Ibid., p. 266.   [46] Ibid., p. 265.

opportunity it is seeking to take advantage of. This awareness of emotions in turn allows us consciously to advert to them and to consider various modes of action that might respond to the perceived problem or opportunity. Consciousness and self-consciousness give us greater degrees of freedom from biological conditions. The behavioral flexibility[47] and self-awareness provided by self-consciousness give some control over how to act in light of induced emotions. The ability to understand another person's feelings also makes it possible for us to experience emotions such as compassion and pity.

As supremely cultural animals, we are profoundly shaped by learning. The emotion of fear responds to what is perceived as dangerous, and we usually identify what is dangerous through a long process of learning and not just as a matter of instinct. Some experiences of immediate danger are "visceral" – perhaps fear of snakes – but the content of what we feel is dangerous is profoundly influenced by socialization, instruction, imitation, cultural transmission, and personal life experience. What is true of fear is also true of the other basic emotions such as sadness, happiness, anger, surprise, and disgust. Emotions are ways of responding to situations that confront us, and since we can only respond to our emotions by interpreting them, it makes little sense to think of emotions or feelings as noncognitive, irrational impulses or as any more "natural" than our cognitive capacities. As Charles Taylor points out, feelings implicitly include judgments. A description of and reflection on feelings can help to shape particular emotions and the place we give them in our lives.[48]

Human choices, while often considered primarily cognitive, are profoundly affected by emotions. Damasio maintains that life experiences give neuro-chemical "tags" – or "somatic markers" in the pre-frontal cortices – to particular questions, stories, images, and the like, that interpret these objects either in negative terms, as a kind of "automated alarm signal," or in positive terms, as an indication of expected pleasurable or beneficial outcomes. "Somatic markers" are shaped by the contingencies of each particular emotional life-history and might be understood in terms of Aristotelian "habituation." They shape memory and imagination, influence attentiveness to sense experience and social interactions, and provide an emotional framework that helps to guide decision-making by predisposing the agent toward or against a given object, but they do not take the place of decision-making.

---

[47] See Damasio, *The Feeling of What Happens*.   [48] See Taylor, *Human Agency*, pp. 100–101.

Christian ethicists have recently stressed the importance of emotions for the moral life.[49] Unlike all other animals, we are not endowed with a strong set of instincts but rather need to be taught how to live. Behavioral plasticity is one of the most important features of human nature, and emotions can affect actions in ways that are either beneficent or maleficent. Moral training brings a more or less settled set of patterns to how the agent responds to natural emotional capacities and it empowers the mature adult to transcend the appeal of immediate gratification of desire for the sake of more worthwhile ends. It allows one, for example, to resist the emotional appeal of quick retaliation as a response to aggression or betrayal. Reason ideally guides the expression of emotions without suppressing or distorting them.

EMERGENT SOCIAL CAPACITIES

Emergent emotional capacities are closely related to emergent social capacities. We have inherited from our precursors a strong proclivity for sociality – a tendency to live and thrive in groups, to depend upon networks of cooperation, reciprocity, and commitment to one another. The emotional and cognitive capacities already mentioned make possible unprecedented psychological complexity, forms of communication, and cognitive depth and symbolically based reasoning.[50]

If human beings are naturally moved by social emotions, then perhaps pity, empathy, and other pro-social feelings are not simply laid on top of a substrate that is essentially antisocial. If this is the case for human beings, then our first-person experience of caring for others is neither illusory nor derivative from self-concern. In this case social institutions have more to build on than atomistic individualism typically recognizes,[51] and civil

---

[49] In Christian ethics, see Diana Fritz Cates, "The Religious Dimension of Ordinary Human Emotions: Working with Gustafson and Nussbaum," *Journal of the Society of Christian Ethics* 25.1 (2005): 35–53 and "The Virtue of Temperance," in Stephen J. Pope, ed., *The Ethics of Aquinas* (Washington, DC: Georgetown University Press, 2002), pp. 321–339; Paul Lauritzen, "Emotion and Religious Ethics," *Journal of Religious Ethics* 16.2 (fall 1988): 307–324; Edward Collins Vacek, *Love Human and Divine: The Heart of Christian Ethics* (Washington, DC: Georgetown University Press, 1998). In moral philosophy, see, among others, Robert Solomon, *The Passions: Emotions and the Meaning of Life* (Indianapolis, IN: Hackett Publishing Company; repr. 1993 [1976]); Martha Nussbaum, *The Therapy of Desire: Theory and Practice in Hellenistic Ethics* (Princeton, NJ: Princeton University Press, 1994); Michael Stocker, *Valuing Emotions* (New York: Cambridge University Press, 1996).
[50] See Sussman and Chapman, eds., *Origins and Nature of Sociality*.
[51] See Charles Taylor, "Atomism," *Philosophy and the Human Sciences: Philosophical Papers*, vol. II (New York: Cambridge University Press, 1985), pp. 187–210.

society is more than a cease-fire called between antagonists. Society is a network of communities that make a positive contribution to human well-being. Social ethics takes its bearings from an account of what contributes to the common good, and politics ought to work for its promotion through laws and public policy.[52] Certain kinds of social conditions can increase the likelihood of aggression and make necessary the engagement of reconciliatory behaviors. Some of these may simply be "friendly and affiliative" behaviors rather than "reconciliatory."[53] One interesting question this poses is whether competition and agonistic behavior are only generated from the breakdown of social order rather than its underlying basis.[54]

Human beings are, under the proper circumstances, capable of remarkable degrees of self-control and emotional refinement. Certainly human emotional capacities, like all other human capacities, are themselves the product of our long evolutionary past. Human motivations are mixed, ambiguous, and wide-ranging. They are only possible because of the evolution of the "limbic system" and neurotransmitters. Cognitive and emotional human capacities should thus not be separated from one another or assumed to be inherently mutually antagonistic. The kinds of complex reasoning usually associated exclusively with the recently evolved neocortex of the higher brain does not just "ride on top of" the much older and more primitive lower brain associated with the emotions and physical states. Both parts of the brain are intimately connected in reasoning and in emotional experience. "Nature appears to have built the apparatus of rationality," Damasio writes, "not just on top of the apparatus of biological regulation but also from it and with it."[55]

The major counter-argument to the position developed here regards pro-social behavior as derivative from a more primary egoistic and competitive set of motivations. One cluster of related emotions that were especially significant to Darwin was sympathy, compassion, and pity. The term "sympathy" was more current in Darwin's background than compassion, and it carries a slightly different connotation. "Sympathy" suggests a "feeling sorry for" another person who suffers, whereas compassion involves a "suffering with" a victim. Historically, the term "compassion"

---

[52] See Hollenbach, *The Common Good and Christian Ethics*.
[53] See Augustin Fuentes, "Revisiting Conflict Resolution: Is There a Role for Emphasizing Negotiation and Cooperation instead of Conflict and Reconciliation?," in Sussman and Chapman, eds., *Origins and Nature of Sociality*, ch. 10.
[54] Robert W. Sussman and Paul A. Garber, "Rethinking Sociality: Cooperation and Aggression among Primates," in Sussman and Chapman, eds., *Origins and Nature of Sociality*, ch. 8.
[55] Antonio Damasio, *Descartes' Error: Emotion, Reason, and the Human Brain* (New York: Avon Books, 1994), p. 128.

had a certain resonance with the emotion of pity. In antiquity a person was said to have "pity" when witnessing the undeserved suffering of another person. It was made possible by the natural imaginative and affective capacities conjoined to our social nature. The modern period introduced a new egoistic way of thinking about pity. Hobbes defined pity, compassion, and "fellow-feeling" alike as "grief, for the calamity of another ... [which] ariseth from the imagination that the like calamity may befall himself."[56] In this definition, the most powerful have no compassion for the most miserable because they have no possibility of ever experiencing such a state. Hobbes thought that the experience of compassion creates in us a feeling of our own vulnerability, but that it is always directed to those who suffer from pains that the agent could also suffer. When a person acts out of compassion, Hobbes thought, he or she does so not only with the intention of benefiting himself or herself but also to be removed from the experience of the pain associated with pity. He believed that the good Samaritan acted in order to relieve his own anxiety or emotional discomfort, not directly to help the victim for his own sake.

Jean-Jacques Rousseau, a philosopher usually depicted as Hobbes' intellectual opponent, also developed an egoistic account of compassion. Rousseau, like Hobbes, held that we care about others to the extent that doing so contributes to our own self-interest. The truly happy person, Rousseau held, is the solitary person, completely alone: "It is man's weakness which makes him sociable."[57] Pity is thus "sweet" because we feel the pleasure of not suffering what the other has gone through. Moreover, Rousseau confessed,

when the strength of an expansive soul makes me identify myself with my fellow and I feel I am, so to speak, in him, it is in order not to suffer that I do not want him to suffer; I am interested in him for love of myself, and the reason for the precept [of the Golden Rule] is in nature itself, which inspires in me the desire of my well-being in whatever place I feel my existence ... Love of men derived from love of self is the principle of human nature.[58]

---

[56] Hobbes, *Leviathan: Or the Matter, Forms and Power of a Commonwealth Ecclesiasticall and Civil*, ed. Michael Oakeshott (New York: Collier, 1977), p. 53. Aristotle defined pity as "a feeling of pain caused by the sight of some evil, destructive or painful, which befalls one who does not deserve it, and which we might expect to befall ourselves or some friend of ours, and moreover to befall us soon"; "The Rhetoric," 1385b15, in *The Basic Works of Aristotle*, ed. Richard McKeon (New York: Random House, 1941), "The Rhetoric," trans. W. Rhys Roberts, Book II, ch. 8, p. 1396. Hobbes transforms Aristotle's definition in a way that gives a new prominence to fear for self.

[57] Jean-Jacques Rousseau, *Emile or On Education*, trans. Alan Bloom (New York: Basic Books, 1979 [1762]), p. 221.

[58] Ibid., p. 235.

The same egoistic view of compassion and other pro-social emotions has been advocated by neo-Darwinians. Behavioral ecologist Richard Alexander argues in *The Biology of Moral Systems* that "the most general principle of human behavior" is that "people in general follow what they perceive to be their own interests."[59] According to Alexander, societies are simply "collections of individuals seeking their own self-interests."[60] People do not constantly choose to act on the basis of extensive analyses of self-interest – we are not computers, after all – but nature provides emotional predispositions that perform the rough calculations for us.

Pro-social emotions evolved because they promote fitness. The costs of group living are extensive but they are worthwhile in the face of between-group competition. For Alexander, the antisocial relations between groups fuel the tendency of human beings to cooperate within groups and to compete within groups in a reasonably controlled manner. Antagonism between groups thus helped to provide fuel for human evolution: "In no other species do social groups have as their main jeopardy other social groups of the same species – therefore the unending selective race toward greater social complexity, intelligence, and cleverness in dealing with one another."[61]

Aggression and fear drove the expansion of social complexity primarily as a system of "indirect reciprocity."[62] Within systems of indirect reciprocity, individuals often engage in "social hustling" to gain maximum benefits while paying minimally acceptable costs. "Good Samaritanism" works, Alexander argues, when it provides a net benefit to the agent. Individuals want to appear to be willing to engage in indiscriminate social beneficence so that they can extract such beneficence from others.[63] Benevolence is socially rewarded, so "it can pay to give the impression of being an altruist even if you are not."[64]

Alexander explicitly rejects moral universalism whereby natural sympathies are extended from loved ones "to the men of all nations and races."[65] On the contrary, we have evolved with an emotional preference for family members and to behave ethically toward reciprocators within the group, but to be immoral, or at best indifferent, toward out-groups and their members. Because of our animal nature, Alexander maintains, we are strongly inclined to "in-group amity" and "out-group enmity."[66]

---

[59] Richard D. Alexander, *The Biology of Moral Systems* (New York: Aldine de Gruyter, 1987), p. 34.
[60] Ibid., p. 3.   [61] Ibid., p. 80.   [62] Ibid., pp. 81, 85ff.   [63] Ibid., p. 102.   [64] Ibid., p. 114.
[65] Cited from *The Descent of Man*, in Alexander, *Biology of Moral Systems*, p. 173.
[66] Alexander, *Biology of Moral Systems*, p. 95.

The depth to which this egoistic presupposition has penetrated our culture is revealed in the extent to which many people would take these citations from Rousseau and Alexander to be self-evidentially true. As we will see in chapter nine, the claims of "kin altruism" and "reciprocal altruism" are neo-Darwinized versions of what were already proposed by Hobbes and Rousseau: cooperation and affiliation justified in egoistic terms. Fear plays a large role in compassion because we have compassion only when we fear that what happened to the victim could happen to us. This explains, they would say, why people sometimes can experience simultaneously both pity for the sufferer and relief that it is not they who suffer.

CHRISTIAN ETHICS AND HUMAN NATURE IN LIGHT OF EMERGENT COMPLEXITY

I will now examine the relevance of this general account of emergent human complexity for our understanding of human flourishing and for our theological interpretation of human nature.

Human goodness in the most comprehensive sense lies in the full development of our human capacities and powers. A human being who suffers deprivation and other harms is less "good" in this comprehensive sense (but not necessarily in the specifically moral sense of goodness). One egregious case of such harm is seen in the famous Nova story of "Genie" the "wild child," who spent her childhood locked in her bedroom and subjected to such severe long-term deprivation by her parents that she was subsequently unable to develop more than minimal social connections.[67] "Goodness" in this context signifies an ability to attain the ends natural to a human being — to learn to speak, to form bonds of affection, to develop social competence, and so forth — and it obviously carries no suggestion of moral judgment. A person who is unable to speak more than a few words thus "participates" less fully in what it means to be human. This does not mean that the person is less human, but, because language use and social communication are essential attributes of being human, it does mean that she suffers a very serious defect.

This view of human nature presumes an account of what is essentially human and what is incidentally or "accidentally" human. What is "natural"

---

[67] See www.pbs.org/wgbh/nova/transcripts/2112gchild.html (accessed August 19, 2005). See also Susan Curtiss, *Genie: A Psycholinguistic Study of a Modern-Day "Wild Child"* (New York: Academic Press, 1977).

to human beings is ultimately rooted in the distinctive traits that make each member of the species a human person. Thomas Aquinas believed that the Creator individually designed each species at the beginning of creation, and that creation brought into existence unchanging essences.[68] He also believed that creatures continue to exist throughout time in the form in which they had been originally created, the archetype of which exists in the mind of God. Knowledge of evolution replaces this static notion of species with one that is more open to change, is less stable, and allows considerably more diversity.

Our humanity gives us certain characteristic inclinations. We can distinguish our own felt desires from the inclinations that are proper to human nature. People for all sorts of reasons – traumatic experiences, chemical imbalances, genes, socialization, and suchlike – can develop desires that are not oriented to properly human ends. We possess a capacity to make intelligent and free decisions regarding our actions and ends. The morally good person makes these choices in accord with what is most human in us (in accord with what is the normatively human in us), and the person who acts badly makes choices that violate what is most human in us. The effects of these choices accumulate over the course of a lifetime to shape our characters. Our ultimate end is the beatific vision, but we attain some degree of limited but real flourishing in this life by living virtuously.

Knowledge of human evolution calls into question the notion of permanently fixed essences. Thomas thought that the form of a species was permanent and uniform and that departures from the norm were due to "mistakes" or "accidents" in nature. He was not completely wrong here, since genetic mutations do fail to copy genetic material accurately, but he was mistaken to believe that reproduction tends to duplicate the male form in his offspring, that is, that the active force in the male seed is ordered to produce a perfect likeness of itself.[69] It was not until Mendel that we knew that reproduction naturally produces new combinations of genes and that species benefit from diversity.

Darwin showed that species are not immutable, that what were thought to be "mistakes" can constitute productive variations, and that concretely existing individuals who depart from the norm can be the forerunners of beneficial new adaptations. We now know that species are characterized by diversity, not uniformity. The Creator uses the interaction of law-like regularities and chance events as secondary causes to produce a vast pattern of events within which God's providence works. The divine will works

---

[68] See *ST* I, 9,2 ad 3. [69] See *ST* I, 92,1 ad 1.

through the diversity and heterogeneity of the evolutionary process, not through a precisely scripted set of stages or through a permanent set of essences.

If traditional moral theology was limited by a static view of human nature and the goods proper to it, the evolutionists' view of human nature is morally crippled by their truncated account of the human good. From the point of view of Christian ethics, biology considers a range of goods that compose part of what Thomas called "temporal happiness," but this end is, at its best, radically incomplete since the biological does not encompass psychological, social, or cultural goods, let alone moral and religious goods. Evolutionary theory, then, will always fail to satisfy those who seek in it a complete account of the natural law. Indeed, evolutionists need to be subjected to critical scrutiny when they present a kind of quasi-natural-law argument suggesting that the values of our own particular culture are best suited to address our natural needs as human beings.[70]

The human good also includes not only "external goods" and "goods of the body" but also "goods of the soul" (such as religion and true friendship) that are not reducible to other goods and that are more difficult to achieve than the others. Since what is "natural" for the human person is not simply what is "biological" or "organic" or "genetic," the comprehensive attempt to explain or justify natural law exclusively in terms of evolutionary theory is bound to fail. The "natural" includes the full range of inclinations identified by Thomas, including those desires common to rational beings: for knowledge, for life in political community, and for union with God. These distinctively human orientations point to the highest human good: the knowledge and love of God.

Biologically based emotional proclivities and motivations are objects of deliberate moral choices and behaviorally developed habits. The "phenotypic plasticity" of "open programs"[71] allows for, and even requires, choices that accumulate to shape our more or less persisting habits. Yet human nature is not infinitely plastic. Human passions and emotions require formation and habituation. If there is an evolved innate capacity to resist being cheated, then the content of many moral codes that prohibit cheating does have a basis in human nature. Communities tutor what evolutionary thinkers have called our natural sense of fairness – a natural orientation to

---

[70] E.g. E. G. Beckstrom, *Darwinism Applied: Evolutionary Paths to Social Goals* (Westport, CT: Praeger, 1993), and David M. Buss, *The Evolution of Desire* (New York: Basic Books, 1994).

[71] Ernst Mayr, *Towards a New Philosophy of Biology* (Cambridge, MA: Harvard University Press, 1988), p. 68.

compare cases, to recognize cheating and bias, and to insist that similar cases be treated similarly.[72] Various communities differ in their sense of what is fairly asked of self, others, and the community as a whole. Because agents are tempted to cheat under various circumstances, communities throughout the world have to indoctrinate the habit of setting aside the desire for personal advantage that can be gained through cheating in order to promote the good of others.

The perspective of human evolution underscores the heterogeneous, mixed, and conflicted nature of human inclinations and the goods toward which they move. Evolutionists overstate their case when they see nature as nothing but the scene of conflict and therefore dismiss cooperation as derivative and altruism as illusory. As chapter nine will point out, nature itself is pervaded by symbiosis, interdependence, and cooperation.

One way of dealing with this complexity ranks kinds of goods in terms of their inherent excellence, registers the existential priority under some circumstances of more urgent goods (e.g. physical safety over poetry), and regards lower-level goods as in the service of higher-level goods. Thomas ranked objects of desire on a scale of excellence whereby spiritual goods are more important than external goods, intellectual goods than bodily goods, and so on. The "order of love" provided ethical criteria for distinguishing more important priorities from those that are less so.[73] Caring for one child with regard to one kind of good may mean subordinating the well-being of another child with regard to another good, perhaps temporarily but sometimes permanently. Moral hierarchy here justified a policy of sacrificing lesser for greater goods and doing lesser rather than greater evils. Thomas did not propose a universally applicable scale of goods such that one should always pursue, say, goods of the mind prior to goods of the body, but instead knew that at best we negotiate concrete decisions by exercising the virtue of prudence in concrete social contexts.

Evolutionists stress the extent to which human nature is at odds with itself, divided by conflicting inclinations. Christian ethics recognizes that the disharmony built into the human condition by evolution is aggravated by sin. Disharmony is not entirely the product of sinful human choices and

---

[72] See, for example, Paul Seabright, "The Evolution of Fairness Norms," *Politics, Philosophy & Economics* 5.1 (2006): 33–50; Gretchen Vogel, "The Evolution of the Golden Rule," *Science* 303 (February 2004): 1128–1131; and Sarah F. Brosnan, Hillary C. Schiff, and Frans B. M. de Waal, "Tolerance for Social Inequity may Increase with Social Closeness in Chimpanzees," *Proceedings of the Royal Society B: Biological Sciences* 272, no. 1560 (February 2005): 253–258.

[73] See *ST* II-II, 26; Stephen J. Pope, *The Evolution of Altruism and the Ordering of Love* (Washington, DC: Georgetown University Press, 1994).

bad luck. Knowledge of human evolution alerts us to the presence of conflict, ambiguity, and confusion within human nature itself. Evolution, and not only free and sinful human acts, is responsible for this conflict. The "only" here is intended to signal the fact that these negatives are brought into the human condition by sinful choices and vicious habits, and that these interact with the complexities of human biological nature in a way that aggravates and adds to the difficulties that it generates.

Human beings inherit not only *potentialities* for harm but even, under certain circumstances, *proclivities* for harming others (e.g. in cases of external threat to one's own community, cases of sexual jealousy). The fact that human beings are so constituted is a reflection of our own long and conflicted evolutionary past, the way in which certain adaptations allowed the species and its forebears to survive, and so forth. This evolutionary view of human nature means that moral goodness is attained by shaping, pruning, integrating, and otherwise ordering divergent and conflicting tendencies. In Christian terms, grace does not simply "perfect" nature. Grace also inhibits, thwarts, suspends, and channels various inclinations of human nature.

The persistence, power, and complexity of human emotional inclinations means that the process of habituation often has results that are unstable, uneven, and incomplete. If people are assumed to be born as "blank slates," the language of virtue can be interpreted to suggest that once they have formed their habits their characters are strongly set. An evolutionary view of human nature, however, suggests that habituation can only go so far and is accomplished with a great deal of effort, and that a permanent disposition to certain kinds of goods, appropriate or not, is an ineradicable part of human nature. There may always be a human tendency to want to cheat to gain an advantage or at least to cheat to counter what one perceives as an unfair advantage held by others. This obdurate part of human nature seems to be especially the case when it comes to our strongest physical appetites, namely for food and drink and especially for sex. This may explain why even exemplars of justice such as Gandhi and Martin Luther King, Jr. had difficulty in this domain. In Thomistic language, the vast majority of people seem to waver between continence and incontinence rather than mount to the summit of perfect virtue or fall into the moral pit of unremitting vice.

The juxtaposition of various kinds of adaptations and motivations leaves the human psyche fraught with moral complexity, ambiguity, and tension. Resentment over unjustified harms testifies to our fundamental human inclination to the good. Nature has made us conflicted beings, but sin amplifies our conflicts and limits profoundly our attempts to negotiate them. The task of the Christian life is to allow grace to transform the self so that one can become both a more decent, just, and morally integrated human being and more Christ-like. Grace works through ordinary as well as extraordinary channels, through schools, neighborhoods, and any other form of human association that shapes the person's character.

Greater attentiveness to our embodiment inspires a deeper awareness of the complexity and significance of sexuality for personal identity, and knowledge of the distinctively human, meaning-constituting character of human sexuality inspires in us a greater reverence for its moral significance. Evolutionary theory gives some support to the Christian tradition's well-known ambivalence toward sexuality as both gift and threat. Whereas the Christian tradition has taught that the gift of sexuality comes from divine creation and the moral threat comes from human sin, evolution suggests that the condition comes from nature and is amplified by sin. Sexual temptation, jealousy, infidelity, and male domination occur because mammalian sexual patterns provide the biological conditions and proclivities that lead, in disordered and sinful human lives, to impulsive acts and harmful consequences. This is not to suggest that we shift all responsibility from human sin to nature, but rather that we see harmful human acts as the interplay between both.

Evolutionary awareness also encourages a greater sensitivity to the impact of physical suffering on human lives, and a greater appreciation for the moral value of science and technology for addressing the biological disorders that thwart human well-being. While compassion has long been the focus of Christian ethics,[74] Christianity has also had an other-worldly strain that tends to diminish the significance of "this worldly" suffering while focusing on the salvation of individuals souls. This kind of piety can lead to quietism, passivity, and fatalism in the face of harms caused by social injustice. It encourages interior conversion but not external action for social change, and leads us to consider sacramental devotion, private prayer, and personal charity to be religiously sufficient for Christian identity.

Archbishop Oscar Romero, in contrast, called the people of El Salvador to the imperative of Christian concern for both souls and bodies. He was sharply critical of two opposite kinds of reductionism in this regard, one

---

[74] See James F. Keenan, SJ, *The Works of Mercy: The Heart of Catholicism* (Lanham: Sheed and Ward, 2005).

that stressed only the "transcendent elements of spirituality and human destiny" and the other that attended only to the "immanent elements of a kingdom of God that ought to be already beginning on this earth."[75] Christian evangelization has to include both aspects, Romero thought, and not focus only on the soul and its highest end. Knowledge of human evolution supports Romero's terrestrial balancing of Christian concern because it underscores our physicality, our common genetic heritage, and our biological interdependence and ecological connections with one another.

CHRISTIAN THEOLOGICAL ANTHROPOLOGY AND HUMAN FLOURISHING

Christian theological anthropology understands the human person as created good, "fallen" through our own personal and collective choices, and redeemed in Christ. The belief that we are created good but "fallen" means, first, that there is an ontological goodness in the very fact that we (like any other creatures) exist and, second, that we have a special capacity, in virtue of our emergent cognitive and emotional capacities, to act as responsible agents ("dominion" in Gen. 1:26) for the well-being of one another and of other creatures.

Neurological and other bodily processes constitute the biological conditions that make possible the spiritual dimension of human nature. Because of the close interdependence of body and spirit, one cannot speak intelligibly about the human spirit outside a body any more than one can speak about a living body without a spirit. Though he attended primarily to the spiritual dynamism moving the human person, Rahner properly rejected the sharp distinction between body and soul that marked popular piety in his day. The activity of the human spirit for him was also an activity of matter – indeed, an expression of its highest capacities as they operate under the influence of grace.[76] A nonreductionist perspective acknowledges that the mind exerts an influence on the brain as well as vice versa. Christian sacramental practices that use bread and wine, water and oil, candles and music, incense and stained glass acknowledge the embodied character of the human spirit. Grace works through bodily realities to make us capable of a kind of communion with God that is not possible for other creatures.

---

[75] Oscar Romero, *Voice of the Voiceless: The Four Pastoral Letters and Other Statements*, trans. Michael J. Walsh (Maryknoll, NY: Orbis Press, 1999), p. 130.
[76] See Karl Rahner, "The Unity of Spirit and Matter in the Christian Understanding of Faith," trans. Karl H. Kruger and Boniface Kruger in *Theological Investigations*, vol. VI: *Concerning Vatican Council II* (New York: Seabury Press, 1974), pp. 153–177.

Human beings of course also have the capacity to thwart God. The Christian doctrine of "original sin" is not concerned with what happened in a mythical garden in the distant past, but is about the universality of sin and the need for salvation in the present.[77] Adam and Eve stand symbolically for the entire human race at all times. The narrative about them discloses how God, both just and merciful, chooses to create, guide, judge, and protect human beings. The later Christian doctrine of "original sin" is in part a theological way of expressing the common experience of human nature as disharmonious, out of order, and a result of defection from God's original purposes (i.e. as a departure from the "original justice" with which we were created).

The affirmation of our created goodness is consistent with our knowledge of the cognitive, emotional, and social capacities that make choice a central feature of human action. The Creator wills a world in which we have the capacity to choose to love, but also to choose not to love, to prefer indifference to caring, and even to hate what is in itself good. God's "permissive will" allows human beings to choose moral evil. There is little doubt that we inherit, both individually and collectively, a deep-seated tendency selfishly and freely to misuse our natural powers. The culpable misuse of our capacity for free choice, which the Christian tradition has called "actual sin," gives rise to "sins of commission" that deliberately intend or actually do engage in harmful acts, and to "sins of omission," in which an agent fails to will what he or she ought to.[78]

The transition from innocence to guilt seems to have begun with a kind of early human innocence that is something like what we see in small children and animals, who manifest what Reinhold Niebuhr calls "animal

---

[77] Evolutionists have responded variously to the doctrine of original sin, depending on how they interpret it. Patricia Williams believes that evolutionary biology requires Christians to abandon it, but this rejection is based on a crudely literalist misreading of the relevant texts. See Patricia A. Williams, *Doing without Adam and Eve: Sociobiology and Original Sin* (Minneapolis: Fortress Press, 2001). Donald Campbell argues that original sin is a religious way of talking about our animal nature and that the only corrective to "bad" biological human nature lies in "good" culture; D. T. Campbell, "On the Conflict between Biological and Social Evolution and the Concept of Original Sin," *Zygon* 10 (1975): 234–249. Finally, Christian theologian Stephen Duffy recognizes that literal interpretations of Adam and Eve fly "in the face of our evolutionary worldview." He interprets original sin in terms of the "sin of the world," in which "The tangle of evil persons with their evil deeds and diseased institutional structures and systems weaves a history which constitutes humanity in its network of interdependence as deaf to the appeal of the good"; Stephen Duffy, "Our Hearts of Darkness: Original Sin Revisited," *Theological Studies* 49 (1988): 608, 598, respectively.

[78] See Daniel J. Harrington, SJ, and James F. Keenan, SJ, *Jesus and Virtue Ethics: Building Bridges between New Testament Studies and Moral Theology* (Lanham: Sheed and Ward, 2002), ch. 7.

serenity."[79] As they are repeated, choices become more and more definitive. In this way, human beings moved from something like animal innocence into a guilty maturity that "knows" (i.e. as experiential knowledge) "good and evil." The process of growing into adulthood replicates in the personal level a truth about the human race that is expressed symbolically by the myth of the Fall. The movement from childhood to adolescence involves the emergence of a multitude of previously unacknowledged desires that leads to emotional complexity and bewildering confusion. Each individual life recapitulates the collective human experience of struggling, and often failing, to achieve a mature harmony from among our wide array of contradictory impulses and wrongful choices.

The emergence of the human capacity to make choices, rather than to be determined to specific courses of action by biological causes, is not immediately accompanied by morally disordered desires. Yet as Rahner explained, "original sin can only be thought of as the first act of man's real, authentic freedom."[80] Primeval human beings, symbolized in Adam, chose to transgress their sense of right relationship with God and one another. In this sense one can speak of a unity of the human race in Adam (see Rom. 5:12), since we all participate in the same pattern of disordered choosing and so in our own way contribute to the further disordering of the world.

Sinful acts led to a new fear of death and to a new preoccupation with avoiding it at all costs, even if the means of doing so involves taking the innocent lives of others.[81] The ultimate evil is often said (though dubiously so) to be death. Death existed well before the emergence of human beings; the first appearance of life on the planet was followed by its demise. Death is a context for bringing forth new life that is finite and vulnerable. Biological death is not imposed by God on a previously immortal human nature as a punishment for the sin of Adam and Eve. As Gustafson points out, "In modern culture few persons with average education any longer believe that biological death is caused by the sins of Adam and Eve, including few who write theology or participate in the Church."[82] Rahner concurs with Gustafson on this point.[83] Yet even if biological death

---

[79] Reinhold Niebuhr, *Beyond Tragedy: Essays on the Christian Interpretation of History* (New York: Charles Scribner's Sons, 1937), p. 294.
[80] Ibid., p. 103.
[81] See Karl Rahner, "The Sin of Adam," trans. David Bourke, in *Theological Investigations*, vol. XI (New York: Crossroad, 1982), pp. 247–262, esp. p. 247. This does not require literal belief in a primal couple, or "monogenism."
[82] "The Sectarian Temptation": 91.
[83] See Karl Rahner, *On the Theology of Death* (New York: Herder and Herder, 1961), pp. 42f.

was not brought into the world as a punishment for sin, there is no reason to doubt the truth of St. Paul's statement that the "wages of sin is death" (Rom. 6:23) as long as "death" means not the literal cessation of biological life but, metaphorically, an "alienation from God," a "spiritual death,"[84] and the "self-destructiveness of sin."[85]

We should note how this view of "original sin" differs from sociobiological pessimism regarding the predominant influence of the "dark side" of human nature on our behavior – what evolutionists see as our evolved proclivity to selfishness and opportunism. Christian ethics, in contrast, distinguishes our natural finitude from the wrongful choices we make and their cumulative effect over time and space on the human condition. It does not confuse our natural inclinations with the vices we develop. In theological terms, sociobiology regards us as created evil, and as naturally aggressive; Christian ethics, on the other hand, regards these as fundamental, self-inflicted distortions of a human nature that was created good but subsequently deformed.

While the source of moral disorder lies in human choices, its solution lies in divine redemption. If sin turns death into the "reality we most feel and fear,"[86] redemption involves the radical reappraisal and even ultimate destruction of the theological and moral power of death over human beings. The greatest evil is human malice, the free and intended commitment to harm another person without justification – an attack not only on others but always also on God. Authentic Christian identity involves a radical process of transformation in Christ grounded in Jesus' fidelity to and obedient love of the Father. Redemption involves a profound transformation of the human person in Christ, a commitment to the realization of the Kingdom of God, the original divine intention for creation. This transformation essentially involves a grace-inspired participation in the life of God. Rowan Williams explains that to be made a "new creation" (2 Cor. 5:17), involves a "far-reaching reconstruction of one's humanity: a liberation from servile, distorted, destructive patterns in the past ... a new identity in a community of reciprocal love and complementary service, whose potential horizons are universal."[87]

---

[84] Joseph A. Fitzmyer, SJ, "The Letter to the Romans," in Raymond E. Brown, SS, Joseph A. Fitzmyer, SJ, and Roland A. Murphy, OCarm., eds., *The Jerome Biblical Commentary* 2 vols. (Englewood Cliffs, NJ: Prentice-Hall, Inc., 1968), vol. II, p. 307.
[85] David Burrell, CSC, and Elena Malits, CSC, *Original Peace: Restoring God's Creation* (Mahwah, NY: Paulist Press, 1997), p. 30.
[86] Ibid.   [87] Rowan Williams, *On Christian Theology* (Malden, MA: Blackwell, 2000), p. 138.

Jesus of Nazareth represents the climax of the ongoing process of spiritual development to which all human beings are called. His suffering, passion, and death attest to the extent and depth of human wickedness, to his unconditionally obedient love for the Father, and to the Father's radical self-giving love for us. The resurrection reveals the transforming power of grace that brings life out of death, love out of hatred, goodness out of evil. This is the absolute heart and soul of Christian morality, which is essentially an ethic of love. While our tendency to sin is inherited from our ancestors, its power is broken by the Cross (Rom. 5) and resurrection (1 Cor. 15:21–22). In evolutionary terms, Jesus of Nazareth, Christ the Redeemer, represents the culmination of the meeting of grace with the natural powers of emergence with which creation has been endowed by the Creator. Redemption restores our desire and potential for goodness, an inclination and capacity that, because of the persistence and power of sin, can only been realized through the grace bestowed by God in Jesus Christ.

God as Spirit strives to engage every person in an ongoing process of redemption. The structure of the biological world provides conditions that facilitate this cooperation, particularly in the gifts of intelligence and will that allow us to cooperate with divine goodness. The "Kingdom of God" is realized to the degree that human beings lovingly and intelligently participate in divine goodness. The Cross reverses the effects of the "Fall," overcomes our alienation from God and one another, and allows us to function as the "image of God" originally intended by the Creator. Redemption neither cancels bodily death nor recovers a lost moral perfection, but it does lead us to God. Living a good life as a Christian does not simply occur spontaneously. As Lash notes, the Christian tradition has long recognized the need for "discipline, *ascesis*, the taming of the violence of the human heart, lifelong education into patterns of attentiveness, and peacefulness, and trust. Humanity, we might almost say, does not come naturally to us."[88]

---

[88] Lash, *The Beginning and the End of "Religion,"* p. 208.

CHAPTER 7

Freedom and responsibility

Ethicists and evolutionists alike have asked whether the intellectual acceptance of human evolution requires us to give up the common assumption that we are responsible for our acts. Morality itself would be fatally compromised if science somehow demonstrated that human freedom and moral responsibility were illusions. For all their disagreements, many evolutionists and Christians concur that we must make a choice between two irreconcilable and mutually exclusive alternatives: either "free will" or evolution. This chapter, on the contrary, argues against this false dichotomy.

The chapter is structured in three steps. First, I will examine critically some of the literature on the "free will and determinism" question produced by various evolutionists, most of whom assume some kind of a reductive fatalism. I argue here that the epistemologically and ontologically reductionistic views of some evolutionists cannot explain away the experience of free choice or demonstrate that it is illusory. Human freedom in its most important senses escapes the purview of science.[1] Sociobiology fails to account for our first-person experience of human action, the starting point for reflection of our capacity for free choice as moral agents. We all have abundant evidence from our own lives of being able to do one thing rather than another and acting on the basis of what we judge to be good reasons. Sociobiology fails in its attempt to explain away this experience as illusory.

Second, I turn to the theme of freedom and responsibility in the thought of John Paul II, who is examined here because of his influence on the general theme of evolution and human free will and because he represents an important strain of thinking about the meaning of freedom and

---

[1] The word "freedom" can of course refer to many things, including Socratic self-mastery, Kantian autonomy, moral license, self-responsibility, unimpeded physical motion, the power of choosing between good and evil, or a nearly endless variety of other things; Isaiah Berlin found over two hundred uses of the term. See Berlin, *Four Essays on Liberty* (London and Oxford: Oxford University Press, 1970).

responsibility. Here I argue that his view of freedom and responsibility is detached from evolution and subject to being understood as a form of dualism because it is so abstracted from the biological dimensions of human existence. I argue that the evolutionary notion of emergent complexity can help us understand the biological basis for the existence of free choice rather than to assume that it is disconnected from the biological dimension of human nature. The capacity for free choice, in other words, fits into a view of innate human predispositions or propensities that incline but do not coerce us to move in certain directions for various kinds of goods.

Third, I will move beyond the diametrically opposite positions discussed in the first two parts of the chapter – sociobiological determinism and theological dualism – by drawing some important philosophical distinctions among various terms employed in the "free will versus determinism" debate. I will propose that our capacities for free choice and moral freedom become more intelligible when placed within the context of the evolution of human emergent complexity.

"FREE WILL VERSUS DETERMINISM"

Sociobiologists were criticized early on for endorsing genetic or biological determinism and rejecting "free will." Yet evolutionary thought has not always been taken to undermine human freedom. John Dewey held that Darwin's rejection of any preordained goals in nature had the effect of leaving human beings free to determine their own fate.[2] C. S. Peirce and William James even credited Darwin with destroying determinism and allowing for human creativity and free choice.[3] Neo-Darwinism, however, is usually taken to deny the existence of human freedom. Thus William Provine argues that according to "modern science," "free will as it is traditionally conceived – the freedom to make uncoerced and unpredictable choices among alternative courses of action – simply does not exist ... There is no way that the evolutionary process as currently conceived can produce a being that is truly free to make moral choices."[4]

---

[2] See John Dewey, *The Impact of Darwin on Philosophy: And Other Essays in Contemporary Thought* (Bloomington, IN: Indiana University Press, 1965 [1910]).
[3] See Philip P. Wiener, *Evolution and the Founders of Pragmatism* (Cambridge, MA: Harvard University Press, 1949).
[4] William Provine, "Evolution and the Foundation of Ethics," *Marine Biological Laboratory Science* 3 (1988): 27–28.

One reason the "free-will versus determinism" debate has been so persistent and unresolved comes from the varied and even equivocal way in which these concepts have been used. "Determinism" is sometimes taken to refer to the belief that events have causes and are so connected with preceding events that the latter would not have existed without the former and that full knowledge of preceding events would allow one to predict the later event.[5] Determinism in this sense is not compatible with free will, where the latter is assumed to be conceived as a kind of "uncaused" cause of the agent's act.

Yet at other times "determinism" simply refers to the fact that nature imposes limits on what is possible, so here it need not rule out all descriptions of agents as "free" in some sense. Agents subjected to a variety of forms of "determination" are not necessarily utterly powerless over their own acts, as if they were completely under the control of external agencies. When "determined" is taken to refer to the state of being placed within certain objective constraints, it is not so easily confused with what Midgley calls "evolutionary fatalism," the claim that action is ultimately controlled not by the agent but by impersonal biological forces.[6] This position recognizes that events have countless "causes" and it speaks of the cause of an event as the most salient factor contributing to it. In this case, agents can be identified as causes of actions without suggesting that they act in arbitrary or uncaused ways.

Fatalism, the belief that moral effort is futile and freedom an illusion, has long been the target of Christian polemics. Augustine's polemic against astrological fatalism focused especially on those who attempted to evade responsibility for their conduct by claiming that they were overcome by bodily passions or fated to act in certain ways by the stars.[7] A similar worry applies to sociobiology and evolutionary psychology today. Augustine, of course, was familiar enough with the power of natural appetite, but he also recognized that beneath all our actions, including those that yield intense physical pleasure, lie choices reflecting the deeper orientation of the human will.[8]

---

[5] See Brand Blanshard, "The Case for Determinism," in Sidney Hook, ed., *Determinism and Freedom in the Age of Modern Science* (New York: Collier Books, 1958), ch. 1. "Determinism" is defined as the belief that "all things in nature behave according to inviolable and unchanging laws of nature"; *Encyclopedia of Philosophy*, vol II, p. 364. It holds that "whatever happens is determined by antecedent conditions, where determination is standardly conceived as causation by antecedent events and circumstances. So construed, determinism implies that at any time the future is already fixed and unique, with no possibility of future development."

[6] See Mary Midgley, *Wickedness: A Philosophical Essay* (London and New York: ARK, 1984).

[7] See *De Civ. Dei* V.   [8] See *De Civ. Dei* XIV.6.

Evolutionists argue, as Robert Wright puts it, that natural selection "reveals the contours of the human mind."[9] If these "contours" have been caused by natural selection and random events, then it seems that individuals ought neither to be blamed for their conduct when it does not meet social standards nor to be praised when their behavior is socially acceptable.[10] Moreover, if genes are the key factor for human development, and humans have no freedom to determine their genetic make-up, it seems unfair to hold people accountable for their acts. Against this view, Lewontin, Rose, and Kamin characterize sociobiology as teaching

> that human lives and actions are inevitable consequences of the biochemical properties of the cells that make up the individual; and these characteristics are in turn uniquely determined by the constituents of the genes possessed by each individual. Ultimately, all human behavior – hence all human society – is governed by a chain of determinants that runs from the gene to the individual to the sum of the behaviors of all individuals. The determinists would have it, then, that human nature is fixed by our genes.[11]

The operative term here is "inevitable." Lewontin and his colleagues are especially concerned about "fatalism," the belief that conscious effort is unable to influence human behavior. Fatalism is often accompanied by "epiphenomenalism," the assumption that consciousness cannot affect behavior but is itself only a neurological consequence of causal forces over which the human agent has no control.[12] Lewontin, Rose, and Kamin want to preserve room for the causal efficacy of human choices in the domain of human behavior, and believe that sociobiology disallows such "free will."

The kind of sociobiological determinism opposed by Lewontin and his collaborators faces a number of well-known objections. One is that it wrongly proposes a direct causal relation between discrete genes and particular traits. Traits, however, are always embedded in particular organisms that are shaped by their own relation to other individuals, populations, and environments. Mayr points out that

> [the] version of Darwinism developed during the evolutionary synthesis was characterized by its balanced emphasis both on natural selection and on stochastic

---

[9] R. Wright, *The Moral Animal* (New York: Vintage, 1995), p. 11.
[10] See ibid., pp. 357, 203.
[11] Richard C. Lewontin, Steven Rose, and Leon J. Kamin, eds., *Not in Our Genes* (New York: Pantheon, 1984), p. 6.
[12] See Midgley, "Consciousness, Fatalism and Science," in Niels Gregerson, William B. Drees, and Ulf Gorman, eds., *The Human Person in Science and Theology* (Edinburgh: T. & T. Clark, 2000), p. 24.

processes; by its belief that neither evolution as a whole, nor natural selection in particular cases, is deterministic but rather that both are probabilistic processes; by its emphasis that the origin of diversity is as important a component of evolution as is adaptation; and by its realization that selection for reproductive success is as important a process in evolution as selection for survival qualities.[13]

Biological determinism has also been criticized for encouraging passivity in the face of injustice and acquiescence in the status quo, no matter how oppressive. Lewontin and his collaborators characterize sociobiologists as arguing that "if men dominate women, it is because they must. If employers exploit their workers, it is because evolution has built into us the genes for entrepreneurial activity. If we kill each other in war, it is the force of our genes for territoriality, xenophobia, tribalism, and aggression."[14] Lewontin's characterization is extreme, yet some sociobiologists give evidence for his charge of determinism. Wright, for example, argues: "Understanding the often unconscious nature of genetic control is the first step toward understanding that – in many realms, not just sex – we're all puppets, and our best hope for even partial liberation is to try to decipher the logic of the puppeteer."[15]

A number of misunderstandings about the relation between freedom and genes have to be avoided. First, journalistic oversimplifications aside, neither sociobiologists nor evolutionary psychologists believe that behavior can be explained by reference to genes *alone*. Evolutionists acknowledge that little major genetic evolution has probably occurred in the human species over the last fifty thousand years – at least in the sense there has been no human speciation in this interval[16] – and that cultural change has had a profound influence on human behavior. Moreover, while members of the human race share the same array of "psychological mechanisms," how and the extent to which a given mechanism is exercised depends on particular environmental factors, especially culture.[17] The language of "evolved psychological mechanisms" is a functional redescription in evolutionary terms of natural powers and capacities with which human beings have been long familiar. The ways in which these capacities are exercised are always shaped by their particular cultural and social contexts. Just as a proneness

---

[13] Mayr, *One Long Argument*, p. 106.
[14] Lewontin, Rose, and Kamin, eds., *Not in Our Genes*, p. 237.
[15] R. Wright, *The Moral Animal*, p. 37.
[16] See Jared Diamond, *The Third Chimpanzee: The Evolution and Future of the Human Animal* (New York: HarperCollins, 1992).
[17] The influence of culture is stressed by Paul Ehrlich, *Human Natures: Genes, Cultures, and the Human Prospect* (New York: Penguin Books, 2002).

to develop callouses on one's feet is only activated when individuals walk in their bare feet, so different psychological mechanisms are activated, or not, in various cultural contexts. Few evolutionary psychologists or sociobiologists think that genes alone can explain human beings in their totality.

Sociobiologists do not think that human behavior is under the "control" of one gene or a collection of genes, nor do they hold that human traits are rigidly predetermined by genetic inheritance. Genes cannot function by themselves, and there is broad agreement that behavior is influenced by a constant interplay of learning and culture with biological predispositions and potentials rather than caused by rigidly determined behavior traits. Genes are simply chemically inert units of DNA. Placed within the appropriate cellular environment, genes direct the synthesis of proteins and these proteins in turn can generate significant effects. The expression of genes is influenced by the physical and chemical condition of the cellular environment in which they exist. Even on the micro-level, it makes no sense to assume a genetic determinism according to which genes by themselves somehow cause behavior. Genes never function as isolated causes of behavior but, as Rose emphasizes, rather as essential components of complex networks. Behavior, moreover, reflects the influence of a multitude of genes (they are "polygenic"). Genes play an important role in the cluster of causes that lie behind behavior, but are not "*the*" cause of behavior.

Second, sociobiologists do not believe that most traits are the result of single genes. Speaking of the "gene for" altruism or a "gene for" cheating is only done for conceptual convenience. The expression "gene for" simply refers to a discernible genetic influence on variation in two populations. Dawkins focuses on genes because evolution works by means of differential genetic replication,[18] but he does not examine human behavior through the lens of genetic fatalism:

> Genetic causes and environmental causes are in principle no different from each other. Some influences of both types may be hard to reverse; others may be easy to reverse. Some may be usually hard to reverse but easy if the right agent is applied. The important point is that there is no general reason for expecting genetic influences to be any more irrevocable than environmental ones.[19]

Third, sociobiologists believe that human beings have been selected to pursue certain kinds of goods. But natural proclivities do not eliminate the

---

[18] See Richard Dawkins, *The Extended Phenotype* (New York: Oxford University Press, 1982), ch. 2.
[19] Ibid., p. 13.

capacity of human beings to make choices between alternatives. The fact that human beings have a natural tendency to mate is not at odds with the ability to choose among potential mates or for an individual to forego mating altogether. In this regard, sociobiology is not proposing an entirely novel idea. Human beings have always been acknowledged to have strong appetites rooted in biology, and sociobiologists argue that there are good Darwinian reasons for these inclinations. The sociobiological difference, however, lies in the relation between biological and other levels of human desire, for example the social, the psychological, the religious. Instead of there being multiple levels of desire with ascending human significance, sociobiologists argue that human desires, even those that do not seem to be concerned with biological goods, are actually indirectly in the service of inclusive fitness.

Most people, of course, do not habitually calculate the relative reproductive profitability of various courses of action. Evolutionists hold that human nature makes unnecessary this deliberate kind of cognitive work, and that the human emotional constitution promotes fitness more efficiently than would an exclusively cognitive system. Wright explains, "A woman doesn't typically size up a man and think: 'He seems like a worthy contributor to my genetic legacy.' She just sizes him up and feels attracted to him – or doesn't. All the 'thinking' has been done – unconsciously, metaphorically – by natural selection."[20]

Many human decisions, evolutionists argue, are made on the basis of priorities already set by evolution and inscribed biochemically in the individual's genotype. Choices are made on the basis of proximate psychological mechanisms that give rise to emotional states such as guilt feelings, sexual attraction, jealousy, the desire for revenge, resentment. Sociobiologists and evolutionary psychologists do not strive to locate the particular "proximate" genetic and developmental pathways that shape the personality of a given individual agent. Interested in discovering and explaining the existence of certain adaptations, they seek to identify the "ultimate" evolutionary reason for the persistence of these adaptations and so are not concerned with how a given adaptation might or might not influence the behavior of a particular individual. Since they want to understand the evolved regularities that underlie human behavior in general, these theories ought to be taken into account by natural-law ethics.

Evolution works through human desires. Ruse expresses this position when he argues, using Frankfurt's terms, that people decide about how to

[20] R. Wright, *The Moral Animal*, p. 37.

respond to "first-order desires" – our immediate wants – according to criteria determined by "hard-wired" "second-order volitions."[21] Family ties provide one example of how these desires operate. The theory of "kin altruism," or "nepotism," holds that parents are more likely to provide resources to family members, especially their direct offspring, than to strangers. Cultures, to be sure, offer different interpretations of the meaning and importance assigned to kin – the status of the extended family is not the same in London as it is in rural Kenya – but all human communities tend to sponsor practices of kin preference of one kind or another. Sociobiologists argue that this pattern reflects the evolutionary influence of kin altruism rather than simply the random choices of countless human agents exercising free will.

Individuals act in ways that diverge from expected behavior patterns, but evolutionary theory tries to explain these departures as well. Even the practice of infanticide, for example, which seems to run directly contrary to kin selection, tends to occur more often in certain contexts rather than others (e.g. in conditions of extreme scarcity, by younger mothers, on severely compromised infants).[22] It is not easy to predict such behavior patterns in advance, sociobiologists concede, but one can explain them retrospectively in light of evolutionary principles. Behavior that is currently regarded as unpredictable is presumed to be influenced by evolutionary factors that one day will be identified and accounted for by science.

The third chapter distinguished between methodological, epistemological, and ontological reductionism. It distinguished the working methodological principle that science can provide explanations of natural phenomena from the epistemological claim that science alone provides genuine knowledge. Cell biologist R. David Cole applies the distinction to the issue of "free will" when he points out that "most scientists do not believe that they are complete slaves to our genes and environmental history. The confusion arises from a failure to distinguish between philosophical determinism and the methodological or operational determinism of science."[23] As noted above, sociobiologists frequently claim that science is naturalistic or deterministic in this methodological sense,[24] but they do not also advert to the

---

[21] Michael Ruse, *Can a Darwinian Be a Christian?* (New York: Cambridge University Press, 2000), p. 215. The notion of "second-order volitions" comes from Harry Frankfurt's "Freedom of the Will and the Concept of the Person," *Journal of Philosophy* 68 (January 1971): 5–20.
[22] See Martin Daly and Margo Wilson, *Homicide* (New York: Aldine de Gruyter, 1988), chs. 3 and 4.
[23] R. David Cole, "Genetic Predestination?," *Dialog* 33:1 (winter 1994): 21.
[24] E. g. Daly and Wilson observe: "As practicing scientists, we embrace determinism; in other roles, as human protagonists, we simply switch to the contrary and seemingly incompatible world view" (*Homicide*, p. 269). They do not provide any rationale for this "switch."

fact that this methodological premise does not on its own justify, let alone require, the further commitment to ontological reductionism.

Sociobiologists proclaim without explanation that we have a power to resist the "tyranny of our genes,"[25] "defy the selfish genes of our birth,"[26] exercise ecological responsibility,[27] avoid violent conflicts,[28] and make conventional sexual mores more compassionate and less hypocritical.[29] We have the "mental equipment" to promote our long-term selfish interests rather than our short-term selfish desires, and we can even "discuss ways of deliberately cultivating and nurturing pure, disinterested altruism – something that has no place in nature, something that has never existed before in the whole history of the world."[30] Dawkins asserts that while we are "gene machines," we still have the "power to turn against our creators. We, alone on earth, can rebel against the tyranny of the selfish replicators.[31]

Evolutionists typically protect "free will" by emphasizing the fact that genes and phenotype always involve an interaction with environment and culture and that behavior is always conditioned by its content. Wilson holds that "genes hold culture on a leash."[32] Evolutionary psychologists John Tooby and Leda Cosmides clearly acknowledge that "every feature of every phenotype is fully and equally codetermined by the interaction of the organism's genes . . . and its ontogenetic environments . . ."[33] Yet substituting environmental determinism for genetic determinism, or adding the former to the latter, does not shed much light on human freedom. Gene-culture co-evolution can amount to a "two-part determinism, not actual freedom."[34] Wilson concurs: "The agent itself is created by the interaction of genes and the environment. It would appear that our freedom is only a self-delusion."[35] "In fact," as William R. Clark and Michael Grunstein point out, "each argues strongly against its existence . . . Combining two

---

[25] Dawkins, *The Selfish Gene*, p. ix.   [26] Ibid., p. 200.
[27] See, for example, Edward O. Wilson, *Biophilia: The Human Bond with Other Species* (Cambridge, MA and London: Harvard University Press, 1984) and *The Diversity of Life* (Cambridge, MA: Harvard University Press, 1992).
[28] Richard Wrangham and Dale Peterson, *Demonic Males: Apes and the Origins of Human Violence* (Boston and New York: Houghton Mifflin, 1996), p. 258.
[29] See, for example, Buss, *The Evolution of Desire*, ch. 10.
[30] Dawkins, *The Selfish Gene*, pp. 200–201.
[31] Ibid.   [32] E. O. Wilson, *On Human Nature*, p. 167.
[33] J. Tooby and L. Cosmides, "The Psychological Foundations of Culture," in J. Barkow, J. Tooby, and L. Cosmides, eds., *The Adapted Mind* (New York: Oxford University Press, 1992), pp. 19–136, at 83; see also pp. 87 and 122.
[34] Ted Peters, *Playing God?: Genetic Determinism and Human Freedom* (New York and London: Routledge, 1997), p. 33.
[35] E. O. Wilson, *On Human Nature*, p. 71.

forms of determinism gives little insight into the origin and meaning of choice in human behavior."[36] There is no doubt, at this point, that every human act is influenced by nurture as well as nature.

The same general difficulty attends Gould's argument against what he takes to be sociobiological genetic determinism. Our species has been lucky enough to be one of the relatively few winners in the natural lottery. A highly unlikely species, "we are an improbable and fragile entity . . . not the predictable result of a global tendency."[37] Gould argues that just as we come into existence by means of a series of accidents, so we could be eliminated as a species by another series. He takes vulnerability to potential catastrophe as a sign of human freedom: "Some find the prospect depressing. I have always regarded it as exhilarating, and a source of both freedom and consequent moral responsibility."[38]

Gould here confuses freedom with indeterminacy. Lack of planning does not mean freedom, at least not in its richest human senses of the term. In fact, in one sense of the word, the person who makes a plan and is then able to act on it displays freedom, and it is the disorganized, scatter-brained person who actually lacks freedom – at least in the sense of effective personal agency to pursue his or her own goals. Freedom is not promoted by lack of direction, and "determinism" is neither identical with fatalism nor necessarily incompatible with human agency. But in order to avoid this nearly inevitable implication, sociobiologists need to propose a constructive philosophical account of human willing and choosing, which they have yet to attempt. Lacking this, readers are left with the impression that sociobiologists hold a self-contradictory position that simultaneously asserts and denies "free will" and responsibility.

A second attempt to address "the problem of free will" emphasizes the "soft-wiring" of the human emotional and cognitive constitution and the "phenotypical plasticity" of human agents. All organisms have some ability to adapt to exigent circumstances, and human beings have a higher degree of behavior plasticity than other species. The "soft-wiring" image allows for more contingency, play, and flexibility in human action than does the image of tighter, mechanistic "hard-wired" behavior, but "soft wiring" is not the same as human freedom.

The notion of "free will" suggests not only an internal locus of control (signified by the term "will") but also a *self-determining*, internal locus of

---

[36] William R. Clark and Michael Grunstein, *Are We Hardwired? The Role of Genes in Human Behavior* (New York: Oxford University Press, 2000), pp. 265, 266.
[37] Gould, *Wonderful Life*, p. 319.   [38] Ibid., p. 291.

control (or autonomy). How this self-determination is to be conceived is the nub of the issue. We seem to be faced with a mutually exclusive choice between autonomy, in which the "free will" is totally independent of natural causation, and mechanistic fatalism, in which the individual is the "puppet" of the genes. If there is no qualitative difference between human behavior and that of other organisms, as Ruse puts it, morality is just "an illusion fobbed off on us by our genes to get us to cooperate."[39]

Sociobiologists hold that core internal "controls" are ultimately set by genes, and that, as Ruse puts it, "second-order volitions" are "hard wired." Unaware of these underlying genetic causes, individuals feel that they exercise "free will" when they experience themselves willing to do what they do. But their willing is itself directed by emotions, and the human emotional constitution has been structured by natural selection. Evolutionary psychologists focus on the evolution of specific mental modules, but these do not operate by themselves. Modules are not a sufficient basis for human action and their operation is sensitive to the individual's biological, social, and cultural context. The environment shapes developmental pathways so profoundly that they are constitutive of the individual organism's natural capacities. In this way sociobiology and evolutionary psychology actually constitute a strong alternative to the view of human choice commonly held in both analytic and existentialist philosophies, according to which the mature individual consciously chooses the values by which he or she lives. Sociobiology argues on the contrary that these choices of values are generated from an emotional constitution whose deep structure is the product of natural selection.

A third attempt to address the problem of determinism proposes a pragmatic dualism that constructs a wall of separation between the scientific knowledge of evolution and the practical affirmation of "free will" in daily life. Contrasting deterministic science with morality, Pinker argues, "The mind is a system of organs of computation, designed by natural selection to solve the kinds of problems that our ancestors faced in their foraging way of life, in particular, understanding and outmaneuvering objects, animals, plants, and other people."[40] Science examines people as material objects, and it seeks to understand the physical "causes" of human behavior.[41] In the absence of scientific explanation, he argues, we ought to understand "free will" as an idealized construct that society had to invent for the sake of maintaining the institution of morality. Morality can only

---

[39] Ruse and E. O. Wilson, "The Evolution of Ethics": 50–52.
[40] Pinker, *How the Mind Works*, p. 21.   [41] See ibid., p. 55.

function if people are viewed as responsible moral agents. The ethics "game" is only "playable" if we accept this moral idealization of people as "free, sentient, rational, equivalent agents whose behavior is uncaused, and its conclusions can be sound and useful even though the world, as seen by science, does not really have uncaused events."[42] Human conduct, he asserts, appears close enough to the idealization to make morality work.

Pinker confesses his sense that it is impossible rationally to show how "free will" has any place within the deterministic world examined by science. Our minds are confronted here with a "computational mismatch"[43] because we have evolved by natural selection to solve practical problems of survival and reproduction, not to do metaphysics. We cannot solve the "free will"–determinism conundrum any more than we can see ultraviolet light with the naked eye or mentally rotate objects in four dimensions. From the point of view of science, human decisions are caused by a system of neural organs designed by natural selection. So Pinker maintains that "free will" cannot be real because there are no events, including neurological events, without physical causes. There is no autonomy but we have to act as if there were in order to maintain some kind of moral order in society.

Pinker claims to value both science and ethics in their own respective spheres, but we do not have to look too hard to see which holds the greatest authority and which makes the most reliable claims about the real world: "The mechanistic stance allows us to understand what makes us tick and how we fit into the physical universe. When those discussions wind down for the day, we go back to talking about each other as free and dignified human beings."[44] Were Pinker's skeptical view of moral knowledge accurate, however, we would have to wonder how long it would take for the truth to set in and for social interactions to take a decided turn for the worse.

Some sociobiological critics might argue that Pinker's practical dualism suffers from an incoherence caused by a failure of nerve that refuses to apply the full implications of his own scientific perspective to morality. It will not do, they might argue, simply to assert the social value of belief in "free will" after having argued that it has no place in a rational account of the world. This view of human nature ultimately implies an amoralism that undercuts the values that Pinker promotes, especially liberal tolerance, respect for diversity, and the open-minded quest for scientific knowledge.[45]

---

[42] Ibid. [43] Ibid., p. 565. [44] Ibid., p. 56.
[45] For a dualistic alternative, see Roger Penrose, *The Emperor's New Mind: Concerning Computers, Minds, and the Laws of Physics* (New York: Oxford University Press, 2002) and *Shadows of the Mind: A Search for the Missing Science of Consciousness* (New York: Oxford University Press reprint, 1996).

FREEDOM AND RESPONSIBILITY IN CHRISTIAN ETHICS

I would now like to turn to Christian ethical discussions of human freedom. I will begin with an analysis of the writings of Pope John Paul II because of the influence and representative quality of his treatment of these matters. His work is complemented by a sympathetic writer, Terence Nichols, but I will argue that both are found wanting in evolutionary terms. The second part of this section proposes that an account of emergent complexity allows us to understand how human free choice, and other aspects of moral agency, have been given to us as part of the evolutionary process rather than specially produced by discrete acts of divine intervention at selected points in the natural history of our pre-human ancestors. A more realistic view of moral agency is more consonant with an evolutionary account of our place in the natural world.

John Paul II

As seen earlier in this book, John Paul II made it clear that he took evolution, properly understood, to be compatible with the Christian view of human beings as responsible moral agents. His major reservation about evolution came from a moral concern – that evolution should not be interpreted in such a way as to downgrade human dignity. Any attack on the power of free choice would constitute an attack on human dignity, he believed, and both are rooted in the reality of the human soul. This conviction led the pope to denounce theories of emergent materialism, according to which higher mental capacities are the emergent properties of matter once it evolves to an appropriate level of complexity and organization. John Paul II also rejected any view of the human soul as somehow evolving from matter. He did not explain what he meant by this notion, but he seemed to believe that any adequate account of human dignity has to rely on belief in the separate, intentional, and direct creation of each human soul by God. The pope did not argue that it is *scientifically* necessary to affirm divine intervention if one is fully to account for biological human nature, but he firmly believed that it is *morally* and *religiously* necessary to do so.

This papal position might mean one of two things. The first option, following the teaching laid out by Pope Pius XII in *Humani generis*, argues that evolution can account for the body and its biological needs, but not for the soul and its higher cognitive, moral, and spiritual inclinations. This view has three advantages. First, it does not reduce spirit to a function of matter, and it clearly leaves room for free choice. Second, it ties human destiny closely to a very personal involvement of the Creator in the creation. Third, it can admit that evolutionary theory accounts for the formation of the human body without need to appeal to divine intervention; this recognizes the proper competence of science without eliminating the separate sphere of free choice. Yet this very strong view of the soul amounts to a form of dualism that detaches mind (or at least some mental capacities) from the human body, thus essentially ignoring the Darwinian project. The second option argues that divine intervention was necessary in the past so that the human race would appear in creation. It also holds that God directly creates each individual soul. This argument suggests that if the course of natural history had not been divinely directed it would not have produced human beings as we exist today; it might hold that predecessor species would have stalled or become extinct at some point.

This divine guidance of natural history of course runs directly against the grain of the neo-Darwinian agenda that explains consciousness, human behavior, and the evolution of *Homo sapiens* as, in principle, subject to explanation in purely scientific terms. It also disagrees with the proponents of evolutionary contingency, such as Gould, who argue that no divine plan could possibly be found in such a haphazard and chancy process as evolution. The pope here suggests that there are moral reasons for recognizing the explanatory limits of science. Because he believed evolutionary theory insufficient for explaining the basic biological underpinnings of human nature, the pope would seem better classified as a "progressive creationist" than as an "evolutionary theist."

Theologian Terence L. Nichols provides theological support for John Paul II's position. Nichols maintains that God bestows the gift of freedom to each human being. He begins his argument with a generalization: "Theistic theories of evolution ... claim that chance, physical laws, and natural selection, by themselves, are inadequate to account for the emergence of highly complex living systems."[46] He makes a very broad claim here about the inadequacy of natural science – not only to explain human nature, but any highly complex living system. Nichols concedes that these factors exist and play a large part in evolutionary processes, but he also believes that God guides the direction and shapes the outcome of the evolutionary process through the "input of information,"[47] including that communicated through "directed mutations" operating on the micro-level of natural selection.[48] God in effect steers the evolutionary process so that it produces human beings – and God has to do so because

---

[46] Nichols, *The Sacred Cosmos*, p. 90.   [47] Ibid., pp. 90, 20f.   [48] Ibid., pp. 110, 171.

the process itself does not have inherent "giftedness" to generate such effects. God guides the evolutionary process by working within indeterminacies to subtly direct the emergence of genetic mutations, but in a way that is neither detectable by scientific investigation nor in violation of the laws of nature. God is thus the cause of human moral capacities.

Nichols by no means denies the presence of contingency and chance in the workings of nature. Christian affirms that God is an active influence in nature, "the constant presence of a field that gently and imperceptibly draws creation toward an ultimate goal of reconciliation with its Creator."[49] This ultimate reconciliation of all things in God is the reason why the universe is structured so as to produce greater and greater complexity and diversity.

Nichols argues that human consciousness is revealed in direct, first-personal experience and that this capacity cannot have arisen by material shaping the human brain through natural selection.[50] If the self emerges from a physical process, then, Nichols argues, emergent materialism also implies that the self will cease to exist once this physical condition reaches its demise. His Christian belief in an afterlife leads him to suspect that material emergentism cannot provide a fully adequate view of the person.[51] Nichols has in mind recent advocates of emergentism such as Nancey Murphy, Philip Clayton, and others,[52] but he would presumably also have reservations about Rahner's belief that matter itself is created in such a way that it is moved by its own nature to give rise to ("transcend itself" into) spirit.[53] Rahner believed that the emergence of the human soul is part of the larger dynamism of an evolving universe that gives rise to beings who become capable of freely accepting God's self-gift. Nichols, on the other hand, and John Paul II are more reserved about the inherent capacities of nature. They tend to believe that we can account for the spiritual capacities of human nature only by taking into account free decisions by God to guide the evolutionary process in very definite ways.

Nichols argues that God establishes a personal relationship with each developing human being that gives him or her a capacity for freedom.[54]

---

[49] Ibid. [50] See ibid., p. 150. [51] See ibid., pp. 154–155.
[52] See ibid. See Clayton, "Neuroscience, the Person, and God: An Emergentist Account," in Robert John Russell *et al.*, eds., *Neuroscience and the Person: Scientific Perspectives on Divine Action* (Vatican City State: Vatican Observatory Publications, and Berkeley, CA: Center for Theology and the Natural Sciences, 2002), pp. 181–214. Murphy's representative work is Warren Brown, Nancey Murphy, and H. Newton Maloney, eds., *Whatever Happened to the Soul?* (Minneapolis: Fortress Press, 1998).
[53] See Rahner, "The Unity of Spirit and Matter."
[54] See Nichols, *The Sacred Cosmos*, pp. 158–159, 174–175.

"God, at some undetermined time, perhaps sometime at conception, perhaps shortly after conception, enters into a relationship with the newly formed person. This personal relationship elevates the capacities and horizon of the developing embodied soul or person, just as grace elevates nature."[55] This is the context in which Nichols interprets Jeremiah, "Before I formed you in the womb, I knew you, and before you were born, I consecrated you" (Jer. 1:5).[56] Material systems are not truly free, he argues, so the only way that a person can be free would be if God supernaturally created this freedom as a personal gift to each developing individual.

At this point Nichols departs from John Paul II's anthropology. Whereas the latter held that God infuses a soul created from nothing into prepared human matter, thus creating a person, Nichols suggests that from the very earliest moments of life each individual has a natural soul, created through the reproductive process.[57] God then enters into a particular relationship with the soul of each individual, animating it with a higher purpose and with a new divinely bestowed power. God, in other words, creates the soul when God creates the person.

The key point here is that, for Nichols, we are each constituted in the most important way by existing in a personal relation to God. This position avoids a dualistic insertion of a soul into a previously existing material substrate. It does however seem to approach emergent materialism. The difficulty it presents revolves around an ambiguity in Nichols' treatment of the meaning of "freedom" and "subjective consciousness." He defines "free choice" as the internal ability to choose to do other than what one actually did in any given situation. His paradigm is the case of someone who chooses not to do something out of moral conviction despite the fact that he or she has a strong desire to do it. The dog feels like running down the street but sits still out of fear of the owner. The person who feels like hitting an annoying peer refrains out of a commitment to benevolence (or at least "non-malevolence"). Nichols refers to "authentic freedom" here as identical to the capacity for "free choice," and he speaks of "free choice" as identical to "free will." Freedom means acting intentionally for moral reasons rather than from internal compulsion, and this capacity, he argues, cannot emerge from a "wholly material system."[58]

Yet while human beings can make choices that are less bound by internal pressures than those of other animals, our acts are often based on unreflective internal desires and spontaneous urges. Moreover, free choice and

---

[55] Ibid., p. 175. [56] Ibid. [57] Ibid., p. 176. [58] Ibid., p. 158.

intentionality are not all-or-nothing capacities. It seems more appropriate, given the phylogenetic heritage we share with higher mammals, to think of these as sitting at various points along a spectrum. In fact, at this point it is hard to resist resurrecting the old image of the ladder, since we can speak of "higher" and "lower" degrees (or none at all) of freedom from genetic programming. Mind has given our species considerable freedom from the genetically structured, ground-up control mechanisms characteristic of lower organisms. It includes cognitive capacities that enable us to respond flexibly and sensitively to the particular circumstances in which we find ourselves immersed at any given time and place. Owen Flanagan thus describes the human mind, using terms from cognitive science, as a "plastic, self-updating representational system."[59]

Social primates do shape their behavior to accord with the habits and organization of their groups. The evolution of human intelligence gives us a range of options, an ability to remember and plan, a facility at communication and human understanding that gives as a vast advantage in considering the context and consequences of various courses of action. The fact that we make decisions by employing our intelligence means, among other things, that we do not habitually proceed directly from biological impulse to behavior. Intelligence gives us the ability to know that we have many different motivations, some of which conflict with one another, and, because of the complexity of our social life, to know that we have many different objects of responsibility.

The capacity to make free choices is neither an abstract power independent of our bodily nature, as dualists suggest, nor a power of control over ourselves given supernaturally. As Midgley points out, "what really is distinctive about human freedom is an individual's ability to act in spite of many inner divisions, as a whole."[60] We do this by "standing back" from the immediacy of the moment, taking stock of our motivations, opportunities, and relationships, and then striving to make balanced judgments in light of relevant wholes. Midgley's "struggle for wholeness" is not so much about a basic ability to decide between two or more real options, though it depends on this capacity, but rather about moral freedom, or moral goodness. Nichols would thus do well to distinguish between free choice and moral freedom. This distinction draws our attention to at least three kinds

---

[59] Owen Flanagan, *The Science of the Mind* (second edn., London and Cambridge, MA: MIT Press, 1991), ch. 7, "Minds, Genes, and Morals: The Case of E. O. Wilson's Sociobiology," p. 304.

[60] Mary Midgley, *The Ethical Primate: Humans, Freedom and Morality* (London and New York: Routledge, 1994), p. 7.

of "freedom" found in Christian ethics: first, the power of choice that comes naturally to us as human, particularly as possessing important cognitive, affective, and social capacities; second, the moral accomplishment that constitutes a truly virtuous character; and third, the gift of sanctity, found in the person who has so cooperated with God's grace that he or she experiences the height of moral freedom.

One could argue that the consciousness of different species of animal also runs along a spectrum, and that, again, the fact that we reside on one end of the spectrum does not imply that we are the only ones on it. If many higher animals have consciousness, and some even self-consciousness, there seems no reason to postulate a special divine initiative to account for this feature of human nature or to think that God has to provide any missing power that is not already a potency of nature itself.

The difficulty with Nichols' position is that it seems to lean in the direction of a "God of the gaps" argument. I use the term "lean" because it does not fit the description of an argument that attempts to show that God has "work" to do in the natural order that would otherwise have no explanation. Nichols does not attempt to demonstrate that God exists on the grounds that some empirically observable regularity in nature could not otherwise function in the way it does. At the same time, he does maintain that if God had not taken deliberate steps to create it, human freedom could not have evolved from the evolutionary process. We cannot explain rationally or scientifically how human freedom came to exist, Nichols argues, and therefore the existence of human freedom constitutes evidence of divine influence on the course of natural history and in the lives of individual human agents today.

Another difficulty with Nichols' position is that, like "God of the gaps" arguments, it is open to being discredited by the expansion of scientific understanding. In Nichols' favor, the issue of human freedom is probably not going to be resolved by some unforeseen empirical discovery or theoretical breakthrough. The issue is more philosophical than it is scientific. At the same time, we ought to be wary of moving too quickly from our lack of knowledge to special divine activity. As Richard H. Bube points out, "The list of phenomena invoked by Christians to defend the God-of-the-Gaps is very long and still with us."[61] The alternative position held by Van Till and others maintains that God does not have to "intervene" (I have already noted the problems with this term) or otherwise "guide" or "direct" the

---

[61] Richard H. Bube, "The Failure of the God-of-the-Gaps," in James E. Huchingson, ed., *Religion and the Natural Sciences: The Range of Engagement* (Fort Worth: Harcourt Brace, 1993), p. 135.

evolutionary process because creation has already been given the potential for producing intelligent and loving beings.[62] This potentiality is actualized as nature is allowed to manifest its potentialities over the available cosmic scales of time and place. The universe is organized in such a way as to generate life, and then to generate more and more complex forms of life, and then to generate more and more intelligent forms of life. God sustains the creation in being but does not need to guide the evolutionary process. This is the "fully gifted" cosmology of Van Till, among others. God uses the structures of nature and contingent events to produce creatures, among whom are human beings. In this view God can interact with nature in such a way as to guide its unfolding, but need not do so. There is no need to refer to supernatural agency in an explanation of natural history, and no need for a "God of the gaps."

THE MEANINGS OF FREEDOM

The meaning of human freedom must be addressed in a way that takes seriously each person's experience as a self-directing moral agent while at the same time acknowledging that human beings function within contexts that are to a large extent not of their own making. The notion of human emergent complexity acknowledges some degree of self-determination within the natural limits of human behavior. We do not experience freedom directly, of course, and it is a mistake to think about it as a noun, as if it existed somewhere on its own rather than as a property of acting human beings.

Ayala rightly believes that it is intuitively obvious to most people that the ability "to choose between alternatives is genuine rather than only apparent."[63] He is not alone in this claim. We can advert to its existence as a dimension of our acts as we perform them. Every person has the daily experience of deciding to act in one way rather than another on the basis of good reasons, or moral principle, or an overriding loyalty or a desired goal. To dismiss all of these motives as illusory, or as "slaves" of deeper genetically driven motivations, runs too profoundly against the experience of what it means to be human, to be persuasive.

Freedom and related notions are central to moral reflection, but not much has been written lately on the issue of "free will versus determinism"

---

[62] Van Till *et al.*, *Portraits of Creation*.
[63] Francisco J. Ayala, "The Difference of Being Human," in Holmes Rolston III, ed., *Biology, Ethics, and the Origins of Life* (Boston and London: Jones and Bartlett Publishers, 1995), p. 122.

by Christian ethicists. Writers such as Josef Fuchs[64] and Karl Rahner[65] focus on transcendental freedom and the freedom of the fundamental option; social ethicist David Hollenbach, SJ,[66] and liberation theologians Gustavo Gutierrez[67] and Jon Sobrino, SJ,[68] are concerned with political freedom and human rights; legal scholar John Noonan[69] is focused on freedom of religion; the more psychologically concerned theologian Patrick McCormick[70] writes about freedom from addictions; moral theologian Charles Curran[71] writes about academic freedom and freedom of inquiry within the church. Major texts in Christian ethics that are written by Richard Gula,[72] Russell Conors and Patrick McCormick,[73] and others take for granted the existence of freedom of choice but do not explicate its meaning for an understanding of humanity influenced by the biological sciences. Christian ethics, however, does provide some key notions and distinctions that help us grasp the significance of evolutionary treatments of human behavior for our understanding of moral agency.

Theologians usually approach the issue of freedom by appealing to the ordinary experience of making choices between alternatives. They do not attempt to prove the human capacity to make choices, but rather take it as a background condition of everyday life or as a transcendental condition of human action. Rahner had valid reservations about the prospect of anyone ever constructing a demonstrative philosophical "proof" of freedom: "Freedom is not the datum of any empirical psychology," he wrote. Our knowledge of freedom is always "from within" and "grasped as a form of

---

[64] See Josef Fuchs, SJ, *Personal Responsibility and Christian Morality*, trans. William Cleves *et al.* (Washington, DC: Georgetown University Press and Dublin: Gill and Macmillan, 1983).

[65] See Karl Rahner, SJ, "The Theology of Freedom," trans. Karl-H. Kruger and Boniface Kruger in *Theological Investigations*, vol. VI: *Concerning Vatican Council II* (New York: Seabury Press, 1974), pp. 178–196.

[66] See Hollenbach, *The Common Good and Christian Ethics* and *The Global Face of Public Faith: Politics, Human Rights, and Christian Ethics* (Washington, DC: Georgetown University Press, 2003).

[67] See Gustavo Gutierrez, *A Theology of Liberation*, trans. Sister Caidad Inda and John Eagleson (revised edn., Maryknoll, NY: Orbis Press, 1988).

[68] Jon Sobrino, *Witnesses to the Kingdom: The Martyrs of El Salvador and the Crucified Peoples* (Maryknoll, NY: Orbis Press, 2003).

[69] See John T. Noonan, Jr., *The Lustre of Our Country: The American Experience of Religious Freedom* (Berkeley, CA: University of California Press, 1998).

[70] See Patrick McCormick, *Sin as Addiction* (Mahwah, NY: Paulist Press, 1989).

[71] See Charles E. Curran, *Catholic Higher Education, Theology, and Academic Freedom* (Notre Dame, IN: University of Notre Dame Press, 1990).

[72] See Richard M. Gula, SS, *Reason Informed by Faith: Foundations of Catholic Morality* (Mahwah, NY: Paulist Press, 1989), ch. 6.

[73] See Russell B. Conors, Jr., and Patrick T. McCormick, *Character, Choices, and Community: Three Faces of Christian Ethics* (Mahwah, NY: Paulist Press, 1998), ch. 3.

transcendental experience as such by the fact that the subject knows itself to be incapable of being turned into an object and thus knows itself to be free."[74] This appeal to experience might be dismissed by some critics as a monumental evasion, yet the critics' own appeal for readers to question their subjective experience of freedom is itself only intelligible on the grounds that we have a genuine choice to examine the truth of their claim. Denial of this choice, and the kind of freedom it involves, undermines any serious approach to the matter.

Sociobiologists do not deny that people experience themselves and others making choices, but they argue that the search for objective knowledge by scientific means must prescind from subjective experience. The most philosophically adequate perspective, for them, views human beings as "objects" of scientific analysis. They insist that personal experience is not something that has any objective status for science, so it must be relegated to the status of a merely subjective phenomenon that is more of a distraction than a guide for understanding human agency. The subjective experience of making choices is caused by emotions that have been assembled according to genetic instructions that in turn have been designed by the evolutionary process to promote fitness.[75]

In contrast to sociobiology, Christian ethics affirms that we have a natural orientation to the good, a central inclination of our created nature, that is a particularly important part of our human nature. As theologian Philip Clayton argues, agential freedom in which the agent's own choice is the sufficient cause for some of his or her actions is made possible by an emergent capacity that is "both a product of evolution and a sign of the image of God in humanity."[76] Emergent cognitive, emotional, and social capacities allow us to know and love things for their own sakes, and not only as instrumentally valuable to us or our genetic "interests." Evolution has given animals an evolved orientation to the ends proper to them, and their behavior can be understood in terms of natural selection. Nature has given animals varying degrees of capacity for voluntary action – the bird voluntarily flies overhead, the squirrel voluntarily climbs the tree, and so on – and evolution has given human beings an enhanced capacity for voluntary action and a capacity to select what we think are the best means to the end that we identify as good for us. Animals tend to select from among

---

[74] Rahner, "The Theology of Freedom," pp. 190–191.
[75] See E. O. Wilson, *On Human Nature*, ch. 4.
[76] Philip Clayton, "Biology and Purpose," in P. Clayton and J. Schloss, eds., *Evolution and Ethics: Human Morality in Biological and Religious Perspective* (Grand Rapids, MI: Eerdmans, 2004), pp. 332–333.

various courses of action in accord with their fitness interests, but we make our decisions in light of our more fundamental orientation to what is good for us. Human beings, as Aristotle observed, have no option but to seek what they consider to be good – "every action and pursuit is considered to aim at some good."[77] Our primary moral challenge lies in properly identifying the content of the "good life."

As much as we are an integral part of the natural world and share a common animality, we also stand apart from the rest of the animal world. We have a natural inclination not only to this or that class of goods, say, food or drink or sex, but also, and more importantly, to what is good in general or what is comprehensively good. We pursue this good by exercising our capacity to make practical choices. Each person makes choices about how to act in particular circumstances in light of his or her sense of what is good in general.[78]

We can understand better the significance of human agency, including freedom, by distinguishing between five terms that frequently become muddled or confused with one another: choice, will, free choice, freedom, and responsibility. Before going further, though, I would like to suggest that the phrase "free will" is unhelpful because it often implies a dualistic separation of the "will" from the rest of the human being within which it "resides." To be "free" is best not construed as a quality that modifies only the "part" of the person called the "will." Freedom is a quality of the whole person, not of the will alone. This having been said, distinguishing these terms allows us to relate the evolutionary context of human behavior to the exercise of human agency without distorting one or the other.

First, the term "choice" refers to the ability of the agent to decide upon a course of action. Most broadly, choice enables an organism to address challenges from its environment. Social animals identify costs and benefits of various options within contexts of social exchange. Primatologist Frans de Waal shows that primates make choices in the social domain by entering into advantageous alliances, enforcing agreements to cooperate, detecting cheaters, selecting the best means of dealing with cheaters, minimizing costs and maximizing benefits in various cooperative arrangements, and so forth. If the term "choice" means simply selecting one option rather than another, then chimpanzees and bonobos, and even Skinnerian pigeons

---

[77] Aristotle, *Nicomachean Ethics* 1094a1.
[78] According to Thomist Yves Simon, it is the "necessity of the adherent of the will to the comprehensive good [that] entails the indifference of the practical judgment"; Simon, *Freedom of Choice*, ed. Peter Wolff (New York: Fordham University Press, 1969), p. 105.

make hundreds of choices a day. This use of the term "choice" is, of course, radically different from its meaning in Aristotle and Thomas Aquinas, for whom choice flows from deliberation.[79]

Sociobiologists liken choices to decisions simulated in the calculations of game theory, yet it makes sense to speak of accurate and inaccurate calculations but not of "free calculations." In evolutionary theory, an organism that chooses to act in a way that runs against its fitness interests will be judged to have made an error as a result of compulsion, ignorance, or manipulation by more clever competitors.[80] This is why evolutionists tend to regard human acts of genuine moral altruism for non-kin and nonreciprocators as aberrations and, in evolutionary terms, as mistakes.

Evolutionary theory strives to provide useful information about the evolved mechanisms that lie beneath many human choices. Human choices reflect the "behavioral plasticity" proper to the species, however, and are not just the result of machine-like procedures. Human nature is composed of propensities, proclivities, tendencies, or inclinations that are more likely to be elicited in some circumstances than others and that make it easier to learn some things than others. Thus people are more likely to be protective of their own children than those of strangers, to be trusting toward friends and cautious around strangers.

The very notion of "human nature" suggests that human beings will be *inclined* to respond to what they perceive to be objects that fulfill some kind of essential needs. But it does not, in and of itself, mean that this response is automatically preordained by genes or the evolutionary process. Statistical probabilities do not contradict the fact that individuals can depart from norms.

Second, the term "will" refers to the human person as a deciding, desiring, striving being. In classical Christian thought, the term "will" signifies the *whole person* as desiring and choosing; it is not some special "part" of the person, a kind of "moral organ," namely the "free will" that directs the movement of the body. Willing is both a cognitive and an appetitive activity joined in unison; it is not helpfully identified with desire acting independently of intelligence. Willing relates the whole person to this or that particular action. It also, and more fundamentally, leads the agent reflexively and as a whole to a deeper attachment to what he or she

---

[79] See *Nicomachean Ethics* 1113a and *ST* I-II, 13. See also Simon, *Freedom of Choice*.
[80] This line of thinking was developed by Richard D. Alexander, *Darwinism and Human Affairs* (Seattle and London: University of Washington Press, 1979) and *Biology of Moral Systems*, esp. pp. 102–103.

judges to be good (or evil) depending on the choice made. When we choose to do something, we do it because we believe it to be, in some sense, good for us *as a whole* and not only for our senses of touch, our wallets, our genes, or any other part of what we are. Frankfurt's "second-order volitions" are themselves either explicitly or implicitly willed and acted upon in light of the person's determination of what is good. They reflect the deeper priorities of the person that guide concrete choices about how to respond to immediate "first-order desires" here and now.

Sociobiologists sometimes speak of "free will" as a *part* of the person that exercises a special kind of force on other parts of the person, usually the body (popular language reinforces this image when it speaks of "will power" controlling appetite). Whereas evolutionists such as Pinker, E. O. Wilson, and Dawkins typically discuss "free will" in order to debunk it, the phrase itself is not found in the writings of classical authorities such as Aristotle and Augustine. Philosopher Vernon Bourke points out that Thomas rarely refers to "free will'" (*libera voluntas*) and "practically always discusses man's freedom in terms of free choice (*liberum arbitrium*) of means."[81] In fact, the phrase was seldom used before the modern period.

The notion of "will" refers to a unified center of personhood. Pinker, Wilson, and Dawkins maintain that the common-sense affirmation of unified selfhood is a fiction. Just as there is no one controlling part of the brain, Pinker argues, there is no unified governing psychic center that oversees and controls the subsidiary parts of the self. The mind, he maintains, is better conceived as "a congeries of parts that operate asynchronously." This is not to say that the brain is just a collection of completely autonomous "modules" – since the ability to make any choice, even a trivial one, depends on neural coordination – but it is to call into question the stable psychic unity implied in the phrase "free will."

Modern science, and especially psycho-pharmacology, challenges, "our sense of what is consistent with the self and what is mutable."[82] Yet it is not necessary to maintain that each person has a metaphysically immutable or completely unified core in order to acknowledge the existence of stable personal identities that are structured in large part according to what

---

[81] Vernon Bourke, *Will in Western Thought: An Historico-Critical Survey* (New York: Sheed and Ward, 1964), p. 67. See, more recently, David Gallagher, "The Will," in Stephen J. Pope, ed., *The Ethics of Aquinas* (Washington, DC: Georgetown University Press, 2002). Many medieval philosophers, such as Aquinas, did not attempt to provide a positive demonstration of the existence of free choice, but they were not working within a mechanistic theory of human behavior that required its explicit philosophical justification.

[82] Kramer, *Listening to Prozac*, p. 21.

MacIntyre calls "narratives" and Charles Taylor "articulations." Loss of memory, and especially of personal life, leaves individuals without a sense of self.[83] Thus Rose's claim that "Nothing in biology makes sense except in the light of history"[84] also applies to human life. The insufficient sense of historicity on the part of sociobiologists undermines their ability to appreciate the dependence of effective agency on narrativity. Instead of simply the product of narratives, of course, we also interact with these stories as we appropriate them to become, in some sense, the active authors of our own lives. The capacity for willing is thus exercised and developed to the extent to which we come to greater self-knowledge and "self-articulacy."[85]

Third, the phrase "free choice" refers to the human ability to decide upon a course of action that is neither necessitated nor restrained by external coercion. It signifies that the agent has a degree of control over how, and even whether, he or she acts. Free choice is displayed in acts produced by reflection, deliberation, and judgment. Most human choices are made out of routine or habit rather than freely in this sense, but they remain voluntary.

Pinker seems to assume a notion of "free will" that regards a free choice as essentially uncaused. This suggests to him that free choice exists only if "free will" somehow suspends the normal ordering of nature. Pinker assumes that the "will" is completely independent of all physical causation and that it gives rise to behavior that is capricious and unpredictable. He adopts this view from Gilbert Ryle's famous criticism of the Cartesian bifurcation of the person into a physical body, the "machine," and a purely non-physical entity, the "ghost," which somehow mysteriously operates on the body from inside. Pinker accepts Ryle's repudiation of the view of the mind as a kind of uncaused cause, able mentally to "jump start" itself and act independently of the body and our evolutionary heritage.

The classical Christian tradition, particularly as expressed in the work of Thomas Aquinas, uses the term "cause" in a much more nuanced and complex way than one finds in most evolutionists. Free choices have causes: both the agent, as the subject of the act (as efficient and formal causes), and the human good (as final cause), as the object of the act and that to which the agent is naturally drawn. From the point of view of Christian ethics, the agent as a whole is the secondary cause of his or her action because God is

---

[83] See Stephen G. Post, *The Moral Challenge of Alzheimer's Disease: Ethical Issues from Diagnosis to Dying* (Baltimore: Johns Hopkins University Press, 2000).
[84] Rose, *Lifelines*, p. 157.   [85] Taylor, *Human Agency*, vol. I, p. 71.

the first cause of all events, though not in a way that is detrimental to human freedom.[86]

Sociobiologists self-consciously adopt the modern philosophical rejection of final causality,[87] but they implicitly offer a new, Darwinized understanding of teleology, according to which living things are driven by genetic self-replication. "Free will" is usually conceived by sociobiology as working only in the mode of efficient causality because this alone would leave it truly free. But efficient causality can be interpreted as disruptive to "free will" and there is no reason to think, as dualists do, that the exercise of free choice requires a suspension of preceding efficient causes. Peacocke, for instance, argues: "Free choice is only possible at all in a milieu in which natural processes follow 'lawlike' regularities so that 'free' choices have broadly predictable outcomes. Choice would be illusory if all were totally unpredictable."[88] Free choice is thus better construed as a capacity to select from influencing factors rather than as a complete transcendence of them. It is a modification of the ordering of nature, not a suspension of it. As philosopher Philip Kitcher puts it, "Freedom does not consist in the absence of determination but in the way in which the action is determined."[89] Agents thus exercise free choice in order deliberately to determine their acts.

The fourth term, "freedom," concerns moral power. There are a number of helpful ways of characterizing moral power. The moral "struggle for wholeness" is made possible in evolutionary terms by the fact that that we are "*predisposed* biologically to make certain choices"[90] but have the power "to conform to some drives of human nature and to suppress others."[91] The power to make these choices and to act on some influences and not on others is quite variable among individuals. Exercising this power in one act depends on the capacity for free choice, but being able to do so in a consistently good way is moral freedom. We gradually approach greater moral freedom through engaging in Midgley's "struggle for wholeness" in the midst of various components provided by our particular biological constitution, culture, and life history. The context for training natural capacities for self-direction, self-organization, and self-determination is found in one's particular family of origin, friendships, and communities.

---

[86] See *ST* I q. 83, a. 1.
[87] See Francis Bacon, *The Advancement of Learning, Book I*, ed. Hugh G. Dick (New York: Random House, 1955), p. 193.
[88] Peacocke, *Theology for a Scientific Age*, p. 75.
[89] Kitcher, *Vaulting Ambition*, p. 407. [90] E. O. Wilson, *Consilience*, p. 250.
[91] Ibid., p. 251; also E. O. Wilson, *On Human Nature*, p. 97.

The source of moral freedom is virtue, the habitual disposition to do what is right and to embrace what is good. One of the meanings of the Latin root of virtue, *vir*, is "power." Virtue is the basis for the most profound sense of freedom. This is the sense in which "the more one does what is good, the freer one becomes."[92] Conversely, vices such as envy, sloth, and cowardice restrict and damage human freedom. It is not the evolutionary process that limits this sense of freedom but our own choices and cultural contexts. The *potential* for virtue is provided by genotypes and biological constitutions and it is developed by choices made in proper social and cultural environments. It is *actualized* by the process of moral development in which the agent is gradually enabled to assume responsibility for cultivating his or her own moral freedom. This sense of freedom is neither expanded nor even maintained by isolated subjects. It is extended and strengthened, or retarded and weakened, by habits, friends, work, and other social and cultural circumstances.

Greater understanding of the human predicament contributes to moral freedom. Sociobiological writings can be interpreted as offering a warning about the danger that lies in "unexamined emotions,"[93] a theme echoing much of traditional Christian ethics. A significant part of moral education amounts to learning to identify in oneself operative emotions and to make judgments about the role they ought to play in one's life. Being able to engage in this kind of "evaluative self-reflection" is certainly one of the capacities distinguishing us most sharply from other primates. Sociobiologists and evolutionary psychologists might be able to contribute to self-reflection by identifying psychological mechanisms whose operation or influence upon us can constitute either moral liabilities or positive moral possibilities. To date, though, the plausible components of what sociobiology has claimed to be a unique way of viewing human nature is already known from common sense, for example the fact that people are naturally self-centered, biased in favor of their own groups, and prone to favor reciprocators. Evolutionists argue that they can account for *why* we have these propensities, and that this knowledge might enable us to be more vigilant and to take more effective corrective steps to avoid unhelpful behavior than we have in the past.

Capacities for emotional self-control – the classical virtues of temperance and courage – enable the agent consistently to live in accord with

---

[92] *Catechism of the Catholic Church* (second edn., Vatican City: Libreria Editrice Vaticana, Ligouri Publications, 2000), pars. 1731 and 1733.
[93] Wrangham and Peterson, *Demonic Males*, p. 197.

appropriate "second-order volitions." These choices are based on the agent's interpretation, judgment, and evaluation of information provided, in part, by "first-order desires" that in some cases might indeed be influenced by evolved psychological mechanisms. But acknowledging their influence on "first-order desires" is different from reducing choices to epiphenomenal manifestations of evolved psychological mechanisms. Similarly, recognizing that some evolved mechanisms have a role in our spontaneous emotional responses to experience does not mean that a person's entire character is nothing but an epiphenomenon of his or her genes. In contrast, the notion of moral freedom reflects the ancient view of character as formed by the habitual choice of acts over which the agent has some control.[94]

The final term, "responsibility," usually refers to the agent's accountability to relevant parties for his or her conduct, that is, his or her blameworthiness or praiseworthiness. It can also refer to the agent's responsiveness to others, sense of moral duty, and role expectations.[95] It can require accepting responsibility for oneself, one's neighbors, and one's community. Evolutionists argue that human emotions, far from opposing moral responsibility, can be understood as constituting capacities for moral responsibility that we ought to train and tutor in appropriate ways.

The ability to exercise moral responsibility was made possible by the evolution of the human brain and the kind of higher cognitive and emotional capacities examined in the previous chapter. Consciousness gives various animals some ability to create representations of other beings in their own minds. The inability of monkeys to recognize themselves in a mirror gives some indication of the limitations of this species' consciousness. These kinds of representations are often accompanied by emotional associations, for example the image of predator with the emotion of fear, the mother's image of an offspring with the emotion of care. Damasio differentiates the "core consciousness" of the present found in other animals from the distinctly human capacity for an "extended consciousness" of the past and future. Clearly "extended consciousness" is mediated by emotions as well as reasoning, and these emotions are themselves influenced by the "old brain" we inherited from our distant forebears. There is no doubt that the affective significance of powerful symbols draws

---

[94] See Aristotle, *Nicomachean Ethics* 1114b.
[95] Helpful works on the theme of responsibility include H. Richard Niebuhr, *The Responsible Self* (San Francisco: Harper and Row, 1963), and Hans Jonas, *The Imperative of Responsibility: In Search of an Ethics for the Technological Age* (Chicago and London: University of Chicago Press, 1984).

on the human capacity for "extended consciousness" and can contribute to the integration of the self. As Tattersall puts it, "while every other organism we know about lives in the world as presented to them by Nature, human beings live in a world that they consciously symbolize and re-create in their own minds."[96] While scientists can pinpoint specific regions or pathways of the brain involved in particular functions, it is the whole person who is conscious, who decides, and who acts.

As argued above, human nature need not be conceived in ontologically reductionistic terms. In fact, adhering single-mindedly to a reductionistic perspective leads one unreasonably to obscure the dynamic character of organic nature in general and human life in particular. Identifying the serious shortcomings of the reductionistic program, and especially the scientism and scientific materialism that often accompany it, shows that we need not grant sociobiologists and evolutionary psychologists the privilege of framing the unanswerable question: how can "free will" exist in a mechanistic evolutionary world? – a question that amounts to asking: what is the place of free will in a world in which no wills are free?

The broad view of human emergence developed here, on the other hand, calls attention to ways in which wholes influence the functioning of their constituent parts. The nonreductionistic, holistic perspective suggested by the notion of emergence enables us to recognize ways in which higher or emergent capacities reshape and reorder lower capacities. The notion of "sublation" accommodates the proper functioning of lower capacities in their own right, but also acknowledges ways in which new capacities direct them to higher goods. Speech, for example, can serve not only to share information vital to cooperative living organized for instrumental purposes, but also to communicate knowledge and share experience within a genuine community of friends. Sex, similarly, can function not only in "mating behavior" but also in "making love" to serve the unitive goods of mutuality, companionship, and affective intimacy as well as the procreative end of sex.[97] In specifically religious terms, the transformative symbolism of the Eucharist offers a powerful depiction of the sacramental project that is human existence. The sacramental imagination promotes the transformation and reordering of these evolved proclivities and mechanisms so that they are able to serve more and more fully the Christian moral life.

In a Christian ethical perspective, *the* human responsibility is to engage, as deeply and seriously as possible, in lives of ethical "sublation": to employ the capacities given to us by God through the evolutionary process for the

---

[96] Tattersall, *Monkey in the Mirror*, p. 78.   [97] *Humanae Vitae*, par. 12.

highest ends for which we can strive, to use the capacity for free choice in ways that move us to closer union with the ultimate good, and to live in such a way that we are increasingly drawn to goodness and virtue. From the perspective of Christian ethics, evolution is the means employed by the Creator to produce creatures, including human beings, and human freedom is best understood in terms of evolved human nature and not as a bizarre exception to everything else known scientifically about human beings. Whereas the capacities that make us distinctively human emerged through the blind working of the evolutionary process, virtues never appear without the active cooperation of human agents.

In moral terms, the best outcome of human evolution is obtained through the engagement and extension of human freedom. As Peacocke explains, "God has made human beings thus with their genetically constrained behavior – but, through the freedom God has allowed to evolve in such creatures, he has also opened up new possibilities of self-fulfillment, creativity, and openness to the future that requires a language other than of genetics to elaborate and express."[98] Rather than constituting a dichotomy, then, the practice of human responsibility shows that our evolved human nature can be actualized in a way that radiates freedom in its most profound sense.

---

[98] Peacocke, *God and the New Biology*, pp. 110–111.

CHAPTER 8

Human dignity and common descent

This chapter is concerned with the dignity of the person in light of human evolution. It focuses on intrinsic human dignity as opposed to worth based on particular traits, such as social status, racial identity, income, or talent. The question here concerns whether the notion of human dignity is any longer viable, given our common descent from animals and, if so, how common descent ought to influence our interpretation of the meaning of human dignity and the ethical implications we draw from it. Widespread moral disagreement that characterizes modern pluralistic societies inclines us to take individual rights as our major moral reference point, and rights are usually said to have their moral justification in human dignity. Thus the Universal Declaration of Human Rights (1948) states: "All human beings are born free and equal in dignity and rights," and all members of the "human family" share an "inherent dignity."[1] The "Universal Declaration of the Human Genome and Human Rights" approved by UNESCO in 1997 depended heavily on the notion of human dignity. The Preamble insisted that "the human genome underlies the fundamental unity of all members of the human family, as well as the recognition of their inherent dignity and diversity."[2] Article six of the Declaration denounced any discrimination on the basis of genetic traits as a violation of human dignity. It also forbade reproductive cloning and germ-line manipulation on the grounds that they are contrary to human dignity.

Appeals to human dignity affirm that each person has an intrinsic value and therefore ought to be protected from certain kinds of harm. Because people have dignity, they have rights that shield them from certain kinds of

---

[1] Universal Declaration of Human Rights, Preface, in Henry J. Steiner and Philip Alston, *International Human Rights in Context: Law, Politics, Morals* (second edn, New York: Oxford University Press, 2000), p. 1376.
[2] UNESCO, Gen. Conf. Res. 29 C/Res. 16, "Universal Declaration on the Human Genome and Human Rights," available at www1.umn.edu/humanrts/instree/Udhrhg.htm (accessed August 24, 2005).

harms (e.g. being subjected to arbitrary arrest, unjust seizure of property) and that require that they be given certain kinds of benefits (e.g. education, health). The notion is thus of vital concern to bioethical discussions surrounding, for example, genetic research, artificial reproduction, cloning, as well as to ethical reflection on social justice and the ethics of war.

Darwinism was from its beginning regarded by some of its sharpest critics as undermining our sense of human dignity. In this chapter I argue that human dignity is congruent with our growing knowledge of common descent, even if this notion cannot be given a "foundation" in science itself.

THE DARWINIAN CHALLENGE

Darwin was not sure about the mechanism for the beginning of life itself, but he supposed that all the organic beings on the planet descended from one primordial form. The theory of common descent functioned as a unifying principle that, as Mayr points out, "at once gave meaning to the Linnean hierarchy, the archetypes of the idealistic morphologists, the history of biota, and many other biological phenomena."[3] It also made Darwin alert to the commonalities and continuities across species lines, including the one that separates humans from members of other species. The third chapter of *The Descent of Man*, "Comparison of the Mental Powers of Man and the Lower Animals,"[4] argued that we can trace descent not only through a study of comparative anatomy and physiology but also through comparing the cognitive abilities of various higher animals. Darwin was convinced that there is no "fundamental difference between man and the higher mammals in their mental faculties."[5] Higher animals share with humans intellectual capacities such as attentiveness, memory, imagination, and even reason, which we see in their ability to "pause, deliberate, and resolve."[6]

Darwin's abandonment of orthodox Christian belief was due in part to what he thought was an anthropocentric conceit that regards all of nature as centered on the human species. According to historian John Brooke, "For Darwin the very enterprise of natural theology had become incurably anthropocentric, reflecting man's arrogance in believing himself the

---

[3] Mayr, *One Long Argument*, p. 95. See Philip Kitcher, "Darwin's Achievements," in N. Rescher, ed., *Reason and Rationality in Natural Science* (New York: University Press of America, 1985), pp. 127–189, especially pp. 171, 184–185.
[4] In *Darwin: A Norton Critical Edition*, ed. Philip Appleman (second edn., New York: W. W. Norton, 1979), pp. 176–186.
[5] *Descent of Man*, ch. 3, in ibid., p. 177.   [6] Ibid., p. 183.

product of special creation."⁷ At the same time, Darwin tried to show ways in which distinctively human traits – cognitive, emotional, social, linguistic, and even moral – are also held, albeit in more rudimentary ways, by other animals.

Darwinism from its inception was taken to present a major challenge to human dignity. In fact, observers long before Darwin had regarded claims of human–animal commonality to be a threat to morality. René Descartes, for example, noted in the *Discourse on Method* that "there is nothing that puts weak minds at a greater distance from the straight road of virtue than imagining that the soul of animals is of the same nature as ours."⁸ Descartes believed that assumptions of commonality would undermine belief in the afterlife and destroy the hope, or fear, that it instills in believers.

Despite apprehension in some circles regarding the moral implications of common descent, Darwin himself saw no inherent conflict between evolution and human dignity. The closing observations of *The Descent of Man* argued:

As love, sympathy and self-command become strengthened by habit, and as the power of reasoning becomes clearer, so that man can value justly the judgments of his fellows, he will feel himself impelled, apart from any transitory pleasure or pain, to certain lines of conduct. He might then declare ... in the words of Kant, I will not in my own person violate the dignity of humanity.⁹

Darwin's detractors felt otherwise. Samuel Wilberforce, Bishop of Oxford, registered one of the early worries about Darwinism in his review of *The Origin of Species*: "Man's derived supremacy over the earth; man's power of articulate speech; man's gift of reason; man's free will and responsibility ... – all are equally and utterly irreconcilable with the degrading notion of the brute origin of him who was created in the image of God ..."¹⁰ Nineteenth-century critics contrasted Darwinian degradation with the biblical message that the human race is the crown of creation. P. R. Russel, for example, lamented that Darwin's view of evolution,

casts us all down from this elevated platform, and herds us all with four-footed beasts and creeping things. It tears the crown from our heads; it treats us as

---

⁷ Brooke, *Science and Religion*, p. 305.
⁸ René Descartes, *Discourse on Method and Meditations on First Philosophy*, third edn., trans. Donald A. Cress (Indianapolis and Cambridge: Hackett Publishing Co., 1993), p. 33.
⁹ *Descent of Man*, in *Darwin*, ed. Appleman, p. 310.
¹⁰ Samuel Wilberforce, "On the Origin of Species," *Quarterly Review* (1860), p. 258.

bastards and not sons, and reveals the degrading fact that man in his best estate – even Mr. Darwin – is but a civilized, dressed up, educated monkey, who has lost his tail.[11]

While Darwin himself did not construct an ethical theory on the basis of nature, some of his peers were less hesitant. These included most famously Herbert Spencer, whose social theory actually antedated the publication of Darwin's work. He thought that *laissez-faire* capitalism allows nature to take its proper course and that civil leaders ought to let the most "fit" lead the way of progress for society. He denounced social programs intended to assist the poor on the grounds that they interfere with the "laws of nature," enable the "unfit" to reproduce unnaturally, and thereby actually increase the misery of the poor as well as harm society as a whole. He insisted that the state ought to protect private property and the liberties of the citizens, and that it ought not to provide care for the indigent or aid for the poor. Spencer was opposed to private philanthropy for the same reason. In the global context, the alleged natural superiority of the British was invoked to legitimate the domination of colonized natives and to support their sense of responsibility to "civilize" "savages" dwelling on lower stages of the evolutionary ladder.[12] Darwin was invoked to justify British dignity but not necessarily human dignity. In Germany, biologist Ernst Haeckel developed an interpretation of evolution (though not a very Darwinian account of it) that regarded "higher races of men" (Europeans) as more valuable than "lower races of men" (Africans).[13]

Social Darwinism was used to give scientific justification to the accumulation of vast amounts of individual wealth in the midst of great urban poverty. Steel magnate Andrew Carnegie found solace in this social philosophy: "I remember that light came as in a flood and all was clear ... Not only had I got rid of theology and the supernatural, but I had found the truth of evolution. 'All is well since all grows better' became my motto, my true source of comfort."[14] Though he accepted Darwin's view of morality as the product of the evolutionary process, T. H. Huxley repudiated Spencer's

---

[11] P. R. Russel, "Darwinism Examined," *Advent Review and Sabbath Herald* 47 (1876): 153, cited in Numbers, *The Creationists*, p. 5.
[12] See Robert J. Richards, *Darwin and the Emergence of Evolutionary Theories of Mind and Behavior* (Chicago and London: University of Chicago Press, 1987), chs. 6 and 7, and Bowler, *Evolution*, ch. 10.
[13] Richards, *Darwin*, p. 596.
[14] Andrew Carnegie, *The Autobiography of Andrew Carnegie* (Lebanon, NH: Northeastern University Press, 1986). See Herbert Spencer, *The Data of Ethics* (New York: D. Appleton and Co., 1895). For a historical treatment, see James R. Moore, *The Post-Darwinian Controversies* (Cambridge: Cambridge University Press, 1979).

efforts to derive an ethical and political philosophy from Darwinian biology. Whereas nature teaches a lesson of "survival of the fittest," morality needs to resist the workings of the inhumane "cosmic process."[15] Dignity for Huxley is created in the human fight against base instincts rooted in evolution.

Opponents of Darwinism worried that it would inevitably lead to a degradation of human worth. William Jennings Bryan, the famous Presbyterian anti-evolution crusader of the Scopes "Monkey Trial," left no doubt regarding his judgment:

> I do not mean to find fault with you if you want to accept the theory; all I mean to say is that while you may trace your ancestry back to the monkey if you find pleasure or pride in doing so, you shall not connect me with your family tree without more evidence than has yet been provided.[16]

Bryan was convinced that common descent could not but lead to personal and social immorality: "How can teachers tell students that they came from monkeys and not expect them to act like monkeys?"[17] He argued that claims of "monkey ancestry" would loosen the ethical restraints on the taking of life and the waging of war, and he even connected Darwinism to the horrors of the First World War: "The same science that manufactured poisonous gases to suffocate soldiers is preaching that man has a brute ancestry . . ."[18] Others insinuated a connection between common descent and the practice of abortion: "Why worry about disposing of 'unwanted' babies in the womb . . . since they are simply baby primates?"[19]

If one accepts the premise that animals have only instrumental worth, then to regard humans as animals is naturally to lead to the conclusion that humans have only instrumental worth. Because "brutes" have no souls, no rationality, and no moral agency, they have no intrinsic moral worth. If we are descended from animals, and share their essential traits as animals, how can we resist the implication that we have no intrinsic worth? One way of opposing this conclusion was used by the utilitarians: instead of lowering humans to the level of moral consideration given to the animals, the utilitarians tried to raise animals up (or closer) to the level of moral consideration given to humans; they argued that animals should be given moral consideration on the grounds that they too can suffer.[20] Escalating the value of animals, however, can be accompanied by lowering that of humans.

---

[15] T. H. Huxley, *Evolution and Ethics*, p. 91.
[16] Cited in Larson, *Summer for the Gods*, p. 20.   [17] Cited in ibid., p. 116.
[18] Cited in Numbers, *The Creationists*, p. 41.   [19] Cited in ibid., p. 337.
[20] See Michael Bradie, "The Moral Status of Animals in Eighteenth Century British Philosophy," in Jane Maienschein and Michael Ruse, eds., *Biology and the Foundations of Ethics* (New York: Cambridge University Press, 1999), pp. 32–51.

Since Darwin's time the thesis of common descent has received an enormous amount of supporting evidence from many different disciplines within the natural sciences. The "molecular clock" of evolution, which measures the degree of divergence between organisms in terms of amino-acid or nucleotide sequences, provides especially compelling evidence for commonality. It gives a biochemical basis for reconstructing the "tree of life," a pattern that reveals both commonality with, and the splitting of various lineages from, precursors. Mayr observed that "there is probably no biologist left today who would question that all organisms now found on the earth have descended from a single origin of life."[21] Recent studies of DNA sequencing showing that chimpanzees share more in common with humans than they do with orangutans or monkeys suggest that there is no radical biological gulf between humans and other animals, especially the primates.[22]

Sociobiologists offer the most stridently demeaning view of human dignity found among contemporary followers of Darwin. The reduction of the human person to nothing but an expendable "robot vehicle" for genetic replication does not provide a helpful ground for affirming human dignity. E. O. Wilson and Dawkins are not social Darwinians, as mentioned already, but it is hard to see how their explicit endorsement of democracy, toleration, decency, and individual rights is consistent with their extended campaign to convince us that we are nothing more than "clever brutes,"[23] and that our moral ideals are really only "enabling mechanisms for survival"[24] and ideational tools for manipulating others.

Whatever its moral relevance, knowledge of common descent is well established. Mayr explains: "The discovery that prokaryotes have the same genetic code as the higher organisms was the most decisive contribution of Darwin's hypothesis. A historical unity in the entire living world cannot help but have a deep meaning for any thinking person and for his feeling toward fellow organisms."[25] If "monkey ancestors" troubled the Victorians, how much more threatened they would have been by "prokaryote ancestors." G. G. Simpson put the point with characteristic clarity when he wrote that

---

[21] Mayr, *One Long Argument*, p. 24.   [22] Tattersall, *Becoming Human*, pp. 109ff.
[23] Michael Ruse, "Evolutionary Ethics: Healthy Prospect or Last Infirmity?" *Canadian Journal of Philosophy*, Supplementary Volume 14 (1998): 29.
[24] E. O. Wilson, *On Human Nature*, p. 3.   [25] Mayr, *One Long Argument*, p. 163.

in the world of Darwin man has no special status other than his definition as a distinct species of animal: He is in the fullest sense a part of nature and not apart from it. He is akin, not figuratively but literally, to every living thing, be it an amoeba, a tapeworm, a flea, a seaweed, an oak tree, or a monkey – even though degrees of relationship are different and we may feel less empathy for forty-second cousins like a tapeworm, than for, comparatively speaking, brothers like the monkeys. This is togetherness and brotherhood with a vengeance...[26]

And the central moral question put to this description is well expressed by philosopher Stephen R. L. Clark: "If the young are taught that they are only animals," he asks, "will they not behave as they think 'animals' behave? Why *should we not* be alarmed?"[27]

PHILOSOPHICAL NATURALISM: RACHELS

Philosopher James Rachels presents a typical neo-Darwinian assessment of human dignity. Rejecting in no uncertain terms the notion of human beings as created in the "image and likeness of God," he argues: "After Darwin, we can no longer think of ourselves as occupying a special place in creation – instead, we must realize that we are products of the same evolutionary forces, working blindly and without purpose, that shaped the rest of the animal kingdom."[28] His *Created from Animals: The Moral Implications of Darwinism* offers a philosophical elaboration of Darwin's view of our place in nature. The natural world is "not made exclusively for our habitation" and human life does not have "a special, unique worth."[29]

Rachels applauds the fact that these claims contradict what he calls "traditional morality" – including Thomistic and Kantian ethics – because doing so clears the ground for a more productive ethic that he calls "moral individualism." Moral individualism maintains that, "How an individual should be treated depends on his or her own particular characteristics, rather than on whether he or she is a member of some preferred group – even the 'group' of human beings."[30] Rachels deliberately devalues human

---

[26] G. G. Simpson, *This View of Life: The World of an Evolutionist* (New York: Harcourt, Brace and World, 1963), pp. 12–13. Other texts exploring aspects of common descent include Martin J. S. Rudwick, *The Meaning of Fossils* (Chicago: University of Chicago, 1985); Robert L. Carroll, *Patterns and Processes of Vertebrate Evolution* (Cambridge and New York: Cambridge University Press, 1997); and, in a more popular vein, Ian Tattersall, *The Fossil Trail* (New York: Oxford University Press, 1995).
[27] Clark, *Biology and Christian Ethics*, p. 39; emphasis in original.
[28] *Created from Animals: The Moral Implications of Darwinism* (New York: Oxford University Press, 1990), p. 1.
[29] Ibid., pp. 3, 4.   [30] Ibid., p. 5.

beings in order to correct our tendency to disregard the well-being of animals. Undoing our prohibitions of suicide and euthanasia is of a conceptual piece with improving our treatment of animals because both hinge on a proper understanding of the (very ordinary) place of humans in nature. According to Rachels, "Darwin simply denied the whole idea that there is something special about man's intellectual capacities ... [all] are the result of natural selection."[31] The differences that do exist are in degree, not kind. Since intelligence is common to animals, Rachels argues, there is no need to propose, as Christians have done in the past, that God directly intervenes to create a "rational soul" for each new person.

Rachels thus believes that one of the most important implications of Darwinism is its rejection of special human dignity, a "basic idea that forms the core of Western morals."[32] The *imago Dei* text carries a fundamental message: human life is sacred and animal life is not; human life deserves special consideration and animal life does not. The doctrine was amplified by Thomas Aquinas' assimilation of Aristotle's "rationality thesis," which holds that we are special because we alone are rational. Rachels quotes Thomas Aquinas: "Of all parts of the universe, intellectual creatures hold the highest place, because they approach nearest the divine likeness. Therefore the divine providence provides for the intellectual nature for its own sake, and for all others for its sake."[33] Darwin, on the contrary, concluded: "Few persons any longer dispute that animals possess some power of reasoning. Animals may constantly seem to pause, deliberate and resolve. It is a significant fact, that the more the habits of any particular animal are studied by the naturalist, the more he attributes to reason and the less to unlearnt instincts."[34] Recent evolutionary studies of animal intelligence, Rachels argues, have significantly extended Darwin's doubts that human beings have unique mental powers.

Rachels also thinks that the Christian belief that God created human beings with certain features ignores the fact that the motors of the evolutionary process are randomness and natural selection. Evolution, not the Creator, caused us to have the exact traits that we have. Finally, Darwin undermined the Christian claim that human beings alone are moral when he proposed that the human "moral sense" is an expansion of social instincts, kinship, and cooperation common to many social animals, and when he showed that human sympathy is not vastly superior to the range of concern exhibited in animal altruism. (Whether Darwin really "showed" this to be the case is of course a matter of debate.)

---

[31] Ibid., p. 57. [32] Ibid., p. 86. [33] Ibid., p. 88. [34] Cited in ibid., p. 133.

Rachels builds an alternative approach to ethics on the basis of a thoroughly naturalistic view of humanity. While not wanting to extract norms from evolution, he uses biological descriptions of human nature to undermine belief in the "rationality thesis" and the *imago Dei* doctrine, which he dismisses as unjustifiably exceptionalist and immorally anthropocentric. Rachels again concurs with Darwin when he wrote: "Man in his arrogance thinks himself a great work worthy of the interposition of a deity. More humble and I think truer to consider him created from animals."[35] Thomas Aquinas' interpretation reveals a fundamental flaw of Christianity, Rachels argues, which includes a willingness to have animals killed without consideration for their own needs. The doctrine of the *imago Dei* must be jettisoned and replaced by the practice of judging the dignity of each being, animals as well as humans, on a case-by-case basis: "how an individual may be treated is to be determined, not by considering his group memberships, but by considering his own particular characteristics."[36]

THOMISM: JOHN PAUL II

One prominent response to the neo-Darwinian "downgrade" attempts to retrieve the Thomistic tradition in light of a critical appropriation of scientific knowledge regarding common descent. Pope John Paul II provides a good example of this enterprise. As we saw in chapter four, for him, "Evolution presupposes creation: creation in the light of evolution is a temporally expansive event – *creatio continua* – in which God is visible to the eyes of faith as the 'creator of the heavens and the earth.'"[37] His famous 1996 speech before the Pontifical Academy of Sciences acknowledged that the theory of evolution "has been progressively accepted by researchers, following a series of discoveries in various fields of knowledge. The convergence, neither sought nor fabricated, of the results of work that was conducted independently is in itself a significant argument in favor of this theory."[38] The pope made it clear that he takes evolution, properly understood, to be compatible with the Christian view of the universe as the product of divine creation.

The pope's major reservation about the theory of evolution came from a moral concern – that nature not be interpreted in such a way as to deny or

---

[35] Cited in ibid., p. 1. [36] Ibid., p. 173.
[37] Comments made by John Paul II on April 26, 1985 to participants of a symposium on "Christian Faith and the Theory of Evolution," published in *Insegnamenti* 8.1 (1985): 1132.
[38] John Paul II, "Message to Pontifical Academy of Sciences on Evolution": 352, at par. 5.

downgrade human dignity. Because of this concern, as we have seen, he did not accept the possibility that the human soul "evolves" from matter.[39] He regarded belief in a special act of divine creation for each soul as a safeguard for human dignity. Following Pope Pius XII, John Paul II noted that the body, but not the soul, can be understood in terms of evolution: "If the human body takes its origin from pre-existent living matter, the spiritual soul is immediately created by God."[40] Evolution accounts only for the precursors to *Homo sapiens* and the structure of ontogenetic development in early life, not for the creation of individual souls.

The marks of the image of God in us – rationality and freedom – enable us to enter into communion with God and one another. Human dignity is expressed in art, science, and work. The pope concurred with Thomas Aquinas' belief that our likeness to God resides especially in the "speculative intellect." "But even more," he maintained, humanity is "called to enter into a relationship of knowledge and love with God himself, a relationship which will find its complete fulfillment beyond time, in eternity."[41]

The doctrine of the *imago Dei* carries powerful moral implications. It provides the basis of a strong ethic of human dignity,[42] from which flow a set of human rights, both positive entitlements and negative immunities, that protect and promote human well-being. The justice of any society can be judged according to its respect for the rights that protect human dignity. It is violated in racism, xenophobia, and economic exploitation. The *imago Dei* has special significance for sexual ethics. John Paul II believed that the image of God is disclosed not only in the mind but in the human body. Genesis 1:27 – "male and female he created them" – refers to the interpersonal complementarity of sexual attraction and the call to interpersonal communion, especially when read in light of Genesis 2:18, "it is not good for the man to be alone." "Male and female" are meant for each other and they are, moreover, equally created in the image of God. Evolution has endowed us with a critically important human difference. While functioning in animals only for the sake of reproduction, sexuality in the human person is an avenue for an intense interpersonal communion through

---

[39] See above, chapter seven, pp. 217–218.
[40] John Paul II, "Message to Pontifical Academy of Sciences on Evolution": 352, par. 5.
[41] Ibid.
[42] Human dignity is also based on redemption in Christ. *The Catechism of the Catholic Church* grounds human dignity both in creation in the image of God and in redemption by Christ: "Created in the image of the one God and equally endowed with rational souls, all men have the same nature and the same origin. Redeemed by the sacrifice of Christ, all are called to participate in the same divine beatitude: all therefore enjoy an equal dignity" (*CCC* par. 1934).

which "man and woman give themselves to one another through the acts which are proper and exclusive to spouses."[43] The pope wrote,

> Because we are made in the image of God, sexuality is *by no means something purely biological*, but concerns the innermost being of the human person as such. It is realized in a truly human way only if it is an integral part of the love by which a man and a woman commit themselves totally to one another until death. The total physical self-giving would be a lie if it were not a sign and fruit of a total personal self-giving, in which the whole person, including the temporal dimension, is present: if the person were to withhold something or reserve the possibility of deciding otherwise in the future, by this very fact he or she would not be giving totally.[44]

Personhood thus incorporates biological functions while placing their meaning in a higher, sacramental and transcendent, frame of reference. The *imago* calls us to move beyond the evolutionary framework without denying the existence of subhuman substrates in our evolved nature.

The *imago* also grounds the sanctity of unborn life. From the moment of conception, the pope writes, "the life of every human being is to be respected in an absolute way because man is the only creature on earth that God has 'wished for himself' and the spiritual soul of each man is 'immediately created by God'; his whole image bears the image of the Creator."[45] Human life is sacred and involves the direct action of God in the creation of the human soul, and no one has the right to destroy innocent life or to treat this unborn life as mere organic matter to be manipulated for human convenience. This rules out, among other things, human cloning, experimentation on embryos, and the destruction of human embryos to obtain stem cells.

John Paul II also drew important lessons from Genesis 1:26–27 regarding ecology. The *imago* reflects God's delegated authority but acts only within the limits imposed by divine sovereignty.[46] We have no dominion over human life, so we are forbidden to kill the innocent. The *imago* does not warrant wanton abuse of nature. It requires responsible stewardship of the earth and of other species, and if we are the "royalty" among all animals,

---

[43] John Paul II, *On the Family* (Apostolic Exhortation *Familiaris consortio*, November 22, 1981) (Washington, DC: United States Catholic Conference, 1982), no. 11, p. 9.
[44] Ibid. (emphasis added).
[45] Vatican Congregation for the Doctrine of the Faith, *Donum Vitae, The Gift of Life, Vatican Instruction of Respect for Human Life in Its Origin and the Dignity of Procreation* (1987), no. 5. See also *"Evangelium Vitae," The Gospel of Life*, no. 53, and the *Catechism*'s interpretation of the fifth commandment (par. 2258).
[46] "Mulieris dignitatem," *Origins* 18/17 (October 6, 1988): 268, par. 10.

this lordship is ministerial rather than tyrannical. Human beings are not absolute and unquestionable masters of nature, but stewards of God's kingdom who are called to continue the Creator's work in the promotion of justice, peace, and responsibility. Our task, described the Book of Wisdom, is to rule "the world in holiness and righteousness" (Wis. 9:3). This positive vision provides a basis for denouncing the evils of ecological irresponsibility and exploitation. The human race must repudiate the habit of acting not as God's steward but as "an autonomous despot, who is finally beginning to understand that he must stop at the edge of the abyss."[47] The *imago Dei* thus demands "ecological conversion" to an ethically responsible relation to the natural world.[48]

The *imago Dei* thus generates significant restrictions on the use of technology. The pope believed that it prohibits cloning, euthanasia, and most reproductive technologies, including in-vitro fertilization with the use of a husband's sperm. Human dignity is protected by self-restraint and respect for the ordering of creation, including its biological ordering. We do not have a mandate to reproduce without restraint; procreation must be responsible but limited to morally licit methods.[49]

The *imago Dei* functioned in the pope's writings as something of an all-purpose moral principle. Though he acknowledged that evolution is "more than a hypothesis," his philosophical phenomenology and biblical interpretation showed no influence of evolutionary thinking of any kind. He often made claims that could be tested by scientific findings. For example, his *Apostolic Letter on the Dignity and Vocation of Women* began with an announced intention to explore "the Creator's decision that the human being should always and only exist as a woman or man."[50] The pope offered a rich analysis of gender but never drew from available scientific discussions of "male and female" or new findings regarding sexual identity, sexual assignment, sexual confusion, and so forth. In fact, neither did the pope's writings show any influence of contemporary scriptural scholarship.

---

[47] John Paul II, "God Made Man the Steward of Creation," Papal Audience January 24, 2001, *L'Osservatore Romano* (2001): 11.
[48] Ibid.
[49] In this "Address to Italian Farmers," November 12, 2000, the pope said "It is against God's will to grow genetically modified organisms to increase farm production – farmers ought to be 'custodians' of the earth and not its 'tyrants.'" Prior to this statement, see GMO Roundup, *Nature Biotechnology* 18 (January 2000): 7. In August of 2003, however, the Vatican Pontifical Council Justice and Peace offered support for GM foods "for the billions of people who go to bed hungry every night." See *The Guardian*, August 24, 2003.
[50] "Mulieris dignitatem": 263, par. 1.

This is in part understandable given the pope's pastoral focus, but it also left open a number of lacunae for his followers to address.

The pope clearly maintained that the essential Christian affirmation that we are created in the image of God need not be undermined by a proper interpretation of our evolutionary origins. He did this by affirming the independent divine creation of each individual human soul, but he did not attempt to relate this claim to the thorny philosophical questions presented by evolutionary views of the relation between mind and body. Many Christians would agree with his claim that a long process of evolutionary development led to the emergence of key human traits that provide the basis for the claim that we are created in a special way in God's image. The animal origins of human nature need not be seen as a degradation of our dignity, any more than the religious and moral aspirations of human transcendence need to be taken as indication that we could not have evolved from other life forms. At the same time, to act as the *imago Dei* today necessarily implies a new and more authentic recognition of our responsibility toward the natural world.

A PROTESTANT RECONSIDERATION: PEACOCKE

Scientist-theologian Arthur Peacocke offers a view of humanity that is thoroughly evolutionary. His project is "to rethink our religious conceptualizations in the light of the perspectives on the world afforded by the sciences . . ."[51] Theology must engage scientific culture and be willing to re-examine the "claimed cognitive content of Christian theology in the light of the new knowledge derivable from the sciences."[52] Instead of operating in an intellectual ghetto, theology must be made believable to modern, scientifically trained minds. God is Creator of a world that is both structured by deterministic laws and pervaded by chance and contingency. Peacocke's emphatically nonreductionistic view of the evolutionary process acknowledges the influence of "top-down causality" and the emergence of new levels of complexity that cannot be reduced to their component parts. Recent discoveries of chaos and complexity encourage a new appreciation of ways in which chance allows for the emergence of novel forms of life and organization, while deterministic laws provide the stability for these new forms to endure. The interaction and combination of law and chance make possible an ordered universe capable of giving rise to new kinds of creatures and new patterns of relationship.

---

[51] Peacocke, *Theology for a Scientific Age*, p. 3.   [52] Ibid., pp. 6–7.

Human dignity and common descent

Creating a universe full of contingencies, God does not determine every occasion in the universe but rather allows for the free play of events. Unlike what seems to be John Paul II's notion of a scripted natural pathway leading up to the evolution of the human body, Peacocke does not believe the Creator planned all along to use the evolutionary process to create us, exactly, but he holds instead that "providing there had been enough time, a complex organism with consciousness, self-consciousness, social and cultural organization would have been likely to have evolved, that is, 'persons' would have appeared on Earth (or on some other planet amenable to the emergence of living organisms), though no doubt with a physical form very different from *homo sapiens*."[53]

What does science tell us about human beings? Like all other animals, we are the product of the lengthy creative interaction of chance and natural selection. Peacocke is highly attuned to the many continuities between humans and other animals, but unlike Rachels he is also keenly aware of the discontinuities.[54] "In humanity," he writes, "'biology' has become 'history' and a new kind of interaction – that of humanity with the rest of the natural order – arises in which the organism, *homo sapiens*, shapes its own environment, and so its future evolution ..."[55] The evolution of capacities to feel pain and pleasure preceded the evolution of animals capable of consciousness. Our own heightened consciousness brings with it "an increase in sensitivity, hence in vulnerability, and consequently in suffering ... in the context of natural selection, pain has an energizing effect and suffering is a goad to action."[56]

Unlike other animals, we have been given special capacities for self-awareness, information processing, symbolic communication (especially language), and creativity. Human beings are drawn to "self-transcendence": "self-awareness and self-consciousness, coupled with our intelligence and imagination, generate a capacity for self-transcendence which is the root from which stems the possibility of a sense of the numinous – and so of the divine."[57] Perhaps among our most important traits is the capacity to experience ourselves as mystery. Peacocke connects this wonder with the *imago Dei*:

If God is the ultimately free and the ultimately good and has created us uniquely with a limited, but real, freedom and capacity for goodness, then we are to this extent potentially *imago Dei*, created "in the image of God." If this is so ... then it is not surprising that we also encounter mystery when we contemplate the

---

[53] Ibid., p. 221.   [54] Ibid., p. 76.   [55] Ibid.   [56] Ibid., p. 68.   [57] Ibid., p. 74.

inherent and inner nature of human personhood as an, admittedly distorted, image of the mystery of the divine.[58]

In humans, nature begins both to understand God's purposes for itself and freely to choose to cooperate with those purposes:

> It is as if the Creator has endowed matter-space-time-energy, the stuff of existence, with a propensity, now actualized in man, of tuning in, as it were, to discern that meaning in the cosmic process which its Creator has written into it ... in man, physicality has become capable of reading those meanings in existence which are the immanence of the transcendent God in the whole cosmic process. The way in which God has made himself heard and understood is by endowing the stuff of the world with the ability to acquire discernment of his meanings and to listen to his word in creation.[59]

Peacocke, like John Paul II, holds that human beings "constitute the highest fulfillment of the Creator's purposes."[60] He, too, claims that we have the potential to be "co-workers" and "co-creators" with God.[61] Peacocke envisions God not only as powerful Creator but also as benevolent, caring, vulnerable and responsible – akin to A. N. Whitehead's "the fellow sufferer who understands."[62] He interprets the First Letter of John's "God is love" (4:16) as coherent with his view of God's relation to the world as one of self-limiting, self-emptying and self-giving love.[63] Indeed, he holds that God undergoes change in and through relationship with human beings.

Three major implications of the *imago Dei* flow from Peacocke's view of God. First, this understanding of the *imago Dei* generates a greater openness to technology than one finds in John Paul II. Knowledge of genetics helps us grasp the processes through which God has created and will continue to create life, and we ought to employ this knowledge in biomedical interventions as long as our choices accord with our role as "co-creators" with God. Second, as created in this image of God, we are to be companions to one another, just as God is a companion to us, and to face the future with creativity, hope, and compassion. God's own creative work will continue into the future, with new possibilities emerging in the

---

[58] Ibid., p. 239. [59] Peacocke, *Creation and the World of Science*, p. 145.
[60] Peacocke, *Theology for a Scientific Age*, p. 127.
[61] Peacocke, *Creation and the World of Science*, pp. 304–306. A similar development of this theme is provided by Philip Hefner, "The Evolution of the Created Co-Creator," in James B. Miller, ed., *An Evolving Dialogue: Scientific, Historical, Philosophical, and Theological Perspectives on Evolution* (Washington, DC: American Association for the Advancement of Science, 1998), pp. 403–420.
[62] Whitehead, *Process and Reality*, p. 413.
[63] Peacocke, *Theology for a Scientific Age*, p. 123.

interchange between God's creativity and creation's response. We should see ourselves not only as "co-creators" but as "co-explorers" of future possibilities for creation, possibilities that are as yet unknown to God. "Man would then, through his science and technology, be exploring with God the creative possibilities within the universe God has brought into being. This is to see man as *co-explorer* with God."[64] Peacocke is even willing to say that the extension of powers through technology also gives God a new way of influencing creation. The danger of such a positive view of technology, of course, is that we will use it to pursue narrowly self-interested ends to the detriment of one another and other species. Yet Peacocke is convinced that religion provides the most powerful means for restraining this tendency and for serving God's purposes.

Third, Peacocke's understanding of the *imago Dei*, like that of John Paul II, backs a strong ecological ethic. Our continuity with other animals grounds a sense of commonality and therein a sense of responsibility. Though he believes that "personhood as instantiated in *homo sapiens* can properly be regarded as an ultimate purpose of God in creation, the existence of the rest of creation must also be regarded as having its own *raison d'être* in God's purposes."[65] Peacocke even goes on to assert: "For although humanity may still be held to be unique and to have a distinct place in the purposes of God, this status now has to be seen as a position of *responsibility for the rest of all that has been created*."[66] The "dominion" of Genesis endorses the responsible exercise of power within creation, not the dominative use of technological power over nature: "But at least one who sees his role as that of co-creator and co-worker with God might have a reasonable hope of avoiding this nemesis [of hubris], by virtue of his recognition of his role ipso facto as auxiliary and co-operative rather than as dominating and exploitative."[67]

PROTESTANT CHASTENING: GUSTAFSON

Gustafson attempts to modify the anthropocentric interpretation of human dignity and to deflate the usual loftiness of the *imago Dei* in order to check human pride and the excesses to which it inevitably leads. He appeals to evolution to support his deflationary hermeneutics. One feature of our current circumstances is the tendency of Christianity to be "increasingly

---

[64] Peacocke, *Creation and the World of Science*, p. 306.
[65] Peacocke, *Theology for a Scientific Age*, p. 247.
[66] Ibid., p. 248.   [67] Peacocke, *Creation and the World of Science*, 305.

advanced as instrumental to subjective temporal human ends."[68] In this regard, Gustafson could have disapprovingly cited Peacocke's proposal that "Christian theology is (or should be) concerned above all else with the consummation of the human and with human flourishing both in this life and *sub specie aeternitatis*."[69] Gustafson's major work, *Ethics from a Theocentric Perspective*, attempts to correct anthropocentrism by shifting our focus from the primary pursuit of our own purposes and glory to serving God's purposes and glory. Gustafson's most radical methodological move argues that theological claims must be indicated by the relevant sciences. He explains: "the 'substantial content' of ideas of God cannot be incongruous ... with well-established data and explanatory principles established by relevant sciences, and must be in some way indicated by these."[70]

What we say about human dignity must be responsive to what we know from evolutionary biology and other relevant sciences. Theologians, he writes, have "not come to grips with the significance of what many developments in modern science have made clear: the minuscule place of human life in the order of nature."[71] Gustafson's sense of interdependence, emergent values, and contingency is similar to Peacocke's, but he would consider extravagant the latter's assertions about God's dependency on humanity and our power over nature.

Most of all, Gustafson believes that current scientific knowledge of our place in the cosmos makes it impossible to assert that God made all of creation for our sake.[72] The extremely long time it took for life on the planet to appear, the contingencies of natural events that led to the evolution of life on this planet (and not elsewhere), and the fact that "one way or another there shall be a finis, a temporal end, to life as we know it [on this planet]" all call into question the belief that creation "exists for the sake of man."[73] It is very difficult to believe the universe is made for human beings, given what we know about the role of countless contingencies of cosmic evolution and the origins of life on earth, and the inevitable if relatively distant "future demise of our universe."[74] Gustafson thus argues that salient features of the evolutionary process call into question the claim that the final purpose of nature, the universe, and the other species on the earth lies in our well-being.

---

[68] Gustafson, *Ethics from a Theocentric Perspective*, vol. 1, p. 18.
[69] Peacocke, *Theology for a Scientific Age*, p. 244.
[70] Gustafson, *Ethics from a Theocentric Perspective*, vol. 1, p. 256. [71] Ibid., p. 41.
[72] Cf. *Gaudium et Spes*, nos. 34 and 39.
[73] Gustafson, *Ethics from a Theocentric Perspective*, p. 83. [74] Ibid., p. 90.

This view of nature provides a perspective from which Gustafson criticizes the anthropocentric use of Genesis 1:26–27 that we see in John Paul II. He points out that the first creation narrative (Gen. 1:1–2:3), unlike the second (Gen. 2:4–25), does not place the creation of humanity at the beginning but rather after the creation of all the conditions upon which we depend – light and air, land and water, animals and plants. These are not described as "good for humanity" but simply as "good." Gustafson agrees with Rachels that evolutionary insights into common descent undercut the standard assertion of radical discontinuity between humans and other animals: nonhuman animals are not simply "irrational creatures."

Gustafson also agrees with Rachels that *imago Dei* language can be ecologically hazardous.[75] Yet, unlike Rachels, Gustafson recognizes "the moral perils in any interpretation of life that in any way weakens our belief in the distinctive dignity of human life."[76] Interpreting the *imago Dei* in light of evolution leads Gustafson to provide a more modest estimate of our distinctiveness and more keen awareness of our responsibility to other species. Yet he acknowledges that our species is "the only one we know that has the capacities to investigate all these matters and to acquire information regarding them"[77] and that we are the only self-determining and morally accountable agents. Moral agency is "grounded in our biological and social natures," and rationality is "expressed in our sorting out our motives and desires, and in directing our actions in a fitting way."[78]

But the *imago* also presents a powerful theological temptation for us to equate God's will with what is good for human beings and to ignore what is good for other species or their habitats. For Gustafson, the kinds of anthropocentrism developed by John Paul II and Peacocke contradict their own stated commitments to ecological responsibility. Gustafson insists that if we want to expand the sphere of what we consider to be valuable or good, we must attend to boundaries, accept our limits, engage in self-restraint, respect natural finitude, and strive to conform to, or at least cooperate with, the natural patterns of interdependence within which we are immersed.

Cultural development must be restrained and populations must be restricted,[79] in part through the use of available artificial birth control. Here of course he is in direct disagreement with John Paul II, despite their moral agreement that we must not be allowed to destroy nature to satisfy unbridled consumer desires. As Gustafson puts it, "If the dignity of the human species is presumed to warrant its exploitation of nature for the sake

---

[75] Ibid., pp. 100ff. [76] Ibid., p. 101. [77] Ibid. [78] Ibid., p. 285. [79] See ibid., vol. II, ch. 7.

of not only human survival and a modicum of comfort but also for the sake of satisfying whimsical desires, humanity is in trouble."[80] And unlike Peacocke, Gustafson does not give the slightest hint of a romantic reverence for nature: "The mosquitoes that carry malaria, parasites that thrive in polluted water and human waste, severe winters in vast regions of the globe, earthquakes, droughts in Africa and elsewhere, and rats that carry bubonic plague are all parts of the natural order."[81] We need to work for human flourishing but this good must be put in a balanced relation to our interdependencies within nature and responsibilities for it.

Gustafson also highlights the interdependency of sexual relations in marriage and the need to give moral guidance on the ways in which we seek to fulfill our social desires and sexual needs. He offers a more complex approach to moral problems than one finds in either John Paul II or the "new natural-law theory" of Grisez and Finnis. Gustafson draws meaning from the traditional ends of marriage – procreation, fidelity, and sacrament – in part because they are based on an understanding of marriage as a social institution that meets biological, emotional, and social needs. The purpose of marriage is not only procreation, the accent in Genesis 1:28, but mutual fulfillment, psychological intimacy, "mutual society, help, and comfort."[82] Gustafson, finally, recognizes the need to use biomedical technology to enhance human lives but only within the limits made possible by a responsible attitude to nature.

ASSESSMENT: HUMAN EVOLUTION AND THE *IMAGO DEI*

This discussion illustrates some of the complexity involved in claims made about human dignity when considered in light of common descent and other aspects of human evolution. Each of these authors responds to and relies on evidence about evolution and our evolved biological nature in different ways.

Some of their differences are methodological while others are substantive. To begin with the methodological issue, Rachels adopts the straightforward position that knowledge of evolution simply undermines the *imago Dei* proclamation of Genesis 1:26–27. All of our theologians present an alternative to Rachels' moral reductionism. John Paul II suggested that human dignity need not be threatened by evolution, properly interpreted, as long as the individual creation of each soul is affirmed. Peacocke and Gustafson agree that evolution requires a significant revision of our

---

[80] Ibid., vol. I, p. 104.   [81] Ibid., p. 106.   [82] Ibid., vol. II, p. 180.

understanding of the place of humanity within nature, but they do not want to see human dignity downgraded, let alone eliminated. They do not, however, invoke the language of "soul," apparently because they find the term to be obsolete, and they believe that sufficient justification for human dignity is given by religious convictions complemented by their reading of our innate human cognitive and moral capacities.

The adoption of an evolutionary perspective influences interpretations of the *imago Dei* in several ways. Rachels is a self-conscious ontological reductionist whose appropriation of sociobiological accounts of evolution coheres with his moral reductionism. John Paul II does not explain what he means by the term "evolutionary worldview" and his comments before the Pontifical Academy blur important distinctions between evolution as fact, as well-grounded scientific theory, and as ideology. Peacocke has the most informed understanding of the various meanings of evolution and he manages to be fairly clear about which meanings he uses when he advances various arguments. Unlike John Paul II, he is informed by contemporary theories of chaos, contingency, and emergent probability. He acknowledges the role of natural selection within natural history, and his awareness of the role of contingency within nature inclines him to view God as an improvising musical conductor:[83] "God creates in the world through what we call 'chance' operating within the created order, each stage of which constitutes the launching pad for the next. The Creator . . . is unfolding the potentialities of the universe . . ."[84] This vision is close to those of Stoeger, Van Till, and Conway Morris examined in chapters four and five above. Gustafson's vision is also deeply informed by knowledge of natural history, including the interdependence of species, the fragile and transitory character of life, and our continuity with other species. He understands the centrality of natural selection without falling into illegitimate reductionism, and he focuses, in ways that the other theologians do not, on the tragic undercurrent of the evolutionary process. His attention to the danger that humanity poses to other species and their habitats provides a theological analogue to Rachels' moral concern for animals.

These views of evolution influence how their authors interpret the *imago Dei* text in Genesis: Rachels rejects it entirely on the grounds that it is both empirically untrue and ethically pernicious; John Paul II interprets the *imago* in terms of the special dignity possessed only by the rational soul and expressed especially in the love for God and neighbor; Peacocke construes the *imago* as a goal of human spiritual growth, as a capacity latent in all

---

[83] Peacocke, *Theology for a Scientific Age*, p. 119.   [84] Ibid.

people but not always actualized; Gustafson critiques our use of *imago* language as cheap support for our hubris, but finally sees the moral importance of viewing human beings as distinctly if by no means supremely valuable. The notion of human dignity functions in John Paul II, and to a lesser extent in Gustafson, to present limits to what human beings ought to do with the new powers of technology at our disposal. John Paul II is very clear that we do not have the right to violate human dignity through the use of biomedical technology in order to direct the evolution of our own species.

We can draw six central points from this analysis. First, the essential Christian affirmation of human dignity, the claim that we are created in the image of God, need not be undermined by evolutionary origins. The evolutionary process generated the development of important and distinctive human capacities, notably to understand and to love, that constitute the natural basis for the affirmation that we are made in God's image. Common descent, then, does not imply identical natures today. The conclusions drawn by sociobiologists about human worth are decisively influenced not by science itself but by their ontological naturalism. On the contrary, an understanding of human nature as a complex of emergent human capacities allows us to see that human evolution comports well with the Christian vision of the person as made in the image of God. This reading of emergent complexity, of course, reflects the theological commitments discussed in the first five chapters of this book and advances a position clearly at odds with both Rachels and the sociobiologists. The animal origins of human nature need not be seen as degrading humanity, any more than the religious and moral aspirations of human transcendence necessarily rule out common descent.

Second, our expanded sense of continuity with other species, our interdependence with ecospheres, and our dependence on the planet as a whole cannot help but heighten our sense of responsibility for the natural world. For all their differences, these authors agree that knowledge of evolution brings, as Peacocke puts it, "pressure for a wider perspective on humanity."[85] Our authors do not all agree on how this wider perspective ought to be related to other concerns or on where the ethical line should be drawn with regard to biomedical issues such as cloning, stem cells, or the uses of reproductive technology. Indeed, John Paul II would consider Rachels to be the apotheosis of the "culture of death."[86] But they do agree that knowledge of evolution helps us to understand both our shared humanity and the

---

[85] Ibid., p. 244. [86] See *Evangelium Vitae*, no. 28.

common bonds we share with other species. The genetic inheritance shared by all members of our species speaks to our deep evolutionary connectedness.

John Paul II, Gustafson, and Peacocke also insist that the *imago Dei* can no longer be interpreted in a way that valorizes human dignity to the exclusion of the well-being of other animals. Dominion must be interpreted to mean stewardship, not domination. The sociobiological view of humans as only one species among others attacks the very notion of human dignity. This having been said, the notion of dignity is a relative term, and most of our authors accord human beings considerably stronger relative dignity than animals. The notion of "intrinsic dignity" captures this sense of hierarchical worth when contrasted with "derived dignity," but the implied reduction of animals to merely instrumental objects derogates from their value independent of human needs. One might speak instead of human beings as having "absolute" dignity and animals merely "relative" dignity, but this formulation fails to note that humanity is merely the *image* of God. Despite this semantic stand-off, the *imago Dei* implies that the human person possesses a higher and special kind of dignity not found in even the most sophisticated of other animal species.

Third, we should not confuse the dignity of the person with the dignity of various human traits. In the past the tradition has identified various qualities that were indications of this dignity, such as the presence of the "intellectual soul" in Thomas Aquinas' theology or the image of the Trinity in the human mind for Augustine. Yet it has also recognized that it is the *person* as a whole who is loved and dignified and not simply specific traits. Human nature has an intrinsic dignity and so do all who partake in it, regardless of the extent to which they instantiate or manifest the various traits that give humanity its special nobility. In Christian morality, mentally handicapped people have the same dignity as geniuses, the lame as Olympic sprinters, and the demented elderly as the most alert young person.[87]

Some contemporary discussions of human dignity value a particular criterion or value, notably rationality or consciousness, more than humanity itself. Some Artificial Intelligence experts value cognitive skills in the same way. One bioethicist argues that we need to distinguish "persons" in the proper sense of the term – those who are self-conscious, rational, and morally autonomous – from "human biological life" in the broader sense, for example as found in fetuses, embryos, comatose patients, and the profoundly mentally

---

[87] See Jean Vanier, *Becoming Human* (Mahwah, NY: Paulist Press, 1998).

handicapped.[88] Christian ethics needs to retain its affirmation of the equally intrinsic value of every member of the human race. In this way Christian ethics remains centrally grounded in the narrative and teachings of Jesus.

Fourth, the notion of intrinsic dignity should not be expected to do too much work in Christian ethics. Although we usually reserve appeals to dignity for the most egregious cases of ethical wrongdoing (e.g. for instances of torture or rape), it is also the case that any violation of justice constitutes an assault on dignity because it essentially communicates a judgment that the victim is not worthy of moral consideration, protection, respect, fairness, and the like. The notion of justice is needed if we are to explicate the ways in which various acts are attacks on human dignity or to show how various policies are the most effective avenues for its protection.

Fifth, the appeal to justice must include not only concern for individual rights, but also for the common good. Christian ethics speaks not of the "group," a collection of "individuals," but of the "community," a network of persons bound together in various forms of friendship in pursuit of the common good. The dignity of the person can only be properly supported by communities and not simply individuals. Appeal to the common good, the dignity of a community, or to "group rights" should not be used in a way that leads to the diminishment of the basic human rights of the person; these rights are "inalienable." At the same time, personal dignity is best understood when the dignity of the community is appreciated as well. This means that the protection of human dignity is best pursued when society and state recognize the legitimacy and value of smaller units of interaction such as families, churches, unions, and other intermediate associations, while the latter also acknowledge their role within the wider society and body politic.

Sixth, the promotion of human dignity and the effective acknowledgment of the human person as the *imago Dei* need to be pursued in both theoretical and practical ways. The practical avenue involves the church in both external and internal contexts. The external agenda pertains to how the church might most effectively promote dignity in pluralistic societies. The full defense of human dignity, as John Paul II held, requires the acknowledgment of both "the spiritual destiny of the human person ... and the moral structure of freedom."[89] Yet the church functions within increasingly secular societies where there is no widespread acknowledgment of this destiny or structure. Modern societies are the scene of great

---

[88] See H. T. Engelhardt, *The Foundations of Ethics* (New York: Oxford University Press, 1996), p. 138.
[89] John Paul II, "Address to the United Nations General Assembly," *Origins* 25 (October 19, 1995): 270.

diversity, and individuals and communities differ over what is consistent with and what violates human dignity. A practice that one tradition will see as a violation of human dignity might be fully acceptable to or even mandated by another (e.g. equal rights for women, equal economic opportunity, equal responsibilities within the household). When different traditions and communities within one pluralistic society converge in their judgments about a particular point regarding human dignity, as, for example, in the American civil-rights movement, they can contribute to the building of a valuable moral consensus with great political significance.

In practical terms, the notion of dignity, whether affirmed for secular or theological reasons, can provide a very helpful moral support for working toward public consensus regarding public policy. In this kind of case, pressure can be brought to bear to shape policies and laws in an appropriate direction. Societies that have perpetrated their own tragic moral errors are poised to seek permanent correctives so that the horrors of the past are not repeated. Because of the Nazi past, the German constitution takes the strongest legal position in Europe against eugenics, euthanasia, and experimentation on human subjects without their consent. The first article of the German constitution announces that, "Human dignity is inviolable. To respect and protect it is a duty of all state authority."[90]

When different traditions do not converge, however, and when they have serious disagreements about the moral status of a particular practice, they must attempt to find a way to comprehend one another, and especially to understand why they disagree and exactly on what matters they disagree. On this basis they can slowly work for possible points of contact, partnership, and cooperation. Moral dialogue involves growing in self-understanding as well as in understanding one's interlocutors; it requires self-awareness, openness, a willingness to consider one's own biases or misunderstandings. The pursuit of this kind of process is appropriate to civil conversation in pluralistic societies and is the only way in which we can advance the cause of human dignity throughout society at large.

This encouragement of public understanding and public consensus-building having been registered, it remains the case that pluralistic societies are the scene of very different views about the nature of the human person, the dignity of the human person, and the extent to which that dignity must be protected. It is a sad fact that even when societies formally acknowledge the moral requirements of human dignity in the abstract, for example

---

[90] "Grundgesetz," "Basic Law," for the Federal Republic of Germany, first article, available at http://library.byu.edu/~rdh/eurodocs/germ/ggeng.html (accessed August 24, 2005).

through the signing of international agreements, they often find it convenient to ignore these acknowledgments when their perceived self-interest is at stake. Here the evolutionary depiction of our tendency to categorize human beings into "us" and "them" seems to find resonance in human experience: "our" dignity is recognized by us and "theirs" is not.

One would be hard pressed to find support for any notion of intrinsic human dignity in the writings of sociobiologists and other ontological naturalists who view the person as an organized ensemble of "selfish genes" and who deny the existence of any spiritual capacity that transcends the demands of inclusive fitness. Evolutionist materialist anthropology regards the human body as a configuration of objects to be technologically manipulated according to the choices of consumers. The growth of human mastery over nature can thus contribute to the mastery of the human body, and this can include the mastery of some humans over others. As C. S. Lewis noted long ago, in this scenario "*Human* nature will be the last part of Nature to surrender to Man."[91]

Evolutionary biology, viewed from a nonreductionist standpoint, offers strong evidence of human emergent complexity and indicates the distinctiveness of human cognitive, emotional, and social capacities. Interpreted from a Christian theological perspective, this evidence can be identified with the natural aptitude that Thomas Aquinas took to be an indication of the *imago Dei* in human nature.

In addition to the external challenge, Christian ethics also faces a very important internal challenge. The church in the twentieth century at times functioned as a great defender of human rights around the world, and particularly of the rights of people suffering under totalitarian regimes; at other times, it failed to meet the challenge of justice. The church does not always treat its own members in a way that accords with their dignity as human beings. Especially prominent here are issues regarding the role of women in the church, the treatment of gay and lesbian people, and the status of divorced and remarried Catholics. This challenge must be met if the church is to live up to its own message that every human person is made in the image of God. The most effective response to ontological naturalists is not to use the negative argument that exposes the conceptual shortcomings of naturalism and the frequent shallowness of its critique of Christianity, but rather to exhibit in a practical way the truthfulness of the Christian message by displaying the love that is its basis. Certainly both theoretical and practical avenues are valuable, but Christian ethics is

---

[91] See C. S. Lewis, *The Abolition of Man* (San Francisco: HarperSanFrancisco, 1974 [1944]), p. 59, emphasis in original.

fundamentally rooted in the latter. Christian ethics supports human dignity when it lives out its implications in everyday life.

This chapter argued that our knowledge of common descent need not threaten our commitment to human dignity and that it can even enhance our awareness of human dignity when understood within the context of a process of emergence that generates distinctively human capacities. The Christian theological justification for affirming that every person has intrinsic dignity, the belief that human beings are made in the image of God, is consistent with appropriately nonreductionistic views of human nature. This sense of dignity cannot be provided by scientific knowledge of human evolution operating outside any ontological or theological framework. In this claim I disagree with Gustafson and Peacocke when they would limit our theological and moral claims to what can be indicated by science. The moral value of the person is "underdetermined" by science generally, certainly by evolutionary biology and its allied disciplines. Ontological readings of evolution may be used to support human dignity but they can be and sometimes have been used to attack it. The point of controversy rests most fundamentally with ontology and theology rather than with science.

CHAPTER 9

Christian love and evolutionary altruism

Darwinism was originally thought to describe human nature as thoroughly individualistic and selfish. Altruism was regarded either as illusory or, at best, as the product of culture alone.[1] The term "altruism" appeared in the nineteenth century as a reaction against more cynical views of human nature as purely self-centered or "egoistic" and, along with the term "compassion," provides an appealing moral counterweight to the pervasive individualism of modern culture and market capitalism.[2] Concern for others can degenerate into what sociologist Robert Bellah and his colleagues call "the subjective feeling of sympathy of one private individual for another,"[3] a form of sentimentality that lacks any conceptually cogent connection to justice. Certainly the general cultural thrust of modern individualism tends to attenuate and dilute calls to develop more altruistic concern.

Economist Robert Frank and his colleagues published a fascinating paper demonstrating the impact of one particularly intense form of social learning in this regard. Asked at the beginning of their respective economics and astronomy courses whether they would return an envelope with one hundred dollars in it, students from both courses responded similarly. Yet when asked the same question at the end of the semester, the economics students made more egoistic decisions than their astronomy peers.[4] The point of the study was not that economics professors explicitly argue against altruism, but that mainstream economists presume a view of human beings as self-interested preference "maximizers" that tacitly discourages altruistic concern.

---

[1] See Donald T. Campbell, "On the Conflicts between Biological and Social Evolution and between Psychology and Moral Tradition," *American Psychologist* 30 (December 1975): 1103–1126.
[2] See Robert Bellah *et al.*, eds., *Habits of the Heart: Individualism and Commitment in American Life* (San Francisco: Harper and Row, 1985).
[3] Ibid., pp. 266–267.
[4] Robert H. Frank, Thomas Golivich, and Dennis T. Regan, "Does Studying Economics Inhibit Cooperation?" *Journal of Economic Perspectives* 7.2 (1993): 168–170.

Some important evolutionary voices, however, have increasingly recognized the cooperative, mutual, and pro-social sides of human nature. This development raises the question of whether the time is ripe for a dialogue between Christian interpretations of the ethics of love and evolutionary theories of altruism. My position is that such a dialogue reveals some common overlap between these two approaches to human pro-social behavior. This chapter begins with a brief description of evolutionary accounts of altruism, offers an assessment of their plausibility, and then examines their relevance to Christian reflection on love. My main point is that Christian ethics of love can build on contemporary accounts of what is naturally human as long as they are purged of improper reductionism. As Cahill puts it, "Biblical compassion builds on and expands a natural capacity for empathic identification with others."[5] While striving to develop, hone, and deepen our evolved natural tendency to care for self, family, and friends, Christian ethics also takes deliberate steps to extend our concern beyond these spheres to encompass what sociobiologists call "non-kin" and "nonreciprocators." For its part, Christian ethics offers a helpful warning to sociobiologists that we should not confuse our natural bent to self with the full scope of our moral responsibilities to others.

EVOLUTIONARY THEORIES OF ALTRUISM

According to evolutionary thinkers, behavior that decreases the fitness of the recipient can have a number of evolutionary bases – kin altruism, reciprocal altruism, indirect altruism, and manipulation. Geneticist William D. Hamilton originally classified social behaviors according to their impact on the fitness of the donor and the recipient of a given act:[6] "selfish" behavior increases the fitness of the donor and decreases the fitness of the recipient; "cooperative" behavior increases the fitness of both; "altruistic" behavior increases the fitness of the recipient while diminishing that of the donor; and, finally, "spiteful" behavior decreases the fitness of both.

The first theory, "kin altruism," holds that organisms can be expected to prefer to aid other organisms to the extent to which they share genetic similarity.[7] Sterile worker castes of insect societies evolved because their

---

[5] Lisa S. Cahill, *Sex, Gender, and Christian Ethics* (New York: Cambridge University Press, 1996), p. 126.
[6] W. D. Hamilton, "The Genetic Evolution of Social Behavior," *Journal of Theoretical Biology* 7:1–2 (1964): 1–16, 17–52.
[7] See W. D. Hamilton, "Innate Social Aptitudes of Man: An Approach from Evolution Genetics," in Robin Fox, ed., *Biosocial Anthropology* (New York: John Wiley and Sons, 1975), pp. 133–135.

organisms could pass on copies of their genes through their fertile relatives within the colony. Applied to human beings, kin altruism theory claims that there are evolutionary reasons why we tend to favor kin over non-kin and why (other things being equal) it might be easier to fear strangers than it is to trust them.[8] Distinguishing kin from non-kin and preferring the former can lead to some aversion toward others, sociobiologists suggest. Whether group allegiance is a form of diluted kin altruism or an extended form of reciprocity, vigilance and aversion toward strangers might yield significant payoffs in situations of intergroup competition. There are reasons for believing that we possess evolved emotional mechanisms to prefer our own kin and our own group, to divide the world into "us" and "them," and to think of others as inferior to members of one's own group.[9]

Yet how cultures regard and keep obligations of kin affiliation reflects the fact that human nature is marked by "open programs" rather than tightly scripted "closed programs."[10] Common human experience tells us that a great deal of human conduct has little to do with reproductive success. In all that follows we have to keep in mind the nonreductionist view developed earlier in this book. Evolution has shaped our emotional capacities and needs, but not in a way that makes culture and personality trivially significant. As Ridley and Dawkins put it, "Civilized human behavior has about as much connection with natural selection as does the behavior of a circus bear on a unicycle."[11] The bear's behavior is an elaboration of some natural capacities but it would not have emerged without extensive and deliberate training. It is certainly not something the bear would have done spontaneously under natural conditions. Natural proclivities do not give us absolutely irresistible inclinations to be concerned with self, friends, and kin, but we can and do learn to care for non-kin. Perhaps altruism beyond the kinship group is analogous to the bear riding the unicycle – it does not come spontaneously but can be deliberately taught.

The second theory, Robert Trivers' "reciprocal altruism," holds that biologically unrelated individuals will trade beneficial acts under certain circumstances.[12] "Reciprocal altruism" adds a factor of time to cooperation:

---

[8] Alexander, *Biology of Moral Systems*, p. 261.
[9] Ian Vine, "Inclusive Fitness and the Self-System: The Roles of Human Nature and Sociocultural Processes in Intergroup Discrimination," in V. Reynolds, V. Falger, and I. Vine, eds., *The Sociobiology of Ethnocentrism* (Athens, GA: The University of Georgia Press, 1987), pp. 60–80.
[10] Mayr, *Towards a New Philosophy of Biology*, pp. 26, 49.
[11] Mark Ridley and Richard Dawkins, "The Natural Selection of Altruism," in J. Philippe Rushton and Richard M. Sorrentino, eds., *Altruism and Helping Behavior: Social, Personality, and Developmental Perspectives* (Hillsdale, NJ: Lawrence Erlbaum Associates, 1981), p. 32.
[12] "The Evolution of Reciprocal Altruism," *Quarterly Review of Biology* 46 (1971): 35–57.

individuals who interact only once will be rewarded for selfish behavior and penalized for altruistic behavior, since there is no opportunity in the future either to reciprocate for benefits or to punish for costs incurred. Over the course of repeated interactions, however, individuals can benefit by engaging in patterns of reciprocity.

Studies of the "prisoners' dilemma" game indicate that in a repeated series the best strategy for winning is neither pure selfishness nor pure unselfishness but a "tit-for-tat" strategy that cooperates with cooperators (do not exploit them) and punishes cheats (do not let them exploit you).[13] The logic of evolution implies that phenotypes engaged in cooperative and caring behavior toward organisms who can reciprocate (as well as those who are close kin) are more likely to leave behind copies of their own genes than are phenotypes who act selfishly, that is, who neither cooperate nor care about others.[14] Cooperators are rewarded with continued assistance, noncooperators ("grudgers") are denied assistance, and defectors ("cheats") are punished. Cheats will exploit suckers but, as cheats come to predominate, their strategy undercuts social order and makes any cooperation unlikely. Suckers are eventually eliminated except when all players are suckers; a single cheat will eventually take over the game. The grudger strategy, which employs the policy of cooperating with cooperators and not cooperating with cheats, proves the most stable evolutionary strategy.[15] Nature thus encourages reciprocity and retaliation rather than the "sucker strategies" proposed by "turning the other cheek."

Cognitive skill is needed because the success of "tit-for-tat" depends on the ability of the parties to identify and distinguish trustworthy from untrustworthy participants. Individuals must be able to recognize those individuals to whom they are indebted and those who refuse to return favors, to know who cheats and who cooperates, and to identify a given act as cheating or cooperating. In human interaction, fear can lead individuals mistakenly to draw back from the risk and vulnerability of the "tit-for-tat" strategy in favor of more defensive, exploitative, or hostile strategies. Reciprocity works best within small groups in which individuals can form lasting impressions and members can accrue reputations. Large anonymous societies comprising millions of individuals can make it easy for cheats to hide, deceive, and manipulate other individuals who tend to

---

[13] See David Axelrod, *The Evolution of Cooperation* (New York: Basic Books, 1984).
[14] See Gerald S. Wilkinson, "Food Sharing in Vampire Bats," *Scientific American* (February 1990): 76–82, and "Reciprocal Food Sharing in the Vampire Bat," *Nature* 308 (1984): 181–184.
[15] Axelrod, *Evolution of Cooperation*, p. 54. See the mention of this axiom in Xenophon, *Memorabilia* 2.6.21; Plato, *Lysis* 221e; Aristotle, *Nicomachean Ethics* 1169b16–22, 1162a16–19.

be more trustworthy and cooperative. "Altruistic cooperation in large groups of unrelated individuals is unlikely to evolve."[16] Perhaps this is why people tend to feel less safe, to be less trusting and to be less trustworthy in large anonymous cities than they are in small towns and villages where face-to-face interactions are the norm.

A third theory of altruism developed by R. D. Alexander begins: "the most general principle of human behavior" is that "people in general follow what they perceive to be their own interests."[17] The premise that human conduct can be understood, "only if societies are seen as collections of individuals seeking their own self-interest,"[18] Alexander argues, allows us to understand morality as a system of "indirect reciprocity." He is not troubled by the fact that people misperceive what is in their own self-interest and that individuals often deliberately act in ways that run counter to their own perceptions of self-interest and without regard to reasonable future payoffs of one kind or another.

Indirect reciprocity is routed through third parties or the group.[19] Rewards can be rendered in a variety of forms, running from the material to the social, but Alexander is most attuned to reputational rewards that enhance the status and perceived trustworthiness of the altruist. Perceived trustworthiness contributes indirectly to an individual's long-term reproductive fitness because it enhances his or her ability to elicit assistance. Individuals with the best reputations as reciprocators are most likely to be valued as reliable partners and therefore more adept at recruiting assistance for attaining their goals. On the basis of social surveillance, individuals within groups assign to one another degrees of trustworthiness for cooperation, coalition formation, and reciprocity. Individuals appearing to be altruistic are rewarded with assistance from others who trust that their good deeds will be reciprocated in the long run. The "altruistic conscience" – regarding oneself as having a duty to aid others – persists because it provides the best advertising for trustworthiness and long-term indirect reciprocity. The evolved physiological reaction of blushing is so convincing precisely because it is "hard to fake."[20] The person known to help others and to feel guilty for failing to help is more likely to elicit the trust and the aid of others than is the individual who has little sense of shame and no tendency to feel appropriate guilt.

---

[16] Robert Boyd and Peter J. Richerson, *Culture and the Evolutionary Process* (Chicago: University of Chicago Press, 1985), pp. 230–231.
[17] Alexander, *Biology of Moral Systems*, p. 34. [18] Ibid. [19] Ibid., p. 153.
[20] See also Robert Frank, *Passions within Reason: The Strategic Role of the Emotions* (London and New York: W. W. Norton, 1988).

A fourth theory regards altruism as a product of manipulation. Animals use deception to protect themselves and pursue prey. A female starling bird who removes an egg from another starling's nest and lays her own egg in its place deceives the other female starling into altruistically raising a nestling that is not her own. Animals are said to be engaged in ongoing constant "arms races" in which the evolutionary improvement of one lineage's traits triggers a corresponding improvement in another lineage's traits.[21] The ability to detect deception evolves to keep up with the enhanced ability of competitors to deceive.[22]

Some evolutionists argue that this communicative "arms race" has been a major factor in driving the increased complexity of the human mind and the evolution of its different levels of consciousness. In this ideology, the self manipulates, deceives, and coerces others in its relentless pursuit of genetic self-interest. Since the most convincing public stance is that of moral sincerity, generosity, and integrity, the most likely strategy for genetic success includes self-deception about one's own deepest motivations. According to Alexander, beneath the conscious experience of every purportedly altruistic agent lurks a deeper unconscious egoist concerned only about its own well-being. People do not always deliberately calculate what is in their immediate self-interest, the argument runs, but even when they want to appear altruistic they usually do what they perceive to be in their own self-interest and this self-interest usually coincides with genetic self-interest.[23]

A fifth theory of altruism has been advanced by philosopher Elliott Sober and biologist David S. Wilson in terms of "group selection." Their project receives a sympathetic reading from many of those who take evolution seriously but resist the narrow focus on the "selfish gene." Sober and Wilson acknowledge the fact that unselfish behavior exists as an important feature of human life, biologically and morally, and they admit that genuine unselfishness, at least in some forms, has evolutionary roots. They understand cultural evolution as a source of "counter-hedonic" acts, but they also argue that there is an evolved biological support for altruism that is the product of group selection.[24]

---

[21] See R. Dawkins and J. R. Krebs, "Arms Races within and between Species," *Proceedings of the Royal Society*, series B, 205 (1979): 489–511.

[22] For a provocative speculative evolutionary psychological hypothesis for the evolution of a "Machiavellian module" that enables us to deceive others and to try to detect deception, see David Livingstone Smith, *Why We Lie: The Evolutionary Roots of Deception and the Unconscious Mind* (New York: St. Martin's Press, 2003).

[23] Ibid., p. 40.

[24] The attempt to understand altruism as a combination of group-selected altruism and cultural evolution is also found in Peter J. Richerson and Robert Boyd, "Complex Societies: The

The main target of this theory is the one-sided focus on individual selection that has dominated evolutionary theory since the 1960s. In its place, Sober and Wilson advocate a reasonably coherent "multilevel selection theory." Group selection does not take place when individuals within a particular group have high degrees of individual variation, or when there is little variation between groups. Sober and Wilson argue, however, that genes for altruism could have been selected if in-group altruism brought benefits to individuals within the group that would be greater than the benefits that would have accrued to individuals living in less altruistic groups. Individuals would sacrifice themselves for the good of the group because living within such a group would on average be more beneficial to individuals than would alternatives. Group-selected altruism, however, can inspire attitudes of hostility to outsiders, suggesting, if true, that human nature is at odds with wider moral concern and, in Christian ethics, the universal love that embraces enemies and strangers.

Rather than replacing individual with group selection, Sober and Wilson advocate a pluralism that accommodates the complexity of evolution. They argue that a group composed of altruistic members can be more successful in evolutionary terms than a group composed of egoists. Their work thus retrieves aspects of Darwin's thought that had fallen out of approval with the pre-eminence of individual selection and gene-level selection in neo-Darwinism. According to Sober and Wilson, morality can contribute to the superiority of one culture over another when it confers a special advantage to in-group organization. To illustrate their point, Sober and Wilson draw on the example of the nineteenth-century struggle for dominance by the Nuer tribe in the Sudan.[25] They argue that the Nuer achieved dominance because of the more effective work of their raiding parties, which were in turn made possible by a lineage system and bride-price obligations that were more costly and that led to a more cohesive tribal order than that found in the practices of other tribes in that region. Greater social organization led to greater military competence.

Sober and Wilson, however, do not engage in their own research on this case, but instead rely on anthropological study done without their own concern for group selection in mind. Their speculative proposal about Nuer social effectiveness does not demonstrate that group selection in fact offers a better account of this success than do alternative culturally focused

---

Evolutionary Origins of a Crude Superorganism," *Human Nature* 10 (1999): 253–290; and Christopher Boehm, *Hierarchy in the Forest: The Evolution of Egalitarian Behavior* (Cambridge, MA: Harvard University Press, 1999).

[25] See Sober and D. S. Wilson, *Unto Others*, pp. 186–191.

hypotheses. A conjectural generalization – "it is likely that much of what people have evolved to do is for the benefit of the group"[26] – is offered without compelling evidence. Sober and Wilson fail to show that behavior that benefits a particular group must be *genetically* based rather than simply retained by the group as a feature of the culture that gives the group a certain advantage, for example in social cohesion. There seems no reason to think that the Nuer have a biological superiority to the Dinka rooted in altruistic genes shared by group members. It is more reasonable to hold that the Nuer's superiority lay in their cultural organization, not in group-selected genetic identity.

Despite its weaknesses, this position at least rejects simple egoism, simple hedonism, and simple pure altruism. There is an intuitive and experiential plausibility to the claim that human beings are characterized by "motivational pluralism." We have a natural tendency to prefer our own benefit to that of most other people but we are also endowed with a natural capacity to care for others for their own sakes, most obviously in the exercise of parental care but also in other forms of love and responsibility.

ASSESSMENT OF EVOLUTIONARY THEORIES OF ALTRUISM

Evolutionary theories of altruism suffer from a number of very significant conceptual shortcomings and confusions that need to be identified and critiqued before we can proceed to a constructive discussion of the positive relevance of evolutionary theories of altruism for Christian ethics. One set of confusions concerns the meaning of the "self" and its "interests." In evolutionary literature, the term "self" is used alternately to refer to the entire person (the phenotype), or to something like the "core self" beneath the self-deceived conscious self, or to the individual's genes, or even to copies of the individual's genes carried in biological relatives. Discussions of "selfishness" carry the same ambiguities. "Selfishness" seems properly attributed to the first subject, less so to the second (which is only "part" of the self), and not at all to the third. Yet the evolutionists surveyed here tend to argue that agents who act in apparently phenotypically altruistic ways are (1) often really acting in phenotypically egoistic ways, though their real motivations are hidden from others and perhaps even from themselves, or (2) really acting in genotypically egoistic ways and, again, doing so without any awareness of their own underlying motivations. Sociobiologists and evolutionary psychologists put themselves in an untenable position when

[26] Ibid., p. 194.

they make bold claims about human motivation while at the same time professing a stance that prescinds from any such kind of claim about internal states.

A second set of confusions concerns altruism. Altruism in ordinary discourse is understood to be action motivated by a predominant concern for the welfare of at least one other person. Social psychologist Daniel Batson, for example, defines it as "a motivational state with the ultimate goal of increasing another's welfare."[27] It seems to me that altruism involves at least four components: an agent must (1) possess a certain kind of psychological *motivation*, namely a motive or desire to act on behalf of others; (2) understand and make a judgment about the worthiness of this motivation, its relation to what is thought to be truly good; (3) decide to act on the basis of this judgment about what is good, which as such becomes the directing *motive* of the altruistic act; and (4) *intend* to use an appropriate means – to pursue a certain course of action – to obtain the desired good.[28]

Some elaboration of the terms "motivation" and "intention" might be helpful. First, the term "motivation" refers to the source of energy that activates behavior. A wide variety of factors influence the identity, impact, persistence, direction, and intensity of motivations.[29] A person's own specific motivational structure is influenced by a complex variety of factors, including physiological, cognitive, emotional, cultural, and social. Psychiatrist Antonio Damasio understands our emotions and feelings as lying between older, more elemental, nonrational, cortical structures and more distinctively human, rational, "neocortical" processes.[30] Neural circuits for drives and instincts help to ensure our physical comfort and survival by directly generating behavior or creating physiological states

---

[27] See C. Daniel Batson, *The Altruism Question: Toward a Social Psychological Answer* (Hillsdale, NJ: Lawrence Erlbaum Associates, 1991), p. 6. Contrast with Dawkins: "An entity, such as a baboon, is said to be altruistic if it behaves in such a way as to increase another such entity's welfare at the expense of its own. Selfish behavior has exactly the opposite effect." Dawkins stresses the fact that "the above definitions of altruism and selfishness are behavioral, not subjective. I am not concerned here with the psychology of motives. I am not going to argue about whether people who behave altruistically are 'really' doing it for secret or subconscious selfish motives" (*The Selfish Gene*, p. 4). Dawkins trivializes the agent's motives by describing all appeals to such considerations as yielding only "subjective" considerations, in contrast to the "objective" point of view provided by scientific analysis of behaviors.
[28] On the meaning of "intention," see Herbert McCabe, *The Good Life: Ethics and the Pursuit of Happiness*, ed. Brian Davies, OP (London: Continuum International Publishing Group, 2005), pp. 69–71, 91.
[29] The meaning of "motivation" used here is also found in Midgley, *Beast and Man* and *Wickedness*.
[30] Damasio, *Descartes' Error*, pp. 128ff.

(e.g. hunger or fear) that induce appropriate behavior. Because our brain circuitry is modified and modifiable by particular experiences in our life histories, Damasio distinguishes early "primary" emotions from later "secondary" emotions. The former are innate, or pre-organized by our evolutionary heritage in the structure of the limbic system, while the latter, secondary emotions, are formed when life history establishes a connection in our minds between certain kinds of experiences and positive or negative emotions. Emotions are changes in the body triggered by the nervous system in response to an interpretation of an experienced event. The human body is a self-regulating system that incorporates various levels from the elementary level of basic metabolism, reflexes, and the immune system to the mid-level capacities oriented to seeking pleasure and avoiding pain to higher-level drives, appetites, desires, motivations, emotions, and conscious feelings. These levels are seamlessly connected but distinctive enough to be identified. In the broadest sense of the term, "motivations" provide an emotional impetus for behavior. Behavior becomes human action in the full sense when it reflects deliberate choices.

Sociobiologists mistakenly suggest that all people share the same genetic and biological motivations, and that these are dominant over and constitute the underlying causes of all other motivations, however much these seem to be chosen consciously by the agent. Individuals have distinctive genotypes and therefore the genetic factors underlying motivation cannot be identical for all people, yet sociobiologists speak at times as if all human beings have the same fixed motivational characteristics. A nonreductionistic reading of motivation, on the other hand, holds that the genetic basis of motivation is one among a variety of factors that can influence an individual's particular motivational structure. The impact of biological factors varies from one person to another, tends to be stronger in some individuals than in others, and functions in a variety of relations to chosen motives.

Psychologically healthy and mature people obviously do not act on every impulse, spontaneous desire, or even valid want, but rather upon choices about consciously identified goods. An agent selects legitimate reasons for acting from a menu of unreflective desires or motivations. Reasonable acts flow from legitimate motives, though of course doing the right thing involves more than simply having a legitimate motive for doing it. Sociobiologists fail to account for the difference that chosen motives make to human acts.

Second, the term "intention" refers to an agent's mental plan for action. Intentions implement the appropriate means for the goals that constitute the motive of an act: the lawyer who helps an unsavory client strictly out of a sense of duty is acting from the motive of professional responsibility. The defense lawyer may also have been moved by other values: to make money, to present an impressive public image, to enhance legal stature, and so on, but these are not the motivations upon which the lawyer chooses to act. The term "altruism" is ordinarily used to describe someone who has a predominant motive of caring for another person. An act need *not* be *purely* or *exclusively* self-sacrificial or involve heroic levels of self-denial on the part of an agent to count as genuinely altruistic.[31] Stated motives are often not all that lie behind behavior, which is often intended for multiple motives. Because motivations only give rise to acts when they are mediated through motives and intentions of agents, any attempt to account for human acts by appealing to biological factors alone is doomed to failure. Evolutionary approaches to altruism are so quickly dismissed by Christian ethicists because of the implausibility of genetic or biological reductionism.

Human acts are often, and perhaps almost always, generated by mixed motives. This fact is ignored by those who want to reduce emotional and psychological complexity to a single egoistic motivation generated by "selfish genes." Attentiveness to the difficulties confronting any efforts to apprehend underlying motivations encourages humility when we make claims about our own motives and reluctance to render negative judgments about the motives and underlying motivations of others. Even admitting the prevalence of selfish predispositions does not mean that we have to acquiesce in a theory of universal "psychological egoism," the claim that people always pursue what they think is in their self-interest,[32] or that we must ignore the presence of genuine altruism when and where it is found. As Jeffrey Schloss points out, there are "widely accepted empirical grounds for seeking explanations of social behavior in terms more extensive than unelaborated 'nepotism' and 'favoritism.'"[33] Referring to those who undertake voluntary poverty, celibate religious vows, and risky rescue behavior, Schloss rightly claims that, "Human beings often manifest radically sacrificial, consequently altruistic behavior that reduces reproductive success without compensatory reciprocation or kinship benefit."[34]

---

[31] See Batson, *The Altruism Question*.
[32] Psychological egoism is distinct from "ethical egoism," the normative ethical theory that holds that one ought always to be motivated by self-interest. See Simon Blackburn, *The Oxford Dictionary of Philosophy* (New York: Oxford University Press, 1994), p. 115.
[33] Jeffrey P. Schloss, "Emerging Accounts of Altruism: 'Love Creation's Final Law?,'" in Stephen G. Post *et al.*, eds., *Altruism and Altruistic Love: Science, Philosophy, and Religion in Dialogue* (New York: Oxford University Press, 2002), p. 221.
[34] Ibid.

There is no reason why an act cannot be altruistic in the ordinary sense and yet at the same time constitute an example of "genotypical altruism" when as a matter of fact it has the effect of promoting the reproductive fitness of another organism at the agent's own reproductive expense. Genotypical egoism thus need not generate psychologically egoistic motives. Philosopher Holmes Rolston argues that the logic of sociobiology requires one to assume that the good Samaritan is "constitutionally (i.e. genetically) unable to act for the victim's sake"[35] and can have only an apparent but not real concern for the victim. Sociobiologists assert that "all appearances to the contrary, there cannot be real altruism here (helping another at one's own genetic expense); there must be a self-interested account."[36] Sociobiologists need to advert more consistently to the fact that genotypical altruism has nothing to do with an organism's motivations, motives, or intentions. It concerns only externally observable and quantifiable aspects of behavior. Because genotypical altruism is purely behavioral, it can be applied not only to people but also to Tasmanian native hens, ground squirrels, and prairie dogs. This kind of similarity is as shallow as it is broad.

The same is true of the attempt to show the biological similarities of different organisms relative to "phenotypic altruism." Knowing only biological traits does not tell us anything about the moral character of the act. To say that a parent who dies saving his own drowning child is both phenotypically altruistic and genotypically egoistic is as ethically trivial as it is emotionally sterile. This complaint underscores the importance of the classical distinction between "human acts" proper and the mere "acts of a person."[37] It is one thing to act on a choice, for example to read a book, run an errand, or talk with someone, and another to do something inadvertently, such as scratching an itch or tripping over a ladder. Truly human action can only be understood when placed in the context of the specific intentions, motives, and beliefs of the agent; mere acts of a person are only incidentally human. The entire sociobiological project of attempting to "explain" human behavior in strictly behavioral terms, then, cripples its analysis of genuinely human altruism.

The key factor in interpreting the meaning of an act lies in the agent's intention. External consequences have an ambiguous relationship to human acts. A well-intentioned boy scout who helps an elderly person

---

[35] Holmes Rolston III, *Genes, Genesis, and God: Values and Their Origins in Natural and Human History* (New York: Cambridge University Press, 1999), p. 252.
[36] Ibid.   [37] See *ST* I-II, 1,1.

cross the street but accidentally injures her is acting with an altruistic intention despite its harmful effect. Yet if in another scenario the boy scout is successful and the grateful elderly person turns out to be a multimillionaire who surprises him with a huge financial reward, we would still describe the act as altruistic despite the vastly disproportionate benefits that came to the scout because of it. An altruistic agent can be the unintentional beneficiary of an act without having to forfeit his or her "altruistic" status. Conversely, an act is described as "egoistic" when performed primarily in order to benefit the self, even when, through accident or poor execution, the action ends up benefiting another more than the self.

Sociobiologists fail clearly and consistently to distinguish these different uses of altruism. Rose rightly complains that sociobiological theories of altruism suffer from a confusing habit of "lumping together many different reified interactions as if they were all exemplars of the one character."[38] When an author describes certain behavior that appears to be altruistic as "really" selfish, it is not clear whether this means morally, phenotypically, or genotypically selfish, or some combination of these different kinds of altruism. The terminology of phenotypical and genotypical altruisms distracts and confuses more than it clarifies. It might make biological sense to say that a rabbit eaten by a hawk has been altruistic, or that a grazing antelope is engaged in egoistic activity, but it makes no moral sense to use that kind of language when speaking about selfish or unselfish acts of human beings.

Conceptual confusion leads sociobiologists to adopt a debunking attitude to altruism. As Trivers puts it, this theory is "designed to take the altruism out of altruism."[39] Wilson, for example, describes Mother Teresa's work with the dying and destitute of Calcutta as self-serving and "cheerfully subordinate" to her "biological imperatives."[40] Yet even a slightly more generous "bear on the unicycle" interpretation of Mother Teresa's altruism regards it as the exception that proves the rule. One does not need sociobiology to provide arguments for the generalization that most people are somewhat selfish or that, for most people, charity not only begins at home but often ends there as well.

Evolution provides an account of why we are emotionally inclined to learn to care for self, our own families, friends, and communities more than for others. Since evolved predispositions are always trained in the context of particular cultures and communities, the specific shapes they take in concrete

---

[38] Rose, *Lifelines*, p. 281.   [39] Trivers, "Reciprocal Altruism": 35.
[40] E. O. Wilson, *On Human Nature*, p. 166.

lives show significant variation. Evolutionists insist that it is hard to learn genuine altruism, that is, altruism for non-kin and nonreciprocators. An inclination for genuine altruism cannot have evolved because it would not have been "selected for": over the span of evolutionary time, organisms that practiced genotypical altruism would have been less likely to reproduce (less "inclusively fit") than their genotypically egoistic counterparts, so they would have been reproductively eliminated from their respective populations. The ultimate genotypical altruists – the parent who feeds other people's children while her children go hungry – would not have left as many descendants as the genotypical egoists. This claim amounts to the uncontroversial statement that human nature cannot have evolved in such a way that individuals would consistently prefer to promote the well-being of others to their own well-being, to prefer to promote the well-being of cheats to reciprocators, or to sacrifice the good of their own community for the good of other communities.

CHRISTIAN LOVE AND THE EVOLUTION OF ALTRUISM

Christian ethicists do not often use the word "altruism," because the term is not morally helpful. A fanatic can be very concerned about the well-being of his or her own community, even to the point of surrendering his or her life, while perpetrating injustices against others. It is unclear whether an act has to be purely other-regarding, with no taint of self-concern, to count as altruistic. Moreover, even trivial acts can count as "altruistic." Acts should not be judged morally good simply in virtue of the fact that they were altruistically motivated; sentimental feelings sometimes do more harm than good. Classical and medieval authorities understood, as theorists of altruism sometimes do not, that moral goodness demands courage, self-restraint, excellence in moral judgment, and especially justice in relation to others and the community at large. Altruism can be destructive of self or others when it is portrayed in mindless and ineffectively self-denying behavior. Other-regarding action, moreover, can be quite immoral, as in the altruistic dedication of the fanatic who cares only about the well-being of his group. Some of the Crusaders and Conquistadors were motivated by idealism and not simply by greed and personal glory.[41] Operatives of death squads sacrifice their own comfort and risk their lives in order to promote

---

[41] For one example, see David Howard, *Conquistador in Chains: Cabeza de Vaca and the Indians of the Americas* (Tuscaloosa, AL: University of Alabama, 1997).

the goals of their group. "The terrorist is fundamentally an altruist,"[42] writes one expert.

The Christian tradition has differentiated uses of the term "love." They include, among others, love as a natural human emotion or "passion" (*amor*), as the product of deliberate human choices (*dilectio*), as the mutual love of friendship (*amicitia*), and as the theological virtue of charity (*caritas*). Love in its most basic sense is an affirmative response to the good, an affective affirmation of what we recognize as good (what Thomas Aquinas called "complacency in good").[43] Love in this mode issues in one of two directions, either as desire for the good that is absent, or joy in the good that is present. The term "care" refers to the response of love to the beloved's need.

The "passion" of love, what we might call our affective or emotional attraction to something that we regard as loveable, might or might not be suitable for us. Human love is distinctive because it is mediated through our understanding and our ability to make choices. We tend to be selfish, biased, and self-deceptive, and our ability to love is only morally reliable to the extent that we have learned and been trained to live virtuously. Instead of focusing primarily on the moral rightness or wrongfulness of particular acts, the concern of altruism, we should also pay attention to the quality of our moral affections, intentions, and motives, to the decisions we make about who and what we should love, and to the connection of love to the virtues on which it depends for its integrity.[44] Altruism cannot function as a stand-alone moral concept.

The nature of human love is complex because it reflects both our animality and our rationality. Christian ethics benefits from evolutionary knowledge of the emotional strata that we share in common with other animals. Evolutionary biology helps us understand why we have so much in common with other animals, and the fact that we share much more with them than the Christian tradition has recognized. Yet our love is also distinctive in that is it shaped by our capacities for knowing and loving.

To make matters even more complicated, we have to recognize that we are prone to care excessively and in the wrong way for ourselves, and to care wrongly and insufficiently for others. Christian language of human nature as "fallen" is complemented by belief in redemptive divine love that forgives, heals, and elevates the sinner.

---

[42] Bruce Hoffmann, *Inside Terrorism* (New York: Columbia University Press, 1999).
[43] See *ST* I-II, 26,1,2; I-II, 28,2.
[44] The best treatment of this set of issues is found in Spohn, *Go and Do Likewise*.

The virtue of charity restores, purifies, and heals our natural human capacity for love, where "natural" means "created" or "as God originally intended." It retains the natural tendency of the person to love most those to whom he or she is most closely bound by marriage, family, or any other kinds of particular connections or friendships. Grace also restores, purifies, and heals our natural ability to recognize in every person the goodness, worth, and dignity that they possess simply in virtue of being human. Charity by its nature inspires love for every human being one encounters. At the very minimum this love means good will (or classically the "love of benevolence") regardless of any other features of the other's conduct, personality, and so on. The virtue of charity thus inspires both a higher quality in the intensity and depth of the most profound kind of particular human friendships and a widening and extension of the range of our moral concern in the direction of universality. This reliance on grace might be taken to suggest that scientifically established knowledge of human nature is irrelevant to Christian ethics, but this is far from being the case because grace acts on and within human nature.

RECENT ANALYSES

Some authors assume an excessively simple definition of Christian love. Theologian Colin Grant's *Altruism and Christian Ethics* defends an interpretation of Christian love as radically self-sacrificial.[45] This kind of position condemns sociobiology for catering to human selfishness, narrow-mindedness, and tribalism, and believes that the stingy concern of the realist constitutes a self-fulfilling prophecy. Sociobiological cynics first describe people as Machiavellian and spend so much time looking for, and finding, egoistic motives that they ignore non-egoistic counter-evidence. The gospel demands not only ethical impartiality, Grant argues, but even a radical other-regard that de-centers the self and its parochial attachments. The model of pure self-sacrifice is Jesus, who taught

---

[45] Grant is largely attempting to revive the interpretation of agape proposed by Nygren and that has subsequently received an enormous amount of criticism from nearly every major commentator on love in the last one hundred years. See John Burnaby, *Amor Dei* (London: Hodder and Stoughton, 1938); Martin C. D'Arcy, *The Mind and Heart of Love, Lion and Unicorn: A Study in Eros and Agape* (Cleveland: World, 1964); C. S. Lewis, *The Four Loves* (New York: Harcourt Brace, 1960); Daniel Day William, *The Spirit and the Forms of Love* (New York: Harper and Row, 1968). What Edward Vacek says of Nygren is also true of Grant: although he "tries to build his arguments from Scripture, his overall procedure [does not maintain] fidelity to the diversity of Scripture"; in *Love, Human and Divine*, p. 192.

unconditional love, exemplified uninhibited compassion, and died on the Cross out of love.

Grant accepts Reinhold Niebuhr's judgment that Jesus teaches a rigorous and perfectionistic ethic that disallows self-love and even "concern for physical existence."[46] In this theological perspective, Christian love seems unnatural and perhaps even anti-natural. Christians must deny themselves, "hate" their family members (Luke 14:26), leave their friends and neighbors – anything less is a half-hearted compromise. The term "hate" here does not refer to strong and settled emotional antipathy, but to priority of loyalty – neither marriage nor family nor any attachments should be allowed to stand in the way of discipleship (for this use of "hate," see also Mal. 1:2–3). According to biblical scholar Robert Tannehill, ancient audiences understood "love" and "hate" to refer less to emotions than to "behavior that either honored or dishonored someone else."[47] Family honor cannot be valued more than discipleship, an injunction captured in Matthew 10:37 when Jesus forbids only loving family "more than me." The univocal interpretation of Jesus' anti-family sayings is impossible to square with the command to honor one's parents (see Exod. 20:12; Eph. 6:1–3; endorsed in Matt. 19:17–19) and with the value of marriage (Gen. 2:24; Mark 10:1–12).

For Grant, though, neighbor-love is unilateral, truly universal and not just apparently altruistic: it renounces self-gratifying ties of friendship, family, and community and embraces radical self-sacrifice as a way of life. E. O. Wilson believes that compassion actually "conforms to the best interests of self, family, and allies of the moment";[48] for Grant, Christian love does the reverse.

Philosopher Michael Ruse would recognize in Grant an exemplification of what he characterizes as the "stronger" interpretation of agape.[49] The "strong interpretation" requires love of neighbor as oneself and commands each Christian to count himself or herself as one and only one person among others. The "weak interpretation," on the other hand, allows for special preferences for self, family, and friends. Unlike Grant, Ruse maintains that Christian ethics ought to be constructed according to what nature makes possible. Ruse holds that kin altruism and reciprocity explain the bulk of altruistic acts. He thinks that there is a correlation between a

---

[46] R. Niebuhr, *An Interpretation of Christian Ethics*, pp. 33 and 25, respectively.
[47] Robert C. Tannehill, *Luke* (Nashville: Abingdon Press, 1996), p. 235.
[48] E. O. Wilson, *On Human Nature*, p. 155.
[49] Michael Ruse, "Evolutionary Theory and Christian Ethics," *Zygon* 29 (1994): 5–24.

sense of moral obligation and a sense of sympathy and/or likelihood of reciprocity. Human beings have evolved such that we are prone to feel moral obligation to people either connected to us through special social bonds (e.g. family, neighborhood) or at least capable of reciprocating good deeds. This does not mean that moral responsibility will be felt only for people who share copies of our genes, for, as just argued, evolution also gives only a proclivity or bias toward reciprocators and those who care most about us.[50]

"Biologically," Ruse explains, "our major concern has to be towards our own kin, then to those at least in some sort of relationship to us (not necessarily a blood relationship) and only finally to complete strangers."[51] Ruse argues that "weak agape" is consistent with our evolved emotional predispositions, and that "strong agape" runs so contrary to our nature that it is unacceptable, irresponsible, and even "morally perverse."[52] De Waal also presents a view of altruism with significant limits: "Altruism is bound by what we can afford. The circle of morality reaches out farther and farther only if the health and survival of the innermost circles are secure."[53] Instead of an "expanding circle," de Waal suggests an image of a "floating pyramid" in which the "force lifting the pyramid out of the water – its buoyancy – is provided by the available resources. Its size above the surface reflects the extent of moral inclusion."[54] Better to trim the ethic down to manageable size, argue Ruse and de Waal, so that it can sustain altruism within the small circle for which it has been adapted.

Grant would simply dismiss this "small circle" as Darwinian propaganda. Agape restrains, if not outright abolishes, other forms of love. Grant repudiates cheap accommodation and calls for loyalty to a distinctive religious community against secular culture and natural ties. Christians receive instruction from the life and teachings of Jesus, not from merely human conventions. Life is to be lived from within the scriptural narrative, not according to the Darwinian "epic of evolution." Christians affirm the saving power of Christ and the power of the Holy Spirit to make each disciple a

---

[50] For further reflection on this point, see the essays in Stephen G. Post *et al.*, *Altruism and Altruistic Love: Science, Philosophy, and Religion in Dialogue* (New York: Oxford University Press, 2002), including Stephen J. Pope, "Relating Self, Others, and Sacrifice in the Ordering of Love," pp. 168–181.

[51] Michael Ruse, *Taking Darwin Seriously: A Naturalistic Approach to Philosophy* (New York: Basil Blackwell, 1986), p. 106.

[52] Ruse, "Evolutionary Theory and Christian Ethics": 17, 19.

[53] Frans B. M. de Waal, *Good Natured: The Origins of Right and Wrong in Humans and Other Animals* (Cambridge, MA and London: Harvard University Press, 1996), p. 213.

[54] Ibid.

"new creature" (2 Cor. 5:17). Disciples graced with the "mind of Christ," as St. Paul puts it (Phil. 2:5–11), are not limited to what is possible for fallen human nature. Those who trust in Christ strive to imitate his unconditional love, including his renunciation of violence and revenge, his special care for the outcast, his unlimited willingness to forgive, and his call to leave family, love enemies, and pick up the Cross.

This polarization of Christian ethics and sociobiology has the virtue of clarity, yet in fact biblical texts and major works from the history of theology present far more complex challenges for interpretation. The "Farewell Discourse" of Jesus to the disciples in chapter 15 of the Gospel of John presents a case in point. In anticipation of his imminent passion and death, Jesus surrounds himself with his disciples, whom he calls "friends" rather than "servants." The command to "love *one another*" (15:17; emphasis added) underscores mutual love within the community of faith,[55] not "unilateralism." None of this denies that Jesus also reminds his disciples to obey his commands, to adhere to his teachings despite pressure to the contrary, to stand firm in the face of persecution, and to remain steadfast in the danger of death. The Johannine passages describing Jesus' love for his own calls into question Grant's dichotomization of friendship and self-sacrifice: the former actually requires the latter. Instead of searching for a single and unitary virtue or norm called "Christian love," it is more appropriate to acknowledge the complexity and richness of its range of meanings.

Neither is there any justification for Ruse's subdivision of these meanings into "weak" and "strong" forms. Ruse never offers a clear explanation of what makes a position "strong" or "weak" in the first place. If "strength" depends on the person making the claim on the agent, we need to keep in mind that demands placed by family and friends are typically much "stronger" than those made by people who are remote. There need be nothing "weak" in the former. If "strength" turns on the extent to which the claim runs against the grain of human nature, then Ruse also needs to recognize that it can be much more difficult to attain reconciliation with a hurtful family member than it is to forgive offensive acquaintances. Ruse and Grant share an unhelpful oversimplification of neighbor-love.

The sharp edges of the conflict between the two extremes represented by Grant and Ruse can be softened. At the same time, Christian ethics must address the claim that persisting moral principles should contribute to, or

---

[55] See Raymond E. Brown, *The Community of the Beloved Disciple* (Ramsey, NY: Paulist Press, 1979), p. 133.

at least not undermine, the genetic fitness of the group that professes to them. This claim is prescriptive, and based on a common-sense ethic that regards morality as intended to allow human beings to live happy and decent lives, an important part of which involves rearing children in a healthy and safe environment that allows them to grow into competent adulthood. The evolutionist position also relies on a descriptive view of human beings as marked by limitations of moral energy and other resources and moved by bonds and affections that are highly particular. In the most charitable interpretation, this does not mean that evolution demands one specific moral and social code for all times and places, for example domestic wives or bread-winning husbands, but it does suggest that cultural practices that run directly contrary to genetic interests will find little support in human biological nature. The best way to promote genetic interests depends on local ecological conditions (for humans this means appropriate social, economic, and cultural as well as physical and biological conditions). This claim of course is almost tautological, but it does have the value of highlighting the real if loose relation between human biological nature and morality.

Evolutionists put Christian ethics in an interesting position. People such as Ruse would claim that when it comes to ethics Christians "talk a good game," especially about love, but that as a matter of fact most Christians live lives that bear little resemblance to the lofty ethic of the New Testament. The Christian response to this observation can go in a number of directions. One insists that no theory holds up to practice, unless the theory itself is not very ambitious. A second (Lutheran) response to Ruse accuses the evolutionists of misunderstanding that moral law was given only to condemn sin and show sinners that we need utterly to rely on God's grace. A third (Augustinian) response argues that the Christian community looks so much like the world because God allows the "wheat" and the "tares" (Matt. 13:24–30) to grow up together and renders judgment at the end of time; the city of God is hidden and its members are known to God alone.

A fourth response, the one endorsed here, agrees that God alone is capable of rendering final judgment on the human heart but disputes the restriction of grace to the elect members of the Christian community predestined from all time to receive grace sufficient for salvation. This view operates with the governing theological principle that since God desires the salvation of the entire world, grace is silently and subtly operative everywhere and in every human life. People respond to this offer of grace in ways made available by their particular cultural and historical contexts. Because of the complexity of history, Christians have been both slave owners and abolitionists, soldiers and pacifists, pro-choice and pro-life. Christians are well prepared to acknowledge the historical limitations imposed on their own judgments. This is precisely the major weakness of the evolutionists – the failure adequately to acknowledge the historical and cultural nature of the human being. Whatever universal species-specific biological traits we have will always bear their moral significance within particular cultural contexts.

LOVE FOR GOD

An adequate interpretation of the relevance of evolution for Christian ethics has to address the three "objects" of love – God, neighbor, and self. "Christian love" is usually identified with neighbor-love, but in fact the Christian tradition has typically spoken of three "objects" of love, three directions in which love is oriented, including God and self as well as neighbor.[56] The development of ecological ethics in the last century stimulated a new attempt to reflect on the natural world as the object of love.[57] The understanding of agape as a form of altruism is an effort to make the former intelligible in humanistic terms, but doing so ignores the fundamental place of love for God in the Christian moral life. Christianity teaches a love for the highest good that transcends love for oneself, for pleasurable activities and good things, and for those near and dear to us. It teaches the reality of a theocentric love that decisively de-centers the self[58] and at the same time turns it toward the common good.[59]

Love for God answers the most important inner desire of the human person for complete happiness and provides the proper perspective from which to view the other recipients of love. Christian love of neighbor flows from a prior and more fundamental reorientation of the agent's entire life from a disordered and wrong-headed egocentric love to a theocentric, all-encompassing love for God. Because God loves all creatures (Ps. 145:9; Wis. 4:10, 7:28), we should love them "in God."

---

[56] See *ST* II-II, 26, and Vacek, *Love, Human and Divine*.
[57] See, for example, James A. Nash, *Loving Nature: Ecological Integrity and Christian Responsibility* (Nashville: Abingdon Press, 1991).
[58] See *ST* II-II, 25,1, on the extension of charity to our neighbor, including "sinners" (aa. 5–6) and "enemies" (aa. 8–9). Ignoring Thomas' theological framework leads social psychologist Batson to make the grossly misleading statement that Thomas represents "the dominant view in Western thought ... that we are, at heart, exclusively self-interested"; in *The Altruism Question*, p. 3.
[59] See Hollenbach, *The Common Good and Christian Ethics*.

The phrase "love of God" is semantically ambiguous. It can be read to refer to God as the subject of love, as in God's love for us – "God's love has been poured into our hearts through the Holy Spirit that has been given to us" (Rom. 5:5). It can also refer to our love for God (as in Ps. 42:2). The former is prior to the latter, as Paul states in Romans 5: "God proves his love for us in that while we were still sinners Christ died for us" (5:8; see also 1 John 4:10). God's love is the cause and sustenance of the creation, and God creates so that the creation, especially human beings, can in turn relate in love to the Creator. The very notion of creation means that God freely created the world out of love, not out of the need to gain anything from it.[60] The same love brings about the covenant with Israel, sends the Son as savior, and gives the Spirit to the church. God's love for human beings is the source of the gifts of grace, forgiveness, and reconciliation. The greatest and unsurpassable revelation of God's love lies in Christ's self-surrender on the Cross. The Christian life, therefore, is shaped in the form of the Cross. The test of love for God lies in love for neighbor, for "those who do not love a brother or sister whom they have seen, cannot love God whom they have not seen" (1 John 4:20; also Matt. 25:31–46).

Our love for God follows as a response to God's prior love for us. Desire for God is natural to human beings and can be discerned as an important factor in the religious yearnings of people around the world. This natural desire for God is answered by the gift of grace and explicitly addressed by Christianity in the life of the church. The Christian life is first and foremost a love for the God revealed definitively in Jesus Christ and carried on in the life of the church. Personal identity is most deeply discovered when life is centered on God, not on the self. What is most fundamental, in other words, is that each person is a child of God, not a child of his or her biological parents. Employing what sociobiologists call "fictive kin relations," the letter to the Ephesians reminds us: "He destined us for adoption as his children through Jesus Christ" (Eph. 1:5). Conversion also leads to engagement with the life and worship of the Christian community. Spirituality is the way in which an individual or community seeks to respond to God's love with a mature, purified love for God. Conversion reshapes and transforms the believer, who turns away from inordinate love for what is not God and toward a proper love for God. This process of conversion is preceded and made possible by God's grace, but works only with the free cooperation of the believer. To cooperate fully with God's grace is to be engaged in a process of sanctification.

---

[60] See *ST* I, 32,1.

Grace reorders the self and his or her relation to all other objects of love – "whoever does the will of my Father in heaven is my brother and sister and mother" (Matt. 12:50). God becomes the center of existence, the greatest good, and the ultimate end of action; relations to self, kin, and reciprocators are all relativized. The center of life is God, not genes or genetically related individuals. Since God's concern for human beings is universal and especially focused on the "least" among us (Matt. 25:31–46), the claims of kin and reciprocators do not always override all other concerns. Lack of gratitude and a willingness to reciprocate are moral failings, but help is not withheld from those who have no ability to return aid. In this sense, Christian love is "unilateral" and not conditioned on reciprocation.

The love of God carries an important implication for the moral worth of human beings. One can argue that human beings, as purely biological entities, have the same status as other "sentient beings." Peter Singer argues that infanticide and euthanasia are morally acceptable and that we are immoral for not taking steps to prevent the extinction of endangered species such as the Bengal tiger or the marsupial mouse (it is of course possible to accept the latter position without embracing the former).[61] Indeed, from a *purely biological* standpoint, it can be argued that individuals belonging to a global population of six billion human beings are less valuable than members of endangered species.

Human beings are regarded as special because of God's love for us and because we are made in God's image (Gen. 1:26), an essential aspect of which involves our capacity to return love for God.[62] This love allows us to cooperate in a special way with God's love for the creation – that is, to exercise "stewardship." All life is valuable, but human beings have a special value that should make us refuse to harm one another without ethical justification. We are called to participate in God's love of creation and to love all human beings – at least in the sense of having good will toward them – not only our kin and reciprocators. Love for God radically alters the contexts envisioned by the Prisoners' Dilemma. Because Christian love is bounded by justice, it does not follow the policy of "tit-for-tat" that encourages the exploitation of "suckers" and retaliates against "cheats." Neither does it follow a simple "sucker" policy of universal cooperation,

---

[61] See Peter Singer, *Animal Liberation* (San Francisco: Harper, 2001), *Practical Ethics* (second edn., New York: Cambridge University Press, 1993), and *A Darwinian Left: Politics, Evolution, and Cooperation* (New Haven: Yale University Press, 1999).

[62] See *ST* I, 93,4, where the image of God is understood in part in terms of our natural aptitude to love and know God.

since justice allows for proper self-regard and proper protection of innocent third parties from exploitation.

Love for God ought to change the *quality* of the believer's love in addition to extending the *range* of his or her appreciation and concern. Love for God transforms the person, and therefore the person's love. This is expressed in St. Paul's notion of Christians as having "the mind of Christ" (Phil. 2:5). Christian love is thus not a "type" of love with specific kinds of objects (e.g. enemies), that makes it morally better, and more Christ-like, than any other types of love. Ruse and Grant, like others, make the mistake of identifying one or two "types" of love as distinctively Christian instead of seeing that God's grace *transforms* all loves, whether for friends or strangers, friends or enemies, self or others.

SELF-LOVE

Self-love is, in different respects, both rejected and encouraged in Christian ethics. The former is more familiar: disciples are commanded to forget themselves and pick up the Cross (Matt. 10:37–39; Luke 14:26, 17:33; Mark 8:34–35). Because of the goodness of creation, self-love is a proper response to the self; proper self-love is legitimate and tacitly praised in the Scriptures, for example, in the Psalms (Ps. 1:1–6) and the Beatitudes (Matt. 5:3–12).

Commands to engage in self-renunciation, to "deny oneself, to "hate oneself," and so forth, are interpreted by Grant and others as issuing a straightforward ban on all forms of self-love. The absence of self-love is often regarded as a kind of litmus test for faith, yet one cannot act on the command to "love your neighbor as yourself" if self-love is forbidden. As Jonathan Edwards observed, "if Christianity did indeed tend to destroy man's love for himself, and his own happiness, it would threaten to destroy the very spirit of humanity."[63] What feminists lament as the destructive effect of an unmitigated ethic of selflessness on women can apply to all people.[64]

Scriptural texts denouncing the evils of disordered self-concern and improper favoritism point to deeper and more complex sources of wrongdoing than do evolutionary analyses. Alexander's analysis of human nature would have been recognized by classical Christian thinkers as a good description of fallen humanity if not the "integral human nature" intended by the Creator. Distorted self-love, is, as Augustine put it, "better called

---

[63] Jonathan Edwards, *Charity and Its Fruits* (London: Banner of Truth Press, 1978), p. 159.
[64] See Barbara Hilkert Andolsen, "Agape in Feminist Ethics," in Lois K. Daly, ed., *Feminist Theological Ethics* (Louisville, KY: Westminster John Knox, 1994), pp. 146–159.

self-hate."[65] We "commit sin to promote our own welfare," he writes, "and it results instead in our misfortune."[66]

Our deeply ingrained tendency to prefer ourselves to others is corrected not by ceasing to love ourselves altogether but rather by loving ourselves in the right way. Instead of completely abolishing self-love, Christian ethics distinguishes between proper and improper forms of it. Scripture often takes for granted natural love of self and family and concentrates instead on correcting their distortions. While holding that inordinate self-love, or pride,[67] is in a sense the cause of every sin, Thomas Aquinas also recognized that self-love is a natural tendency given to us by the Creator and that well-ordered self-love – by which we will the good proper to us, and especially the divine good – is right, natural, and Christian.[68] It is therefore crucially important to distinguish selfishness from proper self-love. Whereas improper self-love makes everything, even God, instrumental to the self, proper self-love thrives when accorded a secondary status to the more encompassing love for God.

Søren Kierkegaard was profoundly suspicious of those who would encourage self-preference. Every person, he observed, "has in himself [or herself] the most dangerous traitor of all."[69] Yet even Kierkegaard believed that this treachery consists not only in the betrayal of others but also in the self-destruction that comes from not willing to love oneself in the right way. He thus encouraged, and did not reject, proper self-love: "To love oneself *in the right way* and to love one's neighbor correspond perfectly to one another... The law is therefore: you shall love yourself in the same way as you love your neighbor when you love him as yourself."[70]

Interpretations of Christian love should not be evaluated positively simply according to the extent to which the self is denigrated or the neighbor elevated in value. Negation of self can be good or bad, depending on circumstances; an obedient Balkan commander, after all, might follow all orders to engage in ethnic cleansing out of selfless obedience. The Christian moral challenge is to move beyond sinful self-love, not to hate oneself or to be oblivious to one's own true good. If it is "more blessed to

---

[65] Augustine, *On Christian Doctrine*, trans. D. W. Robertson, Jr. (Indianapolis: Bobbs-Merrill, 1958), p. 20.
[66] Augustine, *Concerning the City of God against the Pagans*, trans. Henry Bettenson (New York: Penguin, 1984), p. 553.
[67] On pride as the beginning of every sin, see *ST* I-II, 84,2; on improper self-love as the cause of sin, see *ST* I-II, 77,4. See also Stephen J. Pope, "Expressive Individualism and True Self-Love: A Thomistic Perspective," *The Journal of Religion* 71 (1991): 384–398.
[68] See *ST* II-II, 25,4, and 26,4.
[69] Søren Kierkegaard, *Works of Love* (San Francisco: Harper, 1962), p. 39.
[70] Ibid.; emphasis added.

give than to receive" (Acts 20:35), and if "many who are first shall be last, and the last will be first" (Matt. 19:30), then Christian flourishing is found in a way of life focused on caring for others.

Proper self-love acknowledges that each person is more responsible for himself or herself than for other people while also acknowledging the claims of proper neighbor-love. When ordered properly, love for God, self, and neighbor is not fundamentally in competition. Conflict can and does exist, obviously, but here it takes place within the larger scheme of cooperation and caring for the common good.

LOVE OF NEIGHBOR

The most frequently cited formula for love of neighbor is the "golden rule": "do to others as you would have them do to you" (Matt. 7:12; Luke 6:31). Darwin regarded the golden rule as the foundation of morality, but he thought that its support comes from its contribution to the struggle for survival rather than from its religious authority.[71] This formula is often thought to be an injunction to reciprocity – to treat others the way they treat you – and as a religiously backed version of "tit-for-tat" reciprocity based on enlightened self-interest. Psychologist of religion Lee A. Kirkpatrick interprets the golden rule as a religiously sanctioned rule for reciprocity in social exchange.[72] Holmes Rolston regards reciprocity as "a sort of operational version of the golden rule, doing to others as you would have them do to you, while refusing to be taken advantage of."[73] This proviso builds self-protection into the golden rule in a way that echoes efforts in Christian ethics to coordinate love and justice. Rolston elsewhere argues that the golden rule requires that we combine self-love with loving others, ideally in a form of "mutually beneficial reciprocity."[74]

The golden rule, however, differs from reciprocity in a number of fundamental ways. Reciprocity regards other people as solutions to present or future problems faced by the agent. The golden rule is "unilateral" because it requires the agent to act in certain ways regardless of how the agent is treated by the recipient of the agent's act.[75] Because of the love of

---

[71] *Descent*, in *Darwin*, ed. Appleman, p. 319.
[72] See Lee A. Kirkpatrick, *Attachment, Evolution and the Psychology of Religion* (New York: Guilford Press, 2004).
[73] Rolston, *Genes, Genesis, and God*, p. 229.   [74] Ibid., p. 239.
[75] The "unilateral" dimension of agape is emphasized by Gene Outka, "Universal Love and Impartiality," in Edmund N. Santurri and William Werpehowski, eds., *The Love Commandments: Christian Ethics and Moral Philosophy* (Washington, DC: Georgetown University Press, 1992), pp. 1–103.

God, the self does not need to regard others primarily in terms of how they might meet the self's needs or satisfy its desires. It places no conditions of reciprocity on the action of the agent and it makes no mention of limits to what the agent ought to do on behalf of the other. The golden rule does not combine self-concern with concern for the good of the other. The formula is silent about self-regard.

Evolutionists read the golden rule as a means to fitness. "Good Samaritanism" helps to build the altruist's good reputation within a community, Alexander reasons, and thereby indirectly contributes to his or her reproductive fitness. He believes that "whether or not we know it when we speak favorably to our children about good Samaritanism, we are telling them about a behavior that has strong likelihood of being reproductively profitable."[76] This claim, however, misses the point of the story of the good Samaritan (Luke 10:29–37), who, while traveling along the road to Jericho, shows what it means "to be neighbor" by going out of his way to care for a man who had been beaten and left for dead by robbers. The term "neighbor" during Jesus' time meant "near one" but the effect of this story is to "stretch love of neighbor until it becomes love of enemy."[77] The Samaritan is not an "altruist" who helps a member of his own community, but a traveler moved by heartfelt compassion who comes to the aid of a historic enemy.

Alexander's interpretation of the good Samaritan as exemplifying indirect reciprocity is not particularly illuminating; the biblical text, in fact, teaches exactly the opposite of his ethic of self-interest. The poignant strangeness of the parable lies in the historic alienation of Jews and Samaritans. "Being neighbor" means overcoming boundaries, or, in Alexander's language, webs of indirect reciprocity, and extending care to all who are needy, whether they are members of one's own in-group or not. "Neighbor" no longer means "near one," but rather any human being whom one happens to encounter. This universality is "perverse"[78] from a sociobiological standpoint. Evolutionists make the reasonable generalization that when children are taught to act with kindness and compassion, they will be well-respected, more likely to be trusted by others, and make trustworthy friends. Yet even this generalization only holds for some environments and not others. One can think of other, highly stressful social contexts where being cooperative and trusting is taken to be a sign of weakness and invites disdain, exploitation, and ostracism.

---

[76] Alexander, *Biology of Moral Systems*, p. 102.   [77] Tannehill, *Luke*, p. 184.
[78] Ruse, "Evolutionary Theory and Christian Ethics": 19.

Biblical scholar Gerd Theissen focuses on the ways that the Christian ethic of love functions as an "anti-selection" force in cultural evolution that runs counter to biological evolution.[79] Theologian Philip Hefner stresses the need for theology to recognize that biological evolution is always, in human experience, mediated through cultural evolution.[80] Hefner agrees with Theissen that love has a counter-egoistic thrust that encourages Christians to move beyond the restraints of the "selfish gene." The "love-principle" inspires people to extend caring beyond the kin group and the circle of reciprocators – to "cross boundaries and identify with outsiders."[81] It also forms character: "The biological component of a human being faces the developmental challenge to sculpt itself in ways that surmount some of the basic drives or behaviors that are wholly undesirable in a human community."[82] At the same time, Hefner argues, the "love-principle" accords with the deeper meaning of the creation that is best understood in light of its ultimate destiny, the full arrival of the Kingdom of God.

The Samaritan is motivated by compassion rather than enlightened self-interest. As noted previously, sociobiological analyses are crippled – at least in terms of ethics and the structure of human motivations – to the extent to which they restrict their attention to external action and its consequences. Human love, including Christian love, and indeed all love, is inevitably a psychological reality that includes the motivations, motives, and intentions of agents.

Darwin himself argued that interpersonal affections can be extended from the small circle of family, friends, and more intimate groups outward to more remote others. In some way, "nepotism" becomes the basis for more extended care. Yet it is not very clear how such care is possible on neo-Darwinian grounds once behavior moves beyond the circle of kinship and reciprocity. There is no evolutionary incentive for extending responsibility so widely, and no ethical justification forthcoming from nature; even group selection leaves one at the boundaries of one's own group. Evolution does not provide any ethical basis whatsoever for recommending either the extension or the restriction of responsibility, and evolutionists who engage in such moralizing do so outside the bounds, and without the

---

[79] Gerd Theissen, *Biblical Faith: An Evolutionary Approach* (Philadelphia: Fortress Press, 1985), pp. 23, 25.
[80] See Hefner, *The Human Factor: Evolution, Culture, and Religion* (Minneapolis: Fortress Press, 1993). See also "The Evolution of the Created Co-Creator," pp. 403–420, and *Technology and Human Becoming* (Minneapolis: Fortress Press, 2003).
[81] Hefner, "The Evolution of the Created Co-Creator," p. 411.   [82] Ibid.

warrants that function within their own disciplines. Sociobiologists argue that ethical appeals to widespread responsibility are, as a matter of descriptive fact, impossible to fulfill and therefore ethically futile. This is the gist of Ruse's advice that Christians be satisfied with "weak agape." Put in Dawkins' terms, a bear might be trained to ride a unicycle, but no one can teach a bear how to drive a car.

Christian ethics can respond by pointing out that the restrictions discussed by sociobiology have not been demonstrated, only asserted, and by adding that concrete human experience gives us reason for thinking that human beings have greater capacities for goodness than sociobiology admits.[83] Sociobiology misses the fact that the experience of love not only creates a bond between two or more individuals but can create a level of identification with the friend's well-being such that what is good for one is good for the friend and, conversely, what is bad for one is bad for the friend. This kind of mutual identification, called "mutual indwelling" by Thomas Aquinas,[84] makes it difficult to separate what is good for one person from the good of his or her friend. It shows the unhelpfulness and radical inadequacy of examining love in terms of a forced egoism–altruism dichotomy. From a Christian standpoint, sociobiologists ignore the transforming power of grace but also underestimate our natural capacity for goodness. An acknowledgment of emergent complexity works against the moral reductionism that accompanies sociobiological ontological reductionism, and a theology of grace extends our consideration of the love to which we might aspire.

Sociobiologists reduce friendship to mutual service. If no real friendship is possible for ordinary people, then the most that can be done is to "manage" others in order to protect one's perceived long-term self-interest. The elimination of friendship – the quintessentially shared good, a good that cannot be held alone – represents a major assault on the moral underpinnings of the traditional notion of the common good. If goods are replaced by "interests," and if "interests" are always possessed by individuals or groups of individuals, then the notion of the common good is eliminated.[85]

Some Christian ethicists regard love of enemies as the distinctive mark of the gospel. Yet they also need to acknowledge that Scripture provides ample witness not only to the worth but even to the centrality of friendship

---

[83] See Philip Hallie, *Lest Innocent Blood Be Shed* (New York: Harper and Row, 1979).
[84] *ST* I-II, 28,2.
[85] See Hollenbach, *The Common Good and Christian Ethics*.

within the Christian moral life. The Gospels depict the relationship between Jesus and his disciples as one of friendship (John 15:15). Jesus shifted the primary reference away from kin group to the Kingdom, but he too showed a special love for "his own" disciples (John 13:1).

Christian love includes both "unilateralism," the claim that the ethical core of agape flows in one direction, from the agent to the recipient, and "mutualism," which holds that agape consists in bilateral fellowship, friendship, and brotherly love.[86] Christian love includes an attitudinal readiness to serve the neighbor – as seen in the parable of the good Samaritan. But it also sponsors the deep friendship expressed in the table fellowship of the Last Supper – indeed, the Last Supper encompasses both sacrifice and fellowship. The minimal core of agape is a benevolent and steady commitment to serve, but its true fruition is manifested in mutuality where radical "otherness" is ultimately overcome and replaced by genuine friendship. "Christian love," as Stephen Post puts it, "is participatory and mutual within the *koinonia* or fellowship of the faithful, although this circle is by no means 'hermetically sealed.'"[87] In fact, it is precisely this fellowship that inspires and empowers outreach to the "other."

Some Christian ethicists insist that, although it is the product of a particular community, agape (or charity) calls for an unlimited concern for all human beings. This requirement of universal concern is not undercut by the claim that Christians have a responsibility to "be church" and to provide a witness to the truth of the gospel[88] by cultivating an evangelical way of life within the Christian community that also leads to hospitality and mercy to others. Christian love requires response to needy others beyond as well as inside the circle of family, friends, and fellow believers: "If you do good to those who do good to you, what credit is that to you? For even sinners do the same" (Luke 6:33). At the same time, it emphasizes the special commitment incumbent on disciples to build up what St. Paul called the "body of Christ" (1 Cor. 12:12–27) that is the Christian community.

---

[86] The "unilateral" or "non-reciprocal" dimension of agape is given careful examination by Gene Outka in *Agape: An Ethical Analysis* (New Haven and London: Yale University Press, 1972), esp. chs. 1 and 8, and in "Universal Love and Impartiality," pp. 1–103. For an appreciative but critical response, see Stephen J. Pope, "'Equal Regard' versus 'Special Relations'? Reaffirming the Inclusiveness of Agape," *Journal of Religion* 25.1 (1997): 353–379.

[87] Stephen G. Post, *The Theory of Agape: On the Meaning of Christian Love* (Lewisburg, PA: Bucknell University Press, 1990), p. 13.

[88] See Daniel J. Harrington, SJ, *The Church according to the New Testament* (Franklin, WI, and Chicago: Sheed and Ward, 2001), esp ch. 10.

Along with the prophetic call to serve all humanity, Christian ethics also recognizes the need to sustain the special ties that bind those belonging to the Christian community itself. The theological doctrine that all are created in the "image of God" supports the ethical vision that we all belong to the same human family. Alexander argues that "in-group amity" generates "between-group enmity."[89] Yet in a Christian context, the closely bound in-group based on a common faith supports, rather than undercuts, a complementary sense of inclusive openness and generous love for all others with whom we share a common humanity. Both commitments are essential to the Christian ethics, but their unity and difference can only be appreciated within an acknowledgment of the inclusive nature of Christianity. Social oppression is often facilitated by a process of dehumanizing the other. What Aldous Huxley perceived to be happening in the Europe of the late 1930s is a persistent human temptation: "The propagandist's purpose is to make one set of people forget that the other set is human. By robbing them of their personality, he puts them outside the pale of moral obligation."[90] Christian love regards all people as made in God's image regardless of their status vis-à-vis various social groupings; everyone is a member of the most important "group," the human community.

Critics will quickly point out that central streams of biblical ethics support Alexander's divide between "in-group" morality and "out-group" enmity, for example God's command that Israel completely destroy those who occupied the promised land (Josh. 11:20; 1 Sam. 15:3, Ps. 21:9–10, Deut. 20:16–18). These and other biblical texts are more disparate and complex socio-historically, literarily, and theologically than is sometimes recognized. Relevant factors include the commonality of war ideologies in the ancient Near Eastern world, the use of war imagery to build solidarity and group identity, the prevalence of the image of God as the "'divine warrior,'" notions of purity, and theological responses to the royal power of the monarchy. There are, of course, also biblical denunciations of the brutality of war and the emotions that can lead to it (see, e.g., Gen. 49, Amos 1–2, and 1 and 2 Chron.). Yet taking into account these and other factors does not gainsay the many ways in which the Bible supports a strong in-group/out-group division. As one biblical scholar puts it, "God plays favorites and protects his own."[91] The biblical tradition, however, moved beyond

---

[89] Alexander, *Biology of Moral Systems*, p. 95.
[90] Aldous Huxley, *The Olive Tree* (New York: Harper and Brothers, 1937), p. 110.
[91] Susan Niditch, *War in the Hebrew Bible: A Study in the Ethics of Violence* (New York: Oxford University Press, 1993), p. 148.

requiring human sacrifice and the annihilation of the "unclean" enemy, and in the New Testament it came to view God as loving all of humanity.[92]

Christian ethics in the New Testament conceives of every human being as "neighbor" and as a member of one common family, and therefore as encompassed by basic moral protections that cannot be denied for any reason. Instead of focusing primarily on its objects, we should attend to the *transformative* impact of Christian love on the agent. Evolution gives more support to loving one's self, friends, and family than to loving strangers and enemies, yet in some cases the former can be much more demanding than the latter. Friends and family can require degrees of trust, patience, and readiness to forgive that go far beyond what is entailed in good will for strangers and enemies. Wherever this kind of love is directed, it is, as Dostoevsky put it, a "harsh and dreadful thing compared with love in dreams."[93] Indeed, Christian ethics faces the challenge not only of identifying the proper *scope* of love and our resistance to it, such that the range of love encompasses all human beings rather than only the nearest and dearest, but also of cooperating with the ongoing grace-inspired transformation of the *quality* of love so that it more nearly approximates the love revealed, lived, and taught by Jesus. The range of such love expands and deepens as it is permeated by grace.

Several points can be made in response to the alleged dichotomy between familial love and Christian love. First, there are evolutionary reasons for the fact that human beings have a tendency to be selfish, and this evolved proclivity can be highly exaggerated by the individualism and competitive ethos of modern culture. There are also evolutionary reasons for individuals to channel caring activity toward kin and reciprocators and away from strangers and enemies, to form special attachments to their communities and to be less interested in the well-being of communities to which they do not belong. The fact that people are occasionally willing to renounce their own reproductive interests and to expend resources on individuals who are neither kin nor reciprocators might be accounted for in terms of the evolution of group-selected altruism in cases where the needy individuals in question are in fact members of the agent's group. Going outside this boundary, it is also the case that individuals occasionally expend resources on those with whom they share no group membership or kinship

---

[92] The claim that there is a moral development within the biblical tradition should not be taken to imply the "supersessionist" view that the emergence of Christianity superseded Judaism.
[93] Fyodor Dostoevsky, *The Brothers Karamazov*, trans. Richard Pevear and Larissa Volokhonsky (New York: Vintage, 1990), p. 58.

affiliation. According to most evolutionists, such a good Samaritan is merely an illusion because genuinely unselfish acts do not exist. The good Samaritan's action represents evolutionary self-destructiveness, so the tendency could not have become a prevalent part of the human emotional repertoire. One can reply that the good Samaritan reflects his morally conscientious choice, motivated by compassion, to care for the victim of robbery, but evolutionists deny that we have the capacity for such a choice. We might argue that the good Samaritan was moved by the religious culture of his day, and not by biologically based inclinations, to care for the victim, but evolutionists are loath to acknowledge culture as an independent variable, detached altogether from our evolutionary needs. This refusal to admit the possibility of disinterested love exemplifies the limited and cramped imagination of the evolutionary approach to human sociality.

Second, we should not confuse "kin altruism" with familial love. An act can be phenotypically altruistic toward kin without being an expression of love in any sense, for example a parent who gives money to a child in order not to be bothered. Indeed, the very distinction between phenotypical altruism and egoism is undercut by the complexity of familial relations, in which an act can have complex combinations of costs and benefits too intermingled for anyone to be able to make a clear judgment about the matter. There is, moreover, *pace* Ruse, nothing particularly "weak" about love for family. As we see in the challenge faced by parents caring for children with severe Attention Deficit Disorder, or adult children caring for elderly parents afflicted with severe Alzheimer's disease, the quality of love required by the gospel in these spheres can be just as demanding, though in different ways, as the love of strangers. Indeed, because of the emotional intensity and degrees of vulnerability it involves, love of intimates can be more painful, require many kinds of self-denial, and call upon emotional resources that will not characterize expectations generated from more distant relations. Christian ethics can give a special place to marriage and family as religious and social developments of "the natural bases of human love"[94] that in turn provide the emotional and cognitive resources for more extensive altruism. If the circle of altruism is to expand it must do so on the basis of emotional, psychological, and moral resources developed in the intimate spheres of interpersonal life.

Third, we should not identify all forms of familial love with the evil of "nepotism" or "familialism," an unjust bias toward family members. Familial loyalties can be harmful when they become exclusive or myopic.

---

[94] Gustafson, *Ethics from a Theocentric Perspective*, vol. II, p. 160.

Unbalanced familial love can encourage nepotistic favoritism and allow for neglect or abuse of others. The category "neighbor" as it is used in Christian ethics includes every human being, and *not* every human being except our friends and family members. As sociologist Rodney Stark has shown, family ties and other forms of interpersonal attachment, along with an extension of charity "beyond the bonds of family and tribe," played key roles in the rapid growth of early Christianity.[95] Jesus' injunctions to "hate" one's family do not indicate a disdain for the flesh and close personal ties.[96] They should not be interpreted as a reversal of the Fourth Commandment – honor thy father and mother – but simply as a stern warning against disordered "familialism" and caring more for family than for Christ (see Matt. 10:37–39; see also 1 Tim. 5:8).

Moral psychologist and practical theologian Don S. Browning offers a helpful appropriation of evolutionary theories of altruism within a multidimensional view of human nature that is similar to the one developed in this book.[97] He argues that the theory of kin altruism explains the origin of certain common human "premoral goods" involving familial love, reciprocity within marriage, and other aspects of what he calls the natural foundations of love. Browning properly focuses on the value of family life for the development of moral capacities, including those concerning Christian love. I think he is right to argue that evolutionary psychology presents a resource that can help Christian ethics to criticize, refine, and appropriate the moral wisdom of the tradition. Christian ethics, in turn, can supply a basis for drawing an important line between the "pre-moral" valuations discussed in evolutionary theory and more properly moral values and principles.

Browning, a liberal Protestant, finds kin-altruism theory to be a resource that facilitates the retrieval of some of Thomas Aquinas' insights regarding the natural ordering of love.[98] He insists that we recognize the natural and social conditions that are essential for the development of what he calls "a love ethics of equal-regard and mutuality":

kin altruism, the mechanisms of mutual adaptation and attachment in both parent and child, the rise of empathy as the capacity to feel and know the needs of the other

---

[95] Rodney Stark, *The Rise of Christianity: A Sociologist Reconsiders History* (Princeton, NJ: Princeton University Press, 1996), p. 86.
[96] Cf. Owen Flanagan, *Varieties of Moral Personality* (Cambridge, MA: Harvard University Press, 1991), p. 316.
[97] See Don S. Browning, "Science and Religion on the Nature of Love," in Stephen G. Post, *et al.*, eds., *Altruism and Altruistic Love*, pp. 335–345.
[98] Ibid., pp. 340–342.

in analogy to how the child comes to feel and know his or her own needs, the gradual extension of these dynamics to non-kin, and finally, the development of more abstract capacities of cognitive empathy for persons we do not even know ...[99]

Christian love ideally moves toward mutuality and friendship, so it is expressed in a particularly profound way in marriage and the family. It challenges the narrowness of egocentrism and nepotism, and mutuality overcomes the distance of equal regard and the restraints of reciprocity. The virtue of prudence, exercised within the Christian moral tradition, moves to make good decisions in the midst of the conflicts among "premoral" goods that so often characterize the human condition. Christian ethics at its best promotes an appropriately balanced appreciation of the goods and duties of marriage and family along with a wider sense of responsibility to others.

CONCLUSION

This chapter argued that Christian ethics needs to account both for our created capacities, the capacities of emergent complexity, and for the transformative influence of divine grace on the human person. Grant proposes a particularly narrow way of interpreting the latter and ignores the former altogether, which limits the usefulness of his account of agape. Ruse pays attention only to sociobiological accounts of altruism, does not see the severe limitations of such a reductionistic reading of human behavior, and denies the validity of the theological category of creation. He consequently overstates the significance of sociobiology for ethics, obscures some fundamental aspects of human love, oversimplifies Christian teachings on love to the point of distortion, and offers a secular domestication rather than a real understanding of love in Christian ethics.

Christian ethical understanding of love can be enriched by knowledge of the evolved human emergent emotional and cognitive capacities that play various roles in love. The normative core of Christian love comes from revelation – the Christian Scriptures and the ongoing tradition – and its authoritative interpretation. Christian ethics of love cannot then be thought of as derived from or justified by scientific knowledge of human evolution. At the same time, because we are evolved creatures, embedded in nature, possessed of a range of complex evolved capacities, whatever we say about our created capacity to love has in some way to be related to this

---

[99] Ibid., p. 344.

natural basis. Knowledge of evolution makes a modest but real contribution to our understanding of Christian love.

Christian ethics is concerned with the proper development of the human person under the influence of grace, but it is not simply an ethic of grace. Christian ethics can learn from the evolutionary views of altruism studied above when they are purged of their reductionistic presuppositions. They shed light on the evolutionary background of human altruism but they cannot "explain" human action because it is more than an organic event. Nor does evolution provide an ethical reason for acting altruistically in the full sense of the term. In Christian terms, we act for others because this is what Jesus displayed in his life and what he taught his disciples to do. The Christian community encourages us to expand the depth and breadth of what comes to us naturally, and to go beyond the affective and moral limits that typically mark human nature.

One issue concerns universality. Human nature includes predispositions to form affective ties with a number of intimates, to develop real but less intense affective ties with those with whom we have some social connections, and to feel little connection toward those with whom we have little in common. This very elemental evolved set of predispositions makes sense in light of the ordinary experience of most people, even when we realize that there is considerable cultural variability in the meaning found in these connections. It is also given different instantiations in various cultures throughout and across time. As a generalization, it is formal and general enough not to be very exceptional or controversial. We have evolved to form bonds of love with a small number of people (and this is minuscule when contrasted with the population of the human race as a whole or even with the size of a large city today). Because our resources are limited, our caring is channeled to the "nearest and dearest" by evolution. In the absence of deep affection, and to some extent to make up for its limits when extended too far, we speak ethically of the need for justice, a sense of duty, impartiality, and suchlike. Christian ethics challenges our tendency only to see the "nearest and dearest"; it pushes us to regard all people as made in God's image and so to act for their benefit whenever reasonably possible.

CHAPTER 10

The natural roots of morality

Evolutionists have advanced several theories of how morality as a social institution is related to human evolution. They offer three major ways of accounting for morality as a social institution and for the human moral sense: as adaptive, as an evolutionary by-product, and as a product of culture rather than biological evolution. This chapter argues that while each approach has its own weaknesses, each can help Christian ethics understand important aspects of morality.

EVOLUTIONARY THEORIES OF MORALITY

The social institution of morality probably came into use to address certain fundamental human needs rooted in our emergent cognitive, emotional, and social capacities. Morality as such did not evolve biologically, but some human capacities that are essential to morality have evolutionary roots.

Morality as adaptive

Ruse and Wilson argue that the "moral sense" is a biological adaptation, "just like hands and feet."[1] "Belief in morality," they write, "is merely an adaptation put in place to further our reproductive ends."[2] Wilson's answer to Gustafson's question, "What do we value about the human?"[3] is clear: despite what people say, we really value inclusive reproductive fitness above all else. Wilson is convinced that sociobiology can explain the deepest nature and function of morality. Rather than a supernatural code delivered from "on high," or a "spark" of the divine lodged in each person's conscience, moral codes have originated because they serve the fitness interests of their adherents. Wilson even goes so far as to say that specific norms, for example

---
[1] Ruse and E. O. Wilson, "The Evolution of Ethics": 50.
[2] Ibid. [3] Gustafson, *Theology and Christian Ethics*, p. 244.

regarding marriage, property, or truth-telling, are accepted because they yield fitness benefits for those who adhere to them, or at least for those who promote them in others. Membership in dominance hierarchies, for example, "pays off in survival and reproductive success"[4] and compassion "conforms to the best interests of self, family, and allies of the moment."[5]

Morality helps to support a stable and relatively trustworthy social order in which individuals are more likely to cooperate and less likely to defect. In the long run, internalizing norms that resist crude ways of pursuing self-interest actually contributes to one's fitness. (This holds for people living in reasonably stable social orders; in less stable contexts, the optimal strategy might be an aggressive exploitation of available opportunities.)

Wilson believes that moral values reflect the biological "imperatives" encoded in genes: individuals must survive and leave reproductively viable biological kin. Whereas British emotivists such as C. L. Stevenson and atheistic existentialists such as Jean-Paul Sartre maintained that moral values are ultimately the product of the idiosyncratic and arbitrary choices of individual moral agents,[6] sociobiology proposes that in fact there are deep evolutionary reasons for the moral choices people make. Morality in this sense is "natural." Sociobiologists hold that it is in our self-interest to encourage an ethic of self-sacrifice, duty, and honesty because we benefit from living in communities where people act in this manner. If morality is a system of "indirect reciprocity," those who publicly advocate these virtues, and especially those who subjectively believe in them, are more appealing as mates, neighbors, business partners, and the like.

Morality in this view, is essentially self-serving: "People tend to pass on the sorts of moral judgments that help move their genes into the next generation," Wright argues, "and there's definitely no reason to assume that existing moral codes reflect some higher truth apprehended via divine inspiration or detached philosophical inquiry."[7] Ruse and Wilson put this point even more bluntly: "Morality, or more strictly, our belief in morality, is merely an adaptation put in place to further our reproductive ends. Hence the basis of ethics does not lie in God's will . . . [But rather it is] an illusion fobbed off on us by our genes to get us to cooperate."[8] Sociobiology promotes an evolutionary equivalent of Plato's "noble lie."[9] Morality as an essential part of social and individual life can only function

---

[4] E. O. Wilson, *Consilience*, p. 259.   [5] E. O. Wilson, *On Human Nature*, p. 155.
[6] See MacIntyre, *After Virtue*, chs. 2 and 3.   [7] R. Wright, *The Moral Animal*, pp. 146, 147.
[8] Ruse and E. O. Wilson, "The Evolution of Ethics": 50, 52.
[9] See Loyal Rue, *By the Grace of Guile: The Role of Deception in Natural History and Human Affairs* (New York: Oxford University Press, 1994).

properly if it maintains the illusion of objectivity: it "simply does not work unless we believe ... it is objective."[10]

This understanding of morality is developed by evolutionary psychologists in terms of their theory of "Machiavellian intelligence," which crafts behavior to what is thought will be most attractive to associates.[11] When people "advertise" themselves as trustworthy reciprocators who will not cheat, they implicitly invite others to invest in relationships with them. Others pursue a strategy of accepting conventional mores and of acting more aggressively to secure the goods they deem necessary for their happiness; crime pays when, under certain conditions, it is the best means of maximizing fitness. Reciprocity may explain why people naturally resent cheaters and feel indignant in the face of their unfairness. Even brown capuchin monkeys display a sense of fairness when they respond negatively to rewards distributed unequally by a human experimenter.[12]

Monkeys refused to cooperate when they saw a conspecific receive a better reward for the same effort. They responded even more strongly if they saw the conspecific receive the same reward after having done no work at all.[13]

Morality is a means of social control, a way in which individuals living in groups monitor and control one another's behavior. Buss, for example, argues that extramarital sex is prohibited because, and *only* because, it promotes the reproductive interests of individuals.[14] Whereas law controls external behavior, morality is internalized and so produces an even more deep-seated basis for social control. Morality here is nothing but a vehicle for subtle and often masked self-interest, at least genotypical egoism if not crude "what's-in-it-for-me" selfishness. This theory of morality joins moral reductionism to the epistemological and ontological reductionism examined in chapter three.

Self-interest, sociobiologists argue, lies at the root not only of morality in general but of our adherence to particular moral standards as well. As Ruse puts it, "We believe what we believe about morality because it is adaptively useful for us to have such beliefs – that is *all* there is to it."[15] Morality is simply part of a general and flexible human behavioral program. We have

---

[10] Ruse and E. O. Wilson, "The Evolution of Ethics": 52.
[11] See Byrne and Whiten, eds., *Machiavellian Intelligence*.
[12] See Sarah F. Brosnan and Frans B. M. de Waal, "Monkeys Reject Unequal Pay," *Nature* 425 (2003): 297–299.
[13] Ibid., 297–299.   [14] See Buss, *The Evolution of Desire*, p. x.
[15] Ruse, "Evolutionary Ethics: Healthy Prospect or Last Infirmity?": 27–73 at 42; emphasis added.

to cooperate to survive, but this cooperation can run up against a basic evolved behavioral tendency toward selfishness that creates conditions for social conflict. We have a delicate but general decision-making system in which our innate dispositions, the "epigenetic rules," indicate various goals and priorities. Some of these dispositions lie at the basis of moral values. Cooperation is adaptive, even amidst considerable conflict, and we are more efficient cooperators if we believe in cooperation.

Metaphysics and divine commands are not necessary for the functioning of morality. Ruse rejects Kantian ethics because it accords a "necessity to moral imperatives foreign to the modern evolutionist."[16] Moreover, just as Darwinian epistemology requires a "denial of metaphysical reality – the world of the thing-in-itself, not to mention Platonic forms and eternal mathematical truths just waiting to be discovered,"[17] so Darwinian ethics rules out any transcendent basis for morality. There is no objective moral reality over and above the requirements that people impose on one another as members of communities trying to achieve some degree of collective order. To say that someone has acted "immorally" recognizes that he or she unjustifiably harmed someone. This harm typically affects one or more people as individuals (e.g. the victim of theft, lying). But it also inflicts harm, even if in only small or very subtle ways, on the web of relationships constituting the transgressor's community; harms erode social trust, justice, and peace.

It is far from clear, though, exactly what Ruse means by "metaphysical," or how he thinks it even makes sense to suppose that philosophers think it can be "added" to human acts. Ruse's philosophical claims about the general features of matter are themselves metaphysical realities. In fact, the very "being" of human acts constitutes them as inherently metaphysical. Ruse's naturalistic ethic asserts that the effects of human behavior pertain only to the material world, but his position suffers from the same kind of vagueness and ambiguity as Wilson's suggestion that ethics has a purely "material" origin.

Ruse is of course right: to claim that human beings would have a different kind of morality if we were a different species makes perfect sense. Morality enables people to meet some basic human needs and flows from some basic human inclinations, so it would be different if we happened to have had needs and inclinations different from those that we actually do have. This is the point of natural law – the good is relative to human nature. Evolutionary theory can be useful if it accounts for how we

---

[16] Ruse, *Taking Darwin Seriously*, p. 263.   [17] Ibid., p. 269.

came to have the nature we experience today. Showing that morality is appropriate to human nature does not entail its being purely a matter of social convention; on the contrary, it can be taken, as it is by the natural-law tradition, to indicate precisely the opposite.

Wilson's descriptive account of morality is excessively simple and moves too quickly from genes to behavior without taking into sufficient account the impact of culture on conduct. Moral codes are acquired during childhood by various modes of training, imitation, and learning. Because they are the products of contingent human behavior and differ from culture to culture, ethicists who want to understand why various peoples have the codes that they have need to study the historical origins, transmission, and development of these codes.[18]

Lumsden and Wilson's "gene-culture theory" attempts to acknowledge the reciprocal and interdependent relation between culture and genes.[19] They argue that hereditary "epigenetic rules" bias learning and information-processing and incline people to make cultural choices that are generally most likely to increase their fitness. In so doing, they persist in the sociobiological tendency to regard reproductive fitness as the key factor explaining human behavior and thereby continue to undervalue cultural influences.

Personal behavior can only be understood by attending to the particularities of personal history, the flow of experiences most adequately expressed in narratives. To be a person is not only to instantiate the universal category of humanity. It is also to be a unique individual with a distinctive point of view and life story. We can draw connections between what "fitness considerations" would predict and what actually obtains in specific life stories, but in many instances people are motivated by moral considerations to act in ways that run against the grain of what might be predicted statistically on the basis of evolutionary premises. Attempts to explain these away – such as E. O. Wilson's accusing Mother Teresa of selfishness[20] – only discredit sociobiology.

In the same vein, any account of the morality of a particular community has to attend to its history. Human history cannot be usefully analogized to natural history, and Dawkins' analysis of "memes" does not offer any

---

[18] Lumsden and E. O. Wilson, *Genes, Mind, and Culture*, and *Promethean Fire* (Cambridge, MA: Harvard University Press, 1983); L. L. Cavalli-Sforza and M. W. Feldman, *Cultural Transmission and Evolution* (Princeton, NJ: Princeton University Press, 1981); Boyd and Richerson, *Culture and the Evolutionary Process*.
[19] See Lumsden and E. O. Wilson, *Promethean Fire*.
[20] E. O. Wilson, *On Human Nature*, pp. 164–165.

significant new insight into the prevalence of particular symbols in a given culture. Human history is complex not only because of the multiplicity of factors that bear on it and from it – economic, political, legal, linguistic, artistic, military, psycho-social – but also because these factors function concretely within the contingency of human choices.

Wilson himself makes room for human choices when he refers to the individual as only "*predisposed* biologically to make certain choices"[21] and when he recognizes that we need "to conform to some drives of human nature and to suppress others."[22] At their best, evolutionary approaches to human behavior might be able to provide a better grasp of behavioral probabilities and our innate proclivities. Particular choices are not mechanistically determined by the events and processes that precede them, so moral insight will always depend on grasping particularities rather than appealing to abstractions. This is why many humanists are convinced that the depth of moral perception communicated in great novels such as Tolstoy's *Anna Karenina* or Dostoevsky's *Brothers Karamazov* will never be eclipsed by insights generated by science.

Morality as a by-product of evolved capacities

An alternative theory regards morality as a by-product of natural selection rather than as a direct adaptation. Morality is dependent on our biological makeup as its necessary condition (how could it not be, since we are biological beings?), yet, Ayala argues, morality did not come into existence because it was adaptive in itself, but rather as the "indirect outcome of the evolution of eminent intellectual abilities."[23] He holds that three evolved capacities provide the basis of morality: "(i) the ability to anticipate the consequences of one's own actions; (ii) the ability to make value judgments; and (iii) the ability to choose between alternative courses of action."[24] Morality and the capacity to develop a "moral sense" are rooted in the biological makeup of the person, and in our emergent human cognitive, social, and emotional capacities, but they are only the indirect outcome of the evolution of our intellectual capacities for foresight, evaluation, and choice.[25] This strongly nonreductive approach to

---

[21] E. O. Wilson, *Consilience*, p. 250.
[22] Ibid., p. 251; also E. O. Wilson, *On Human Nature*, p. 97.
[23] Francisco J. Ayala, "The Biological Roots of Morality," in Paul Thompson, ed., *Issues in Evolutionary Ethics* (Albany: State University of New York, 1995), p. 302.
[24] Ibid., p. 297.   [25] Ibid., p. 302.

morality employs the less deterministic notion of "basis" and acknowledges that something can be "supported by" (in the weak, nonreductive sense) an evolved base without itself having been directly "caused" (in the reductive sense) by that base.

Ayala's emphasis on the cultural basis of morality seems to be the opposite of Ruse's position. Yet his second condition for morality, the capacity to make value judgments, leaves open the possibility that evolved emotional and cognitive predispositions may incline human beings to adopt some courses of action more easily than others (e.g. to reciprocate). It is difficult to believe, however, that the powerful human inclination to believe and act in terms of overarching normative frameworks is simply a by-product of evolutionary forces. Anthropologist William Irons holds that morality came to exist because it allowed the formation of "better and more unified groups on the basis of indirect reciprocity."[26] Obviously the evolution of basic human cognitive and emotional capacities was a necessary condition for the appearance of morality, but, he argues, the ought-generating "moral sense" was selected in the environments of evolution because it helped its agents obtain their reproductive goals.[27] The fact that this is why it may have originally emerged, of course, does not mean that this is its only significance for us today.

Ayala's approach is supported by Damasio's work. Damasio believes that morality has its origin in the emergence of the capacity for suffering and the growth of human awareness of significant vulnerability to further suffering, which was in turn made possible by an evolved cognitive ability to remember the past, to anticipate the future, and to make and execute plans to bring about desirable consequences. Morality is not proven to be illusory by the sheer fact that it is made possible by brain chemistry, nor is its meaning compromised when it is cultivated by proper social and emotional nurturance in childhood or contributes to a person's survival.

Morality is "natural" but it is not "in the genes," except in the sense that the capacities that allow for morality are based in our biological make-up. Moral codes are transmitted culturally rather than genetically. The body functions in positive ways to support morality. For example, the body's production of the chemical substance oxytocin influences a wide range of behavior, including maternal care-giving and emotional attachment between sexual partners.[28] The performance of altruistic acts can be

---

[26] William Irons, "How Did Morality Evolve?" *Zygon* 26 (March 1991): 49–89, at 67.
[27] Ibid., 60. [28] Damasio, *Descartes' Error*, pp. 122–123.

accompanied by positive emotional states ("positive somatic markers") that express "at any given time, the cumulative preferences we have both received and acquired."[29] Here physiology supports rather than resists emotional bonds and caring behavior.

In the view of Damasio and Ayala, the organic conditions for morality, however, do not dictate any specific moral rules. The "moral sense" should not be conceived on the model of lock-and-key molecular systems or physiological drives such as hunger or thirst. As already stressed in this book, the evolutionary process provides an emotional and cognitive constitution characterized by *general* proclivities, desires, or preferences, not a fixed moral code.

Ethologists understand these proclivities in terms of "prepared learning," that is, as evolved proclivities to learn some things more easily than others. The "open programs" of the human biogram enable the mind to assimilate a tremendous amount of information over a lifetime.[30] A newborn child, for example, learns to recognize her primary caregiver's face more quickly than that of others and to feel attachment to parents more readily than to strangers. Natural proclivities prepare us to learn loyalty to our own group more than to others, to care for close kin more than for strangers, to reward those who cooperate with us and to punish those who violate reciprocity, and to treat others the way they treat us. Life experiences and cultural contexts can diminish or increase the likelihood that these proclivities are actualized.

Evolved predispositions are always mediated to us through a cultural lens. An Ibo chief in Nigeria might experience kin affection very differently from the way a Mormon businessman in Utah experiences it. The flexibility of our "open programming" explains why we should expect such a variety of moral codes throughout history and between cultures. People are not "hard wired" to observe polygamy, or monogamy, or polyandry.

Morality as the product of culture

Holmes Rolston III, in his 1997–98 Gifford Lectures, *Genes, Genesis, and God*, attempts to integrate evolutionary insights into the genesis of human values within a more robust acknowledgment of the cultural and historical basis of morality than one finds in sociobiology. Rolston complains that evolutionary accounts of the genesis of values tend to downplay or even

---

[29] Ibid., p. 199. [30] Mayr, *Towards a New Philosophy of Biology*, p. 26.

ignore the insights that come from "native range experiences, especially our cultural experiences."[31] Intending to offer a theory of values that integrates both biology and culture, he argues that while human values have often been conceived either as purely cultural products or as purely biological products, their creation and transmission has in fact involved both genes and culture. Genes are critical for the evolution of the capacity to value, but genetics does not provide a sufficient explanation of this capacity or of the values that it identifies. Culture thus provides, as he puts it, "a second level of genesis"[32] that cannot be explained by the first.

Since morality is the product of the emergence of mind and the achievements of culture, we must abandon reductionistic efforts to explain moral values in biological terms in a way that amounts to projecting culture onto biology. We must not oversimplify the very complex relation between genes, organisms, species, and their biological environments. Genes do not in and of themselves cause behavior, therefore we cannot think about culture as a domain "controlled" or "explained" by some account of genetics; this does not, after all, even work with human bodies.

Rolston's non-dualistic thesis rejects attempts such as those of E. O. Wilson to explain culture through biology, yet it also repudiates the opposite (and much more common) tendency to regard all human values as completely independent of biology. Human evolution has bequeathed human beings with a flexible, open-ended set of cognitive powers. The significant variation within and between human populations has given rise to the temptation to ignore that we have "dual-inheritance"[33] value systems that rely on genes and culture. Genes constrain behavior in some obvious ways, but they "underdetermine"[34] culture. This point is often missed by the more crude versions of sociobiology that suggest that morality is genetically "blueprinted"[35] onto human nature.

The heart of Rolston's position is that morality, religion, and science rely upon but also "transcend their biological frameworks."[36] This generalization resonates with Ayala's and Damasio's views of evolution of human intelligence as increasingly empowered by emergent capacities for culture, mind, and human intentionality. These emergent capacities enable the moral agent to use his or her biological capacities to think and act in ways that are not dictated, and cannot be predicted, by an analysis of biological or genetic factors.

---

[31] Rolston, *Genes, Genesis, and God*, p. xv.
[32] Ibid., p. xii. [33] Ibid., p. 111. [34] Ibid., p. 121. [35] Ibid., p. 123. [36] Ibid., p. 161.

CHRISTIAN ETHICS

One striking difference between human beings and other animals lies in the domain of morality. Whereas some animals engage in pair bonding, mutual grooming, and other social behaviors, only human beings, Rolston argues, love others for their own sakes and have the ability to act as good Samaritans.[37] Evolutionists like Wilson and Ruse tend to ignore these capacities or treat them as so exceptional as to be of trivial importance. Assuming that people act only out of self-interested motives, reductionists engage in mental gymnastics to argue that all good Samaritans do nothing but pursue their own biological advantage or unknowingly function as dupes for those who do so more efficiently.[38] If we recognize that values are identified and acted upon by human minds, then we are more likely to appreciate their proper status in human personhood and the requirement implanted in our nature that we live out of an "enlarged sense of identity"[39] that identifies with other people and recognizes our responsibility within the natural world.

Christian ethics can profit by recognizing the functional value of morality without presuming that morality is only meaningful for its social functionality. Evolutionists who write about the evolution of morality often do so in a debunking mode, as if morality were nothing but an adaptation or a social convention superimposed on selfish biological human nature. Yet in theory there is no reason why an evolutionary account of the origin of morality should necessarily discredit morality. A scientific explanation of the biochemistry of mother–infant bonding, after all, does not make parental love any less powerful.

Christian ethics can appropriate aspects of these evolutionary perspectives regarding the origin of morality. It is not at all far-fetched to think that evolution has significantly shaped some of the important levels of our emotional and cognitive constitutions as human beings. At what point morality actually emerged from social life is hard to say, and no one has been able to give a convincing argument that completely "explains" the origin of morality. One can speculate that emerging social conventions were based on various kinds of reciprocity, and that these brought a tendency to monitor compliance, to retaliate against cheats, and to reward acceptance of conventions. But, as de Waal points out, it is not possible, at least yet, to identify where fear of punishment and fear of being caught is

---

[37] Ibid., pp. 222, 248, respectively.   [38] Ibid., pp. 252ff.   [39] Ibid., p. 281.

transcended by a more identifiable and distinctive "moral sense" of being obligated by what is right.

Rather than a set of particular normative beliefs or a fixed moral code "engraved" in the psyches of all human beings in all societies, the evolutionary process has created a human emotional and cognitive constitution that includes a set of broad and general proclivities, desires, or "valuational preferences." We inherit evolved proclivities to learn more easily some things rather than others. Natural proclivities play a role in loyalty to one's own group more than to others, readiness for altruism to kin more than to strangers, a willingness to reward those who cooperate and a tendency to punish those who violate reciprocity, and a general desire to treat others the way they treat us. Particular communities at particular times and places attach moral valuation to these preferential tendencies, some channeling parental investment in one direction, toward immediate offspring, others in another direction, toward overlapping care-giving within an extended family.

This accords with the classical position of Aristotle and Thomas Aquinas, who held that each child is born with a range of fairly indeterminate natural abilities, powers, or capacities which are gradually shaped by training, instruction, and habituation to become the adult's "second nature," that is, the virtues or vices that constitute character.[40] Biological predispositions tend to be fairly general in their directionality, as in the notion of "open programs."[41] We are capable of experiencing a variety of basic emotions but they can be subject to human evaluation and direction. Our moral responses to these predispositions can include introspection, criticism, deliberate redirection, and revision of the place they have in our lives.

This generality indicates why we have such a wide variety of moral codes throughout history and between cultures. Our evolved species-wide proclivity to aid closely related kin, for example, takes a wide variety of different expressions in different locales. Moreover, and more disagreeably to sociobiologists, the deeply ingrained species-wide proclivity to maximize inclusive fitness itself is subject not only to delay and redirection but even to abandonment by all sorts of people – missionaries and utopians, artists and poets, prophets and mystics, for example – because of what they consider to be warranted by greater goods.

---

[40] See *Nicomachean Ethics*, 331102b15–1103b25, and *ST* I-II, 49–54.
[41] Mayr, *Towards a New Philosophy of Biology*, p. 26.

The previous evolutionary accounts of the biological roots of morality have five implications that are significant for Christian ethics. The first and most obvious point of contact concerns the natural-law tradition. This tradition holds that the natural inclinations human beings share with other animals are not only biologically significant or interpersonally gratifying, but also morally good when ordered ethically. Evolutionary perspectives underscore the fact that we are psychosomatic unities, not isolated ethereal souls only artificially attached to material bodies. This claim can be developed within a theological understanding of the sacramental significance of ordered human inclinations and an ethical commitment to basic rights of bodily integrity.

This applies to the full range of natural inclinations that underlie our social inclinations to companionship, our desire to know, our sensual desires for food and sex. The good life is guided by reason but allows for the healthy and balanced expression of the full range of our human biological needs and inclinations. Christian ethics can take seriously evolutionary psychological analyses of evolved "psychological mechanisms" and "sexual strategies," both for their descriptive and explanatory insights and for their ability to provide material for ethical reflection on what might constitute positive and negative expressions of human sexuality.

Second, evolved natural emotional and mental proclivities are pervaded by moral ambiguity. They are capable of motivating good or bad character and leading an individual to right or wrong behavior, depending on the agent's intentions and other relevant circumstances. Kin altruism can be good if expressed in ordered parental care and filial loyalty, but can be disordered if it leads to nepotism or xenophobic suspicion of strangers. This moral ambiguity is significantly rooted in the juxtaposition of elementary adaptations in our modern human psyches. The relatively stronger indiscriminate male desire for sexual variety was an evolutionary solution to the crucial challenge of gaining sexual access in ancestral conditions to a variety of women. This evolved sexual tendency may have been adaptive in prehistory but causes a great deal of havoc in modern society. "Men [today] do not always act on this desire," Buss argues, "but it is a motivating force."[42]

Third, understanding more fully the "evolutionary roots of morality" can also serve a critical function. Personal integrity involves evaluating spontaneous objects of desire in light of our comprehensive beliefs

---

[42] Buss, *The Evolution of Desire*, p. 77.

regarding the good life. Evolutionary theory can alert us to obstacles to personal integrity that come from within us by nature (alongside those that come from culture and individual character defects). It alerts us to innate species-wide tendencies to engage in deception, to ignore our own oversights, to minimize our own moral weaknesses and vices, and to justify our biases and those of our friends. "Somatic markers" facilitate decisions in complex social situations but they can be disordered as well, as in uncritical "obedience, conformity, [and] the desire to preserve self-esteem."[43] Bias is felt "in the bones" when a person experiences "unpleasant body states" when she encounters those she finds repulsive, be they mentally ill homeless people asking for aid or an affluent interracial couple on a date. The body's neurally based drive to reduce unpleasant body states can and sometimes does act as a counter-moral force that needs to be held in check.

Evolutionary psychology is particularly adept at attending to ways in which professional advertisers manipulate psychological mechanisms and to ways in which morality can be co-opted to provide ideological support for non-moral ends (e.g. for certain kinds of "mating strategies"). Evolutionary theory strives to get to the roots of many human disappointments and conflicts and thereby to understand more accurately both the depth of the human predicament and the level of commitment required if our conflicts are to be ameliorated more effectively. The conflicted nature of the human emotional constitution reflects the conflicted legacies of the history of natural selection on human nature. Conflicts exist between different biologically based inclinations, between different culturally learned motivations, and between biological and cultural tendencies.

Fourth, understanding the "evolutionary roots of morality" underscores the value of properly training, directing, and tutoring emotions. "Knowledge of the conditions that favor each mating strategy," Buss tells us, "gives us the possibility of choosing which to activate and which to leave dormant."[44] This claim oversimplifies the relation between emotions and behavior, but we can employ knowledge of our emotions to avoid the kinds of conditions that tend to activate what we identify to be undesirable aspects of our evolved "incentive system." Conversely, at least ideally we can deliberately create conditions that elicit desirable kinds of behavior. This is as true of a social ethos and its institutions, or what sociologist Robert Bellah and his colleagues call a community's "social ecology," as it is of personal moral development and individual pursuit of the good life.[45]

---

[43] Damasio, *Descartes' Error*, p. 191.　[44] Buss, *The Evolution of Desire*, p. 209.
[45] Bellah et al., *Habits of the Heart*, pp. 284f.

In Thomas Aquinas' language, then, virtue is not "implanted" in us by nature but formed by habit, and the moral life is a matter of gradually shaping these emotional responses (including what Damasio identifies as their underlying neural machinery) into forms that promote the human good. Damasio's feedback loop involves cognitive processes induced by neurochemical substances, but these are in turn induced by cognitive processes. Habitual action shapes and organizes emotional states and their neurochemical profile.[46] Moral transformation leads not only to a modification of thoughts, words, and deeds, but also, by the repeated action, to reordering neurochemistry, particularly in the prefrontal cortices.[47]

Fifth and finally, understanding the "evolutionary roots of morality" allows us to see more clearly human transcendence of our evolutionary past, a claim more readily appreciated by the "nonreductionist" perspective developed throughout this book. Human beings do not have unlimited freedom, Damasio writes, but we "do have some room for such freedom, for willing and performing actions that may go against the apparent grain of biology and culture."[48] Clearly, accounting for the origin of *some* human values, even central values, does not exhaust the full range of *all* human values.

We have the capacity to act in ways that transcend what accords with our fitness interests. Ethical ideals of universal love, the golden rule, renunciation of violent retaliation and revenge, disinterested regard for others, love of enemies, and solidarity with the poor call people to high standards. Our conduct may reflect the influence of the remote evolutionary past of our species, but we can strive for greater nobility than would be encompassed by natural selection.

At this point it might be worthwhile to consider objections to this proposal. The first objection is that this reading of the "evolutionary roots of morality" is guilty of circularity and a highly selective appropriation of evolutionary theory. Following the path set out in the earlier chapters, it seems to me that one can selectively appropriate neo-Darwinism without also accepting epistemological, ontological, or moral reductionism. I would like to aim at a non-vicious circularity between my reading of evolutionary theory and my appropriation of it within a wider Christian theological perspective.

A second criticism is lodged by those who might argue that since evolutionary theory provides an account of the origin of morality, it cannot have a divine origin. Morality is generated by emotional predispositions

---

[46] Damasio, *Descartes' Error*, pp. 149–150.   [47] Ibid., pp. 182–183.   [48] Ibid., p. 177.

that themselves were caused by natural selection and random variation – and not by God. Ruse claims that the very fact that we feel a powerful desire for transcendent justification for our morality is an "illusion of our genes" that is itself fitness-enhancing. If we seek to know the truth rather than simply justify ourselves, Ruse argues, we ought to employ a scientific understanding of the "evolutionary roots" of morality and throw out the theological understanding of God as the source of morality.[49]

Ruse notes that the natural-law tradition does not regard Christian ethics as flowing from a divine will that is capricious or arbitrary. "God wants what is right, God wills what is right, God demands what is right . . . Morality is part of the nature of things."[50] It seems to me that the claim that evolution has given rise to our "moral sense" – either a direct adaptation or as a by-product – need not compete with "support" from the divine will. Even Ruse admits on this point that "evolution rather clarifies and justifies the Christian position than detracts from it."[51] Ruse goes too far when he suggests that knowledge of evolution might provide ground for either "clarifying" or even "justifying" (his version of) Christian ethics, but he recognizes that the Christian tradition has held that God's will is expressed in the natural order.

Ruse also realizes that the natural-law tradition understands moral claims as warranted not by arbitrary divine dictates but by their contribution to human flourishing. This acknowledgment suggests that we ought to avoid artificially separating divine and natural causality. The set of scientific hypotheses and insights regarding the "evolutionary roots of morality" does not make impossible their religious interpretation, including the claim that God orders the world through the evolutionary process. At times those who insist on understanding human life in terms of nature rather than God naïvely assume a simplistic anthropomorphic image of God. As emphasized in chapter four, this God is an idol, a false god, a mere being in the world. The God of Christian ethics continually sustains the world in being and orders it through the processes and patterns of nature. It is therefore theologically improper to assume that nature is ordered *either* by God *or* by the evolutionary process.

---

[49] See Michael Ruse, "Evolutionary Ethics and the Search for Predecessors: Kant, Hume, and All the Way Back to Aristotle?," *Social Philosophy and Policy* 8 (1990): 65; see also Michael Ruse and E. O. Wilson, "The Evolution of Morals," in Michael Ruse, ed., *The Philosophy of Biology* (New York: Macmillan, 1989), pp. 313–317, and Jeffrie Murphy, *Evolution, Morality, and the Meaning of Life* (Totowa: Roman and Littlefield, 1982).
[50] Ruse, *Can a Darwinian Be a Christian?*, p. 168.   [51] Ibid., p. 202.

From the perspective of Christian ethics, then, it is a mistake to force a choice of either religious or biological "roots." Biological theories compete with other biological theories and not with theology, unless theology offers biological theory (in which case it is no longer theology) or, more likely, relies upon inaccurate biological assumptions. Biblical creationism obscures this distinction in treating Genesis and evolutionary theory as alternate scientific theories. In doing so, creationism ironically joins the skeptics in forcing an unnecessary choice between well-established scientific theory and biblical revelation.

So while it is true that the reductionistic model that equates evolutionary roots with the "essence" of morality will not be acceptable to believers, the "spiritualistic" view that regards God rather than nature as *the* "root" of morality is equally suspect. Neither position allows for the possibility that God works in and through the intrinsic ordering of human nature. A chastened and balanced approach to the "evolutionary roots of morality" examined by scientifically informed sources is not only compatible with Christian ethics, but also helps to illumine the human nature that is divinely created, habituated in the moral life, denigrated by sin, and healed by grace. These theological claims are not and cannot be justified by evolutionary theory on its own terms, at least when it functions in the domain of science with its own proper standards and procedures. The fact that morality emerged as a result of evolution does not necessarily imply that it must be illusory. These kinds of theological claims are apprehended in religious faith, though optimally not by a blind faith but rather one that sincerely appreciates and humbly accepts the insights into the "roots of morality" provided by evolutionary theory.

We can now turn to five implications of this appropriation for Christian ethics. First, Christian ethics has long held that there is a natural basis for morality, one derived from our condition and constitution as creatures, along with an evangelical call that asks the Christian to go beyond the inclinations of nature. Christian ethics, of course, rejects the claim that morality is "nothing but" an evolutionary enabling mechanism. Moral claims sometimes require behavior that runs counter to fitness. Yet grace elicits what is best in human nature while resisting and correcting its historical and existential deformities. The doctrine of redemption concerns the transformation of human nature, not its elimination. Genuine moral commitment, for example, signals trustworthiness and induces cooperation and reciprocity on the part of others. Moreover, it also provides the most stable basis for familial attachments. Even if morality is an enabling mechanism on the functional level, this does not mean that individual moral agents are deceptively using morality to promote their own selfish ends.

Second, Christian ethics does not accept particular moral standards on evolutionary grounds, but the fact that they are at times "fitness enhancing" need not be surprising. Historical awareness of the plurality of moral codes, both within and outside of Christianity, calls into question how they can all be adaptive. Evolutionists, though, hold that different moral codes are adaptive for different cultural settings, that, for example, a relaxed acceptance of sexual promiscuity may be adaptive in some Polynesian communities while a strict monogamy would be more adaptive in Puritan New England. The same could be said of slavery, property rights, or gender relations. This kind of general claim, though, especially in the absence of empirical confirmation, exhibits circular reasoning based on the a priori theoretical presumption that whatever behavior exists must be adaptive.

Third, Christian ethicists can accept the very general statement that morality emerged as a feature of human communities as a result of the evolutionary process. The evolutionary indirect by-product approach is more congenial and persuasive to most Christians because it acknowledges the complexities of morality.

Fourth, Christian ethicists can certainly accept the generalization that morality is often group-serving or even self-serving. The theory that morality constitutes in large part a means for social control for the common good is commonplace in social theory. One of the unique features of morality as a human practice is that it is internalized, made part of the agent's internal motivational system. Sober and Wilson are right to stress the fact of human motivational plurality, a plurality that includes the capacity to take another person's good, or the good of the community, as an end in itself. This capacity evolved as a more successful strategy than one focused narrowly on only the well-being of the self.

This is not something that is discovered for the first time by evolutionary theory, however. It is already seen in ordinary life experience, literature, and religious teachings. Evolutionists claim to offer a special explanation for this narrowness but this claim jumps from the genotypical to the moral dimension of acts without proper justification. "Selfish genes," if interpreted in a properly modest, circumscribed, and metaphorical way, do not necessarily entail selfish people or selfish groups.

Fifth, Christian ethics would seem willing to accept the sociobiological doctrine that people tend to be selfish, both as individuals and as groups. This could be interpreted as one expression of the doctrine of original sin. Yet the Christian tradition distinguishes the condition of finitude and the premoral biological tendencies from the sinful selfishness generated by disordered choices and habits. Christian ethics has always been aware of the widespread human tendency of individuals to prefer their own interests to those of others, to be biased toward members of their own groups and away from outsiders, and to rationalize self-serving behavior through morality. It also recognizes kin bias, in-group favoritism, and out-group prejudice. Human nature as we experience it in ordinary life is much more likely to be expressed in balanced reciprocity than with a willingness to "give without counting the cost." We are spontaneously more willing to care more for our own children and much less for those who are more remote, and willing to subordinate the well-being of other groups to that of our own groups. Christian ethics preaches ethical universalism – that every person is a neighbor – but it seems to be the case that the actual practice of real people, including those who are churchgoing, is closer to the world described by sociobiology than it is to the way of life depicted in the Sermon on the Mount. Yet Christian ethics understands the roots of immorality in religious rather than natural terms – the distortion of the will and intellect by original and personal sin – and its awareness of human fault highlights the importance of transformation and self-discipline for the Christian life.

CHAPTER 11

Natural law in an evolutionary context

Chapter two criticized several major moral theologians from the natural-law tradition for ignoring human evolution. I now turn to four prominent moralists from that same tradition who do take seriously our evolutionary origins, namely Alasdair MacIntyre, Larry Arnhart, Jean Porter, and Lisa Sowle Cahill. After this analysis I argue that knowledge of evolution confirms some aspects of Thomistic natural law and calls for a revision of others.

NATURAL LAW AND HUMAN EVOLUTION: RECENT EFFORTS

Knowledge of human evolution calls for a change in the way we think about biological nature in general and human nature in particular. It brings a heightened awareness of the presence of tension and conflict among particular goods and virtues, and the consequent need for concrete moral decisions that involve trade-offs, denial, and sacrifice of some goods for the sake of others. Human evolution has left us with a heterogeneous, mixed, and conflicted set of human inclinations and the goods toward which they move. Human nature is in some ways at odds with itself, divided by conflicting inclinations that are brought into working order, and not complete harmony, only through intentional effort and discipline.

I have argued in this book that knowledge of human evolution can complement the moral and anthropological commitments found in the Thomistic approach to natural law. This tradition emphasizes our natural inclination to various kinds of goods, from survival and sustenance to sociality and friendship. These inclinations are based in our evolved species-wide emotional and cognitive constitution. At the same time, evolutionary considerations alone account for neither the full range of human goods nor their properly human significance. The genesis of an inclination necessarily accounts neither for what it can mean for human beings now nor for how it ought to function in personal moral decision-making.

Emergent emotional and cognitive capacities allow us to act for the sake of higher-order goods in ways that are neutral or even negative with regard to genetic fitness. The human heart and mind can take on lives of their own and, though we will always need biological goods, we are not driven by a quest for optimum genetic fitness. While acknowledging our need to meet the biological level of human needs, we can and do also strive for higher social, psychological, moral, and religious goods. The human person can struggle to orient his or her life to what is genuinely good in itself, and not only good for his or her phenotype or genotype.

This potential can be described as the capacity for "self-transcendence," which for Rahner means a movement out from the self toward others in understanding and love.[1] We transcend ourselves when we renounce selfish inclinations, feel compassion for a person who is suffering, maintain integrity in the face of temptation, and countless other ways. Theologian Walter Conn holds that self-transcendence can take place in cognitive, moral, and affective ways, and that "every achievement of creative understanding, realistic judgment, responsible decision, and generous love is an instance of self-transcendence."[2]

Evolutionists display cognitive self-transcendence to the extent to which their actual intellectual performance proceeds as a search for truth rather than as an exercise in defensiveness, propaganda, or ideology. Sociobiologists profess to offer a theory of behavior that is true and not simply reproductively useful. If the human mind is "anything but a mechanism set up to perceive the truth for its own sake,"[3] as one sociobiologist puts it, then we can ask on what grounds this statement itself should be taken as truthful. Despite Dawkins' disclaimer, sociobiological discussions of the "selfish gene" are not purely descriptive but rather function in part to shed light on the cause of evil and suffering so that people can come to a better way of treating one another. "Let us try to *teach* generosity and altruism," Dawkins exhorts, "because we are born selfish."[4]

---

[1] See Karl Rahner, "Christology within an Evolutionary View of the World," trans. Karl H. Kruger in *Theological Investigations*, vol. v: *Later Writings* (New York: Crossroad, 1966), pp. 157–192; and "The Unity of Spirit and Matter," pp. 153–177. Rahner's theory of self-transcendence is immensely complicated and its proper explication would require more than can be given here.

[2] Walter Conn, *The Desiring Self* (New York: Paulist Press, 1998), p. 36. Conn relies on Bernard Lonergan but his claim is at this point consistent with Rahner's account of self-transcendence as well.

[3] Michael Ghiselin, *The Economy of Nature and the Evolution of Sex* (Berkeley: University of California Press, 1974), cited in Daly and Wilson, *Homicide*, p. 261.

[4] *The Selfish Gene*, p. 3; emphasis in original text.

Natural law and evolutionary ethics are thus diametrically opposed when it comes to the significance of morality and religion for human life. Whereas evolutionists believe moral appeals and religious ideals serve the goods of survival and reproduction, Christian ethicists regard the latter as important elementary concerns that are subordinate to the love of God and love of neighbor.

The most widely persuasive evidence of our inclination to the good comes not primarily from science but from first-person experience, both in a "reflexive" presence to one's own internal life and, more importantly, the experience of having reverence for, coming to understand, and caring for others. As Rahner points out, the most important knowledge of human existence is not to be acquired by the observations and theories of the natural sciences, as valuable as they are for grasping important dimensions of what we are as members of our species, nor simply by reference to the unique experience of the self in a given concrete moment, but by human life as it is lived in the flow of history and as moving toward the future. In human experience we encounter both the movement of grace and the human desire for the good, which, Rahner and other theologians have long argued, is always also an invitation to respond to the "Absolute Mystery."[5]

The exclusively behaviorist focus of sociobiology and evolutionary psychology claims that "self-reports" and all other first-person approaches to human experience are unreliable, radically defective, or even illusory. The restraint placed on concrete first-person experience might be appropriate for strictly scientific approaches to human behavior, but the evolutionists combine it with a further epistemological and ontological reductionism that rules out other ways of viewing this subject matter.

Awareness of the diversity of human goods allows the natural-law tradition to value disciplinary pluralism. The core of the human conscience calls each of us to orient our lives to what is good and true and as such supports an appropriate sense of humility. An essential aspect of humility includes openness to information about human behavior that can be provided by the sciences about the empirical constituents of human behavior. Knowledge of the human good comes in a privileged way from Scripture and tradition, but it also comes from all the intellectual disciplines that provide a greater grasp of what is real.

---

[5] See Rahner, *Foundations of the Christian Faith*, ch. 2.

Natural law turning to evolution: MacIntyre

MacIntyre's *After Virtue* regards the pre-modern backing given to natural law from Aristotelian "biological teleology" to be indefensible.[6] Since any account of the virtues rests on some account of the human good, MacIntyre argued, concrete communities provide conceptions of the human good to be pursued through the particular virtues that they uphold as central to their notion of morality. Goods internal to practices can be distinguished from the good of a whole life and both of these can be distinguished from the good for human beings as envisioned by a social tradition. The foundational narratives of an ongoing social tradition provide a conception of the human good for those who participate in the defining practices of its way of life.

In *After Virtue*, MacIntyre viewed the moral agent as one who achieves coherence by participating in a personal narrative that is rooted in a rich community narrative and its practices. This kind of coherence would appear to be quite loose when attempting to create a thread of continuity between multiple social roles that characterize the self in modern society. MacIntyre rejected Aristotle's attempt to ground his conception of the virtues in a comprehensive metaphysical account of the human good and replaced it with a "historicized" account of human moral formation. MacIntyre sought to retain a socially teleological aspect of morality, but one based on culture rather than biological nature. A person's orientation to the good comes not through nature but through the narratives of living traditions.

MacIntyre's next major work, *Whose Justice? Which Rationality?* moved beyond Aristotle to Thomas Aquinas and thus to a greater appreciation of natural law and human nature.[7] His more recent writing on *Dependent Rational Animals: Why Human Beings Need the Virtues*[8] employs naturalistic observations about the functions of animals as a context for reflecting on natural purposes. "I now judge that I was in error in supposing an ethics independent of biology to be possible," he confesses.[9] He comes to recognize that

---

[6] MacIntyre, *After Virtue: A Study in Moral Theory* (second edn., Notre Dame, IN: University of Notre Dame Press, 1984 [1980]), p. 58.
[7] Notre Dame, IN: University of Notre Dame Press, 1988.
[8] Chicago and LaSalle, IL: Open Court Press, 1999.
[9] *Dependent Rational Animals*, p. x.

no account of the goods, rules and virtues that are definitive of our moral life can be adequate that does not explain – or at least point us to an explanation – how that form of life is possible for beings who are biologically constituted as we are, by providing us with an account of our development towards and into that form of life.[10]

Education takes place through the development of "language-using powers" that are founded upon various perceptual and cognitive capacities which we share with other species.[11]

Moral training structures our responses to our bodily conditions and allows us to gain a reflective distance from our response to the "goods of our animal nature."[12] At the same time, MacIntyre argues, "what we thereby become are redirected and remade animals and not something else."[13] This biological turn in MacIntyre led to a departure from what he took to be Aristotle's strong emphasis on rationality and linguisticality as what separates humans from all other animals. MacIntyre is convinced that recent scientific discoveries regarding animal intelligence make it more difficult to draw such a sharp line here and lead us to appreciate greater continuities than could be seen by Aristotle. Higher animals, for example, function within a world of significance and with certain degrees of understanding – "Dolphins, gorillas, and members of some other species are no more merely responsive to the inputs of their senses than we are. They too inhabit a world whose salient features can have this or that significance for them."[14]

In this book MacIntyre relates distinctively human rationality to animality and dependence, but he tends to refer to biology only to underscore the reality of human vulnerability and disability in relation to which the "virtues of acknowledged dependence"[15] must be developed and exercised. His interest in animality is usually restricted to various forms of weakness, deficiency, incapacity, and states of suffering some kind of harm. MacIntyre thus begins but does not complete the development of a new teleological view of human nature upon which a revitalized natural law could be built. He recognizes that intelligent animals inhabit worlds of meaning. He also knows that they not only respond to the behavior of other animals but also interpret the actions and purposes of human beings who live in their worlds, and that they respond to their interpretations of these actions and purposes with purposeful behavior of their own. Yet he does not make as much of this commonality as he might.

---

[10] Ibid. [11] Ibid., p. 48. [12] Ibid., p. 49. [13] Ibid. [14] Ibid., p. 60. [15] Ibid., p. 8.

Human beings have the advantage of being able to "stand back" at a reflective distance from their settings, make judgments about goods that have some detachment from their immediate desires, and do so in terms of memory about lessons learned in the past and plans for the imagined future. The natural human life cycle begins with a childish stage of quasi-animal dependence, gradually moves toward mature and independent adulthood marked by practical reason and then, with the aging process, moves back into stages of greater and greater dependence. MacIntyre's reading of this life course moves him to encourage the development of virtues both toward and from the dependent. The former include generosity, justice, patience, and especially a concern for the common good of all members of the community; the latter make it possible to learn to receive what one needs from others in gracious and virtuous ways.

Aristotelian natural right appropriating evolution: Arnhart

Political theorist Larry Arnhart offers the most extensive effort to appropriate evolutionary knowledge for an "Aristotelian Darwinism."[16] Much more explicitly rooted in evolutionary theory than MacIntyre, Arnhart categorizes interpretations of Darwinism into two major philosophical forms: Aristotelian or Hobbesian. Sociobiology, characterized by a largely Hobbesian point of view, holds that because we are naturally egoistic, asocial animals, ethics is most properly understood as an effort to correct for the excesses of dominant egoism and to create conditions that allow for reasonable degrees of social order in spite of the highly competitive nature of human interactions. Arnhart argues that Aristotelian Darwinism is both more faithful to Darwin and more ethically insightful than Hobbesian Darwinism.[17] Darwin's appreciation of human sociality is consistent with Aristotle's view of human beings as "political animals." Arnhart's Darwinian Hobbesianism opposes the tendency of many modern ethicists to regard the moral will as obliged to suppress the animal dimensions of human nature. He is especially critical of Kant's claim that moral goodness lies only in the will, a "supersensible faculty" for moral freedom.[18]

---

[16] Arnhart, *Darwinian Natural Right*, ch. 1.
[17] James Lennox also holds that Aristotle's account of moral agency is consistent with Darwin's account of the origin and function of the moral sense. See his "Aristotle on the Biological Roots of the Virtues: The Natural History of Natural Virtues," in Jane Maienschein and Michael Ruse, eds., *Biology and the Foundations of Ethics* (New York: Cambridge University Press, 1999), pp. 10–31.
[18] Arnhart, *Darwinian Natural Right*, pp. 83f.

Arnhart maintains that the Kantian expectation that reason and culture can liberate us from animality leads to the unfortunate assumption that science, which deals merely with the body, cannot provide any insights of value for ethics. He turns to Darwin's *Descent of Man* to provide an alternative to this bifurcation of nature and culture by arguing that many of our central capacities, and particularly the social aptitude for compassion, cooperation, and sympathy, are rooted in our animal nature.

Aristotelian Darwinism takes its starting point in the fact that all animals desire to meet their needs. Morality is a distinctively human institution that meets some important needs of our species, rooted in our wide-ranging, flexible emotional repertoire and cognitive structure. If the good is the desirable, and the desirable is what human beings generally desire, Arnhart argues, then we can discover the content of the good from learning about what human beings have generally desired in the past. We tend to think of the "past" in terms of history, something like the previous ten thousand years or so, but evolutionary biology suggests that recoverable human history is only a small segment of the relevant past. The most universal and persistent human desires are those we share with prehistoric human beings. Arnhart's catalogue includes desires for social status, wealth, political rule, reciprocity, and (more controversially) war. The collection of these desires in human nature means that our experience will be marked by a wide array of conflicting desires on all levels – internal or intrasubjective, interpersonal, communal, and intercommunal. Pragmatic consensus over common interests can contribute to the resolution of some of these conflicts, but unfortunately evolution does not provide any ethical sources for thinking about how to bring some moral order to these desires, or how to adjudicate these conflicts. If every satisfaction of a natural human desire is good, one critic points out, then Arnhart is not entitled to render a moral condemnation of slavery because it is produced by what he calls our natural human desire for dominance.[19]

Many animals engage in social learning, but human beings alone consciously aspire to shape their lives to reflect their moral ideals. We alone have the kind of cognitive capacities, including foresight, memory, and planning, that allow us not only to identify goals but also to make more encompassing "life plans" and to assess our behavior in light of these goals and plans. These capacities for reflection, deliberation, planning, and

---

[19] See John Hare, "Is There an Evolutionary Foundation for Human Morality?" in Philip Clayton and Jeffrey Shloss, eds., *Evolution and Ethics: Human Morality in Biological and Religious Perspective* (Grand Rapids, MI: Eerdmans, 2004), p. 195.

choice are developments of capacities that we share with some other intelligent animals.

Arnhart finds support for his Aristotelian Darwinism in Frans de Waal's research on primate "proto-morality."[20] As we saw above, de Waal gives convincing evidence that evolution has produced the elemental requisites of morality in primate sociality, sympathy, care-giving, and norms of group living. These traits are precursors of their more sophisticated forms in human evolution, scientific knowledge of which, Arnhart argues, provides an empirical basis for an updated Aristotelian ethic. Aristotle held that ethics and politics allow us to pursue our desires for human goods most effectively through virtue.

Particularly important for Arnhart are Aristotle's doctrine of the virtues and his interpretation of natural right. Fusing Aristotle and Hume, he argues, allows us to consider the virtues as ways of tutoring and expanding our natural moral sense, which includes our sense of fairness, our proneness to become indignant in the presence of injustice, and our tendency to feel guilty after an experience of moral failure. Our natural capacity for sympathy is complemented by a tendency to feel self-love and to develop love for friends and family.

Central to this project is the link drawn between the moral sense tradition in which Darwin was trained, Aristotle's notion of natural right, Humean sympathy, and Aristotelian sociality. Both Darwin and Aristotle held not only that human beings are persuasively social but also that our natural sociality is biologically grounded first in families and then in more extensive social groups. Arnhart argues that we can judge the moralities of various cultures according to how well they support natural sociality and allow for the expression of our native cognitive capacities. He is quick to add the caveat that a negative judgment does not, in and of itself, prescribe a corrective course of action; the virtue of prudence addresses the question of how a particular culture can better meet these goals.

Arnhart also argues that a critical appropriation of Aristotelian Darwinism can complement and update Thomistic natural law by accounting for the origin of our inborn desires and mental capacities.[21] While natural science cannot confirm the biblical basis of Thomas' ethics, it can testify to "the natural truth of biblical religion in its practical morality."[22] Evolutionary

---

[20] Arnhart, *Darwinian Natural Right*, pp. 66–67, 78–79, and *passim*.
[21] See Arnhart, "Thomistic Natural Law": 1–33.
[22] Larry Arnhart, "The Darwinian Moral Sense and Biblical Religion," in Clayton and Schloss, eds., *Evolution and Ethics*, p. 212.

accounts of group-selected altruism, for example, confirm Thomas' claims about the "natural moral truth" of the priority of the common good.[23]

Darwinism offers an account of how these desires came to be part of human biological nature. Some of the desires, such as that for sexual coupling and child-raising, were modified forms of what was inherited from our pre-human primate ancestors. Awareness of the demands of reciprocity and cooperation were transformed over time into a sense of justice and fairness. The basic moral awareness of the distinction between right and wrong became, he writes, "etched into the neural circuitry of the human brain."[24] It is hard to understand whether Arnhart here means our formal sense of right and wrong, the content of which is "filled in" by learning and culture, or whether he intends the stronger claim that our moral sense actually includes substantive content regarding good and bad kinds of acts. In any case, he argues that moral norms governing marriage and family build upon and give direction to our natural desires (Arnhart's "natural desires" may not be the same as Thomas' "natural inclinations," since for the latter you can have no conscious desire for that to which human nature is inclined). In any case, his main point here is that appropriate moral laws satisfy some of the core yearnings of human biological nature. Indeed, Arnhart also believes there is a broad agreement between Thomas Aquinas' view of the moral agent and the sociobiology of the moral sentiments: "much (if not all) of what Aquinas said about the natural inclinations supporting natural law would be confirmed by modern biological research."[25] Arnhart thinks that E. O. Wilson provides the basis through which Thomistic natural-law theory can be further "expanded and deepened."[26]

Arnhart's position faces a major objection from those who would argue that he underestimates the significance of history and the significant differences between the moral codes of various cultures. While moral consensus can be formed within particular communities shaped by their distinctive narratives, this agreement becomes thinner and less helpful as the differences between cultures increase. Moral consensus is achieved in small, tight-knit communities (ideally the size of hunter-gatherer bands but perhaps up to the size of the *polis*). But widespread consensus is impossible to obtain in modern, large-scale, highly populated, pluralistic societies. Pluralistic societies and international agreements between nation-states might be able to share legal agreement on some ethical prohibitions such as those pertaining to slavery, rape, and torture, but their

---

[23] Ibid.   [24] Arnhart, "Thomistic Natural Law": 28.   [25] Ibid., 28.   [26] Ibid., 26.

interpretation and enforcement are by no means easy and uniform. Agreement over purely formal imperatives such as "be loyal," "be just," and "be fair" tell us nothing about exactly what kinds of behavior are required by loyalty, justice, and fairness in specific circumstances. Against Arnhart, critics would emphasize the fact that people across different cultures have varying "moral senses" that tell them different things about justice in male–female relations, property rights, and the responsibility of businesses to their employees. Science cannot provide a basis for moral consensus when disagreements are not over facts but over their moral significance or moral implications.

Arnhart is vulnerable to the criticism that his attempt to "Darwinize" Aquinas ignores the differences between Darwinian and Thomistic-Aristotelian notions of nature. Evolutionary theory takes the term "natural" to mean "happens in nature" with a certain frequency under given conditions. It also takes "natural" to mean "adaptive": it occurs regularly in a species because it serves fitness interests. Species-wide natural inclinations represent adaptive traits – traits that were retained because of their capacity to contribute to fitness in the past environments of evolutionary adaptation. Mothers naturally care for their children, doing so is adaptive, and one finds most mothers in most cultures caring for their children. Human beings are inclined by nature at particular times and places to cooperate, to care for others, to tell the truth, and to protect children. But at other times, at least under some circumstances, it is adaptive to cheat, not to spend energy on others, to lie, and even to commit infanticide. From an evolutionary standpoint, these behaviors are all "natural" in appropriate contexts. Acting "according to nature" in an evolutionary sense may be immoral from a human standpoint.

The Thomistic critique of Arnhart focuses on his own form of biological reductionism. While he is right to point out that Darwin resonates with Aquinas' recognition of the fact that human beings share with other animals a natural ordering to biological kinds of goods, Arnhart departs from Thomas in his belief that these are the most important of human goods or that other goods, including rational goods, can be explained in biological terms. Natural inclinations include but go beyond the biological order. Aquinas considered as rational inclinations our natural desire to know what is true, to live in an ordered political community, and to know God. The human good includes rational as well as natural and sensitive goods, and in fact the latter are ordered to the former.

An elaboration of Thomas' understanding of the person as a "social animal" leads us to appreciate more explicitly the social goods that inspire rich patterns of relationships within communities. Social goods can also be called "goods of social relationship" or "social participation" and they include "gender equality, education, employment, religious membership, and political participation both for individual citizens and for nations and other collective agents."[27] Instead of a "flat" view of human nature characterized only by biological goods and goods of biological significance, it is more appropriate to presume a multilayered and hierarchical view of the human good within which particular goods of the body are subsumed. Craig A. Boyd rightly charges Arnhart with not taking into account Thomas' distinction between natural goods and material goods (a subcategory of natural goods), and between biological goods and rational goods.[28] Arnhart's "naturalization" of ethics ignores completely the Thomistic view of human life as a journey to God. As a result, Arnhart offers moral theory that is closer to Hume than Thomas. Boyd has less confidence than Arnhart does that sociobiology provides a reasonable explanation for either human behavior or morality.

Christian natural law appropriating evolution: Porter

Quite a different approach to natural law, one more attentive to both virtue and the complexity of human nature, has been developed by Jean Porter.[29] Porter appreciates the theological conception of human nature that runs throughout the scholastic tradition of natural law. This theological conception of creation accounted for the links between human and nonhuman animals and for the continuities between rational and nonrational creatures.[30] These inclinations are morally significant even though they are not morally determinative. Porter objects to any theory of natural law that minimizes or even denies the normative significance of nature or that reduces the normative significance of human nature to practical rationality; practical reason does not exhaust the natural law.[31]

---

[27] Lisa S. Cahill, "Toward Global Ethics," *Theological Studies* 63.2 (June 2002): 339.
[28] Craig A. Boyd, "Was Thomas Aquinas a Sociobiologist? Thomistic Natural Law, Rational Goods, and Sociobiology," *Zygon* 39.3 (September 2004): 659–680 at 677. See also his excellent chapter, "Thomistic Natural Law and the Limits of Evolutionary Psychology," in Clayton and Schloss, eds., *Evolution and Ethics*, pp. 221–237.
[29] Porter, *Natural and Divine Law: Reclaiming the Tradition for Christian Ethics* (Grand Rapids, MI: Eerdmans, and Cambridge: Novalis, 1999).
[30] Jean Porter, *Nature as Reason: A Thomistic Theory of the Natural Law* (Grand Rapids, MI: Eerdmans, 2005), p. 13.
[31] See ibid., p. 35.

The Christian doctrinal affirmation of the goodness of creation carries profound consequences for ethics, including, among other things, its recognition of the positive if limited value of possessions and sexuality. Religious affirmation of the creation of every human being in the image of God established the most significant kind of grounding for common human dignity, moral agency, ethical responsibility, and fundamental human equality. Rather than mere doctrinal supplements to an essentially rationalistic theory, these influences profoundly shaped the natural-law tradition. Instead of an abstract, ahistorical conception, as some critics charge, the scholastics employed the notion of natural law in a flexible, context-specific manner that provided ethical insights about developing forms of social life as well as a basis for adjudicating among competing positions lodged by theologians and canonists.

If natural-law principles were flexible and broad enough to be adapted to changing social circumstances in twelfth- and thirteenth-century Christendom, Porter argues, they continue to be flexible and broad enough to be applied fruitfully to contemporary moral problems. She does not replicate the traditionalist assumption that moral arguments can be adequately addressed by simply reminding readers of the moral positions of great authorities of the past. The scholastics believed that human beings have a "species-specific nature" that is both knowable and morally significant; this nature includes the considerable pre-rational, biological roots of human nature. Porter insists that ethicists critically appropriate contemporary scientific insights regarding the animal dimensions of our humanity. Just as Thomas employed Aristotle's notion of nature to update his theory of natural law, so contemporary natural-law ethicists need to incorporate evolutionary accounts of human origins and human behavior. Porter interprets natural law in a way that is both recognizably theological and yet open to scientific and other intellectual perspectives.[32] She argues that "contrary to what is commonly assumed, the scholastics' views do not conflict with the doctrine of evolution, or more generally with the biological sciences."[33] By investigating the basic intelligibilities and forms of ordering found in the natural world, science provides data for theological reflection on the structures of creation.[34]

A moral tradition is intellectually vital to the extent that it engages in this kind of critical extension. Porter is closest to Arnhart when she recognizes the importance of our evolutionary past, but she is farthest from him when she emphasizes the theological and scriptural basis of natural law. Though

---

[32] Ibid., p. 5.   [33] Ibid., p. 51.   [34] Ibid., p. 57.

biology constitutes an important substrate of human nature, the person is more than a biological being. The full range and depth of human aspirations and moral ideals cannot be explained biologically. Like Grisez and Finnis, Porter is alert to the danger of falling into the "naturalistic fallacy" faced by those who seek to derive ethical principles from the evolutionary process. Unlike Wilson and Dawkins, she knows that scientific knowledge of nature is a necessary but not sufficient source for understanding morality and our place in nature.

The virtues give moral order to natural inclinations and desires that are themselves part of God's creation. Porter distances herself from the Stoic axiom that the good life is lived "according to nature," because she wants to distinguish clearly between aspects of nature that ought to be suppressed from those that ought to be encouraged, especially from a Christian perspective. Far from being merely decorative, Scripture provided the scholastic natural lawyers with criteria for determining which aspects of human nature have normative value and which do not.

The Christian belief that each person is addressed by biblical injunctions, free to choose to obey or to disobey moral norms, and directly accountable to God generates a strong emphasis on moral agency and responsibility. The natural-law tradition gives special attention to the intellectual capacity of moral agents to comprehend the basic requirements of the moral law. The scholastics held, in other words, that the norms of the natural law were fundamentally intelligible and believed that people generally are aware of the basic goods of human life.

Porter's historically based account of natural law leads her to claim that there is no one universally credible way of construing the moral significance of human nature. Since morality is "underdetermined" by nature, she argues, "there is no one moral system that can plausibly be presented as the morality that best accords with human nature."[35] This claim would seem difficult to sustain within Christian ethics, since natural law as Porter understands it is most in accord with divine wisdom and best orders us to our final end. Perhaps it should be read as suggesting that every moral system is shaped by its own presuppositions, and that these presuppositions in turn establish the moral framework within which a given "ethical system" is interpreted. Epicurus, Mill, and Nietzsche all proposed alternative ways of basing morality on nature and of identifying what is normative in human nature. Yet though none of the contemporary disciples of these moralists accept the Christian presuppositions that provide the framework for natural law, surely

[35] Porter, *Natural Law and Divine Law*, p. 141.

Porter is committed to holding that natural law grounds the morality that "best accords with human nature."[36]

This contextual approach to Christian ethics should not be taken to imply that we ought to regard the natural law as just one theory among others. Advocates of Christian natural law such as Porter hold that natural law, the golden rule, the prohibitions of the second table of the Decalogue, and the moral protections implied in these injunctions are more reasonable than are contrary positions.[37] This claim can of course also be upheld by positions that accept neither Christian theological premises nor natural-law theory in a broader sense. Porter's main point here is that there is no universal theory standing above history in judgment of all particular moral perspectives. Yet the faith-based perspective of Christian natural law does affirm the Christian interpretation of human flourishing and the means to it.

Rowan Williams affirms: "There is the single authoritative form of human flourishing, liberty before God and full response to God, the Logos made flesh in Jesus ..."[38] The Christocentric interpretation of flourishing involves no "flight from reason" or retreat into particularistic relativity. As Williams explains, "we speak of Jesus' existence as a divine act from first to last because it is recognized as having a potential for bringing together the whole of the world we know in a new unity and intelligibility."[39] One who claims that Christ alone provides the *complete* answer to the question that is human life implicitly proposes this position as the most adequate account of complete human flourishing, at least on Christian terms. The concern with *human* flourishing points to its universal or species-wide dimension and to the necessity of acknowledging the natural law, and the fact that true flourishing ultimately involves becoming "children of God" means that it will be oriented, in some substantive way, to union with Christ (even if not with an explicit identification with the institutional church or even the Christian tradition). (The last part of this chapter will address the question, raised by Lisa Cahill, as to whether interpreting natural law as an expression of cultural and theological particularity can provide the conceptual basis for cross-cultural moral dialogue.)

Porter's project is especially interesting for this book because of the way in which she attempts to incorporate evolutionary theory into natural-law

---

[36] Ibid.
[37] See Lawrence Dewan, OP, "Jean Porter on Natural Law," *The Thomist* 66.2 (April 2002): 275–309, at 293.
[38] R. Williams, *Christian Theology*, p. 173.   [39] Ibid.

ethics. In this regard, her theory of natural law is superior to the other theories considered so far. Her acknowledgment of the theological basis of natural law makes her more theologically adequate (at least from a Christian standpoint) than Arnhart and others who would "naturalize" Thomas. Porter is highly critical of all forms of reductionism and in fact has lately preferred to speak about "biological sciences" rather than evolutionary psychology because the latter is so strongly associated with pernicious forms of reductionism. She is right to regard such reductionism as "scientifically dubious" and based on a "simplistic" account of adaptation and its role in natural selection.[40] She concedes that evolutionary psychology can helpfully account for some of the "regularities" of human nature but she thinks that the disciplines of anthropology and primate studies offer more promising avenues for natural-law reflection. We might note that anthropology and primate studies have their Darwinian as well as anti-Darwinian theorists and that the primatologist cited most often by Porter, Frans de Waal, focuses on the natural selection and functional value of prosocial tendencies in higher social primates. There is simply no way to avoid implicitly accepting evolutionary theory of some kind if one engages the biological sciences, and this includes primatology.

Criticism of Porter might come from two opposite directions. First, some critics will charge that contemporary science has rendered useless all medieval views of the ensemble of creatures ordered to one another and to the whole of the creation. Materialists such as Dawkins and Wilson claim that modern science cannot accept what the medievals took to be axiomatic, namely, that the cosmos as such is ordered to God, and that when this supporting ground is removed, natural-law theory collapses. Earlier chapters of this book, however, argued that this dismissal of purpose from nature is an unacceptable form of reductionism. Science as such does not consider metaphysical questions, and materialistic critics make this claim on the basis of their own philosophical assumptions rather than on scientific grounds *per se*.

Others would charge that Porter glosses over the basic anachronism of the worldview underlying scholastic natural law. Ruse and Beckstrom would argue that Porter is wrong to attempt to rebuild natural law on an evolutionary foundation because evolutionary theory cannot be a "foundation" for anything. According to sociobiology and evolutionary psychology, the universe as simply a cosmic expanse of matter and energy whose order, permeated with chaos and undergirded by indeterministic quantum events, can be adequately and fully accounted for, as Monod put it, by "chance and

---

[40] Porter, *Nature as Reason*, p. 106.

necessity."⁴¹ Human beings are simply the products of what Gould called a "glorious accident."⁴² For him, we are like all other mammals except for some statistically freakish cognitive abilities that are by-products of the evolution of our large frontal cortex. For the evolutionist, attempting to place natural-law ethics on an evolutionary view of human nature is like trying to strap a Conastoga wagon onto the frame of a Formula One race car: either it will not move at all, or if it does it will quickly fall apart.

A second general line of criticism of Porter's project might come from Thomists who reject her entire effort for its abandonment of the Aristotelian teleological philosophy of nature that provided the central metaphysical grounding for natural-law theory. Scholastic natural-law theory was situated within a philosophical, scientific, and cosmological worldview and, contrary to the "new natural-law theory" of Grisez and Finnis, it was never intended to stand as an independent, self-contained ethical theory. This worldview regarded God as the planner of the entire order of the universe, which directs things to the goal of divine goodness, Thomas Aquinas explains, much as a general directs all the soldiers of an army to victory. Central to this teleological worldview is a depiction of God as a designer of the natural order (though not as Paley's "Watchmaker") who creates forms with characteristic inclinations to their proper ends. Aquinas maintained that the order of the universe was devised by God, in whose mind exist the "types of all things" from which things in nature receive their "determinate forms."⁴³ The forms of things are created by a benevolent deity who works through created secondary causes to bring creatures into existence and gives each its own natural constitution, its own potentialities for flourishing, and its own natural inclinations to its appropriate manner of flourishing. God freely chose to create a vast, harmonious, hierarchically ordered universe in which each creature exists both for its own perfection and for the perfection of higher creatures and the cosmos as a whole. Everything is directed to its proper end, animals by instinct and humans by means of intelligent choice.

This view insists on maintaining the entire edifice of Thomistic cosmology as the basis for natural-law ethics.⁴⁴ It repudiates evolutionary theory

---

[41] Jacques Monod, *Chance and Necessity*, trans. Austryn Wainhouse (New York: Vintage, 1972).
[42] See Stephen Jay Gould, *Flamingo's Smile: Reflections in Natural History* (New York: Norton, 1997); see also William Kayzer, *A Glorious Accident: Understanding Our Place in the Cosmic Puzzle* (New York: Freeman and Co., 1999).
[43] *ST* I, 44,3.
[44] Leo Elders, *The Philosophy of Nature of St. Thomas Aquinas: Nature, the Universe and Man* (Frankfurt am Main: Peter Lang, 1997), especially ch. 11.

because it views species as planned by a divine engineer, not as the result of natural selection. It argues that production of new species must have been caused by an "outside cause" beyond the capacities of nature.[45] Some Thomistic critics reject the evolutionary view of nature as wasteful, messy, and disorderly, appetites as inherently conflicted and evolved to aid the individual (and especially the individual's genes) rather than the species, and life as a wildly and oddly shaped branching tree with no preordained *telos*. They also reject the claim that humanity's immaterial "spiritual faculties" can in any way have been the product of evolution.[46] For these Thomists, the attempt by Porter and Arnhart to meld Aquinas with Darwin can only be doomed to failure.

Porter herself allows for a kind of functional teleological analysis in the biological sciences that has been defended by Ayala and Mayr. In this perspective, animal behavior is goal-directed and species pursue characteristic behavior patterns that contribute to their well-being. Porter thus adopts a theory of "natural kinds" that is a recognizable development of aspects of what some scholars find in Thomas Aquinas.[47] The fact that there is no scientific verification of cosmic purpose and no way of scientifically justifying the claim that nature as a whole is purposive means neither that creation as a whole is meaningless nor that nature has only those purposes that are assigned to it by human choices. This is because ontological questions do not fall under the purview of the natural sciences.

A more adequate alternative to this static form of Thomism recognizes that natural law is a living historical tradition that grows in response to challenges presented by contemporary knowledge. Evolutionary theory presents just such a vitalizing challenge for contemporary natural-law ethics. Natural-law ethics placed in an evolutionary cosmology and anthropology must seek the human good amidst an array of patterns of natural ordering rather than one natural order. This ordering, moreover, is construed within a historically minded awareness of the narrative basis of Christian ethics.

Porter's deep sense of historical context and development differs markedly from standard moral theories of the twentieth century, such as that of Alan Gewirth, and resembles the more historically sensitive work of MacIntyre. Porter's concern is clearly not centered on "all rational agents" existing somehow independently of tradition, community, and history (including natural history). At the same time, Porter deals with nature more seriously than do most narrative theologies. Because natural law

---

[45] See ibid., p. 362.   [46] Ibid., p. 363.   [47] See *Nature as Reason*, p. 95.

pertains to human beings as they actually exist in real societies and in their natural environments, the more information we have about these realities, the more likely our ethical reflection will be a source of wisdom about personal and communal lives. Porter thus exemplifies what it takes to keep a tradition vitally engaged with contemporary modes of understanding and explanation while remaining solidly rooted in this moral tradition.

Natural law as the basis for moral dialogue: Cahill

Lisa Sowle Cahill is also significantly influenced by Thomas Aquinas. She believes it possible to achieve reasonable degrees of objectivity[48] in ethical reflection, accords a significant role to the place of Scripture in Christian ethics, and wants to draw resources for contemporary moral reflection from the natural-law tradition. More outward facing than MacIntyre and Porter and more ecclesial than Arnhart, Cahill's writings on sex, gender, and reproductive technology are much more informed by practical moral concerns than are the writings of the three authors I have just examined, and her treatment of moral methodology is honed more by the exercise of practical reason than by a theory about it. Her reading of "nature" is also more informed by dialogue with postmodern authors than are the others examined here. Natural-law moralists always run the risk of projecting cultural presuppositions onto nature and of reading mere biological conditions as determinative of moral decisions that ought to be regarded in more complex ways. Cahill regards the natural law as a historical tradition that goes through developmental processes rather than as a permanent and fixed moral code. Advocates of the natural-law tradition must continually subject themselves to self-criticism and hold themselves accountable to non-Christian sources of knowledge. She is especially resistant to the "physicalism" of the old moral manuals that assumed that the natural law could be "read off" the structures of physical nature, and especially from the structures of human reproductive biology.

Cahill approaches natural law more through experience and inductive reasoning than from ontological speculation. The primary meaning of natural law concerns the "attempt communally, experientially, and reasonably to describe and affirm normative human characteristics."[49] Human

---

[48] Lisa S. Cahill, "Women, Marriage, Parenthood: What Are Their 'Natures'?" *Logos: Philosophical Issues in Christian Perspective* 9 (1988): 11–35, at 14. See also Cahill's *Sex, Gender, and Christian Ethics*, pp. 67ff.
[49] Cahill, "Women, Marriage, Parenthood": 14.

experience discovers the elements of human life that are valuable, worthwhile, and good for human beings. "To say that a human characteristic, relationship, or a form of social organization is 'natural' is to suggest that it is essentially better and more fulfilling for human beings than other alternatives."[50] To claim that a practice or good is "natural" is also to suggest that people from very different perspectives have access to this good simply because they share with us a common human nature. Thus natural law implies that we "commit ourselves to the possibility of shared and even universal discourse about what it is in human life that we should commend."[51]

Cahill, like MacIntyre and Porter, is attuned to the dangers of ahistorical moralizing. Human nature is always a concrete historical and cultural reality, not something to which one can have access outside a particular context. We do have a biological species identity but it is neither comprehensive of the natural law nor ethically determinative of how we ought to live. Cahill adopts something like Mayr's notion of human nature as an "open program" or "basic pattern" rather than as a collection of instincts or programs for behavior. She insists that our knowledge of biological nature and our evolutionary past do not provide much concrete moral assistance in practical moral decision-making. Knowledge of our evolved biological nature informs us about what goods are at stake in a given concrete situation and thus provides data within a process of practical moral reasoning. Biology makes only a limited contribution to ethics, so that, "even to agree, for instance, that the most 'natural' and even the morally best form of procreation is a fertile sexual act between loving spouses committed to share in nurturing their child, is not yet to have determined what is to be done morally when that form is not an option for two people."[52]

Cahill is properly suspicious of attempts to interpret modern western social practices or gender roles as "natural." This concern gives her a critical perspective on the tendency of modern Catholic natural-law ethics to insist on the unexceptional enforcement of absolute moral norms. It also leads her to be suspicious of the writings of sociobiology, such as when its reading of natural male promiscuity and female fidelity is taken to confer on these roles a tacit normative acceptability.[53] Even if these patterns are a "simple de facto legacy of our history as a species,"[54] they should not be taken to have any special moral status on the grounds that they are "natural." Drawing normative conclusions about these matters requires

---

[50] Ibid., 11. [51] Ibid., 12. [52] Ibid., 16.
[53] See Cahill, *Sex, Gender, and Christian Ethics*, p. 93. [54] Ibid.

considerable openness to experiences, careful weighing of different kinds of evidence, and patient and somewhat circumspect moral reasoning. Cahill finds salutary Thomas Aquinas' attempt to make generalizations on the basis of inductive reasoning from particular cases and she appreciates his recognition of the limits of such an exercise.[55] Scientific studies of kinship bonds indicate the evolutionary reason for their significance and they confirm what we already know about their importance for our identity,[56] but this does not mean that there is only one ethically acceptable way of being "kin."

Cahill also emphasizes the public value of natural-law reflection that used to be a central concern of John Courtney Murray, SJ, and continues to be developed by David Hollenbach, SJ. Not given much attention by MacIntyre and Porter, the commitment to public moral discourse is an especially great need given our rapidly expanding social and cultural diversity. It provides language for making arguments without drawing on "confessional" evidence and premises that might not be widely shared in broad pluralistic communities. Because the language of natural law itself has come more and more to be regarded as a distinctive form of Christian theology, and even as a distinctively Catholic form of moral argument that appeals to few people outside the Catholic community, its concerns need to be translated in other, more publicly intelligible language such as rights, dignity, flourishing, well-being, and the like. Certainly both inside and outside this tradition the term "unnatural" is less and less helpfully used as a term of negative moral appraisal. Most of what is "natural" about us is also "cultural," so that we have little access to what is "natural" in itself. Knowledge of natural inclinations is relevant to moral reflection, even if "doing what is natural" is not normative. Christian ethics aims at the human good, as we have said, and biological goods are central to but not all inclusive of the human good. We strive to promote the satisfaction of basic biological, emotional, and social needs, but we do not stop at these.

Natural law moves from the presupposition that human beings everywhere have access to the same basic human goods. Natural-law universality is reinforced by the evolutionary assumption that all human beings are members of one species, share some basic human inclinations and needs, and are likely to thrive only if they have secure possession of certain well-known goods. Christian faith, adherence to doctrine, and membership in

---

[55] Cahill, "Women, Marriage, Parenthood": 13, citing *ST* I-II, 94,2 and 4; see also Cahill, *Sex, Gender, and Christian Ethics*, pp. 47–50.
[56] See Cahill, "Women, Marriage, Parenthood": 24–25; also *Sex, Gender, and Christian Ethics*, p. 93.
[56] Cahill, "Women, Marriage, Parenthood": 14.

the church are not prerequisites for us to understand the importance of basic goods for human well-being. There is a world-wide need for a "global ethic" that can support international commitments to the well-being of the poor. This agenda need not be regarded as a threat to the more distinctive claims of Christian ethics. The convergence of various salient facts pertaining to the basic human needs of the poor in a multitude of different cultural contexts provides the best case for what Cahill calls a "global ethics of the common good."[57] Commenting on the *Millennium Declaration* of the United Nations General Assembly, Cahill notes that differences of commitment over international social justice are focused not on whether there are certain basic human goods that are the critical conditions for any kind of human flourishing, but rather over "who exactly is entitled to flourish."[58]

Cahill's point that certain biological conditions have to be met if we are to flourish finds added support in a recent interdisciplinary study entitled *Hardwired to Connect*.[59] These scholars and scientists found that modern society has failed to provide conditions that allow young people to meet their needs to establish and develop relationships with peers, to put down roots in their neighborhoods, and to feel connected to their communities. Their study proposes that as a society we strive to enhance "authoritative" communities but not "authoritarian" forms of governance. They argue that the social needs of young people are rooted in the evolved nature of their humanity and are genetically "hardwired" into us by the evolutionary process. Failure to establish attachments, incompetence in emotional communication, and inability to form social alliances can lead to serious psychological problems. Young people have a deep natural desire to belong to and feel valued by strong families and communities, and this desire must be honored and satisfied if they and their communities are to flourish.

These needs are not simply expressions of cultural idiosyncrasies. In light of the fact that ten million children under the age of five die every year of preventable diseases in various parts of the globe, UNICEF promotes strategies targeting the unmet needs of children in all developing cultures. The strategies target specific unmet needs, but the goods they promote are universal. The strategies include exclusive breastfeeding in the first six months of life, the extensive use of bed nets to protect children from malaria in mosquito-infested areas (three thousand children die every day

---

[57] Cahill, "Toward Global Ethics": 324–344, at 335.   [58] Ibid., 337.

[59] *Hardwired to Connect: The New Scientific Case for Authoritative Communities* (New York: Institute for American Values, 2003), sponsored by the Dartmouth Medical School, the YMCA of the USA, and the Institute for American Values.

in Africa from malaria – one child every thirty seconds),[60] and the distribution of rehydration salts to combat diarrhea.[61]

More generally, UNICEF's global strategy to help poor children advocates five priorities: early childhood care (including good health care, sound nutrition, clean water, and sound hygiene), sufficient immunization, education for children and especially girls (at the present time, 110 million children are deprived of schooling), effective HIV/AIDS prevention (not only for the survival of the children themselves, but also to prevent them from being orphaned by parental fatalities), and, finally, the protection of children from exploitation, abuse, and violence (particularly in the form of sweatshops, sexual exploitation, and forced conscription into child armies).[62]

UNICEF attends not only to physical survival but also to what it takes for children to thrive. They recognize the evils not only of disease but also of assaults on human dignity and self-respect. Knowledge of evolutionary biology helps scientists in their search for new medicines and vaccines, but it does not shed light on our comprehension of the full range of harms that are done to children by diseases, malnutrition, and violence. Twentieth-century natural-law theory was complemented with an extensive doctrine of human rights, which has now become the closest thing we have to a universal form of moral discourse. These rights are widely intelligible because human beings everywhere – simply in virtue of being members of the same species – share the same basic needs and require the same basic goods if they are to thrive. One does not have to be Roman Catholic to agree with Pope John Paul II that "It is a strict duty of justice and truth not to allow fundamental human needs to remain unsatisfied, and not to allow those burdened by such needs to perish."[63]

Natural-law ethics strives to engage in moral discourse across the boundaries of particular communities about what people in general need if they are to flourish, so it makes sense that much of twentieth-century natural-law social ethics was framed in the language of rights.[64] Because rights have sometimes been misinterpreted to lend support to irresponsible

---

[60] See the April 2003 report of the World Health Organization and UNICEF at http://www.unicef.org/media/media_7701.html, accessed December 20, 2004.
[61] See the "State of the World's Children 2005" issued by UNICEF. Available at http://www.unicef.org/media/media_10958.html; accessed December 20, 2004.
[62] Ibid.
[63] John Paul II, *Centesimus Annus: On the Hundredth Anniversary of Rerum Novarum* (Washington, DC: United States Catholic Conference, 1991), no. 34.
[64] See David Hollenbach, SJ, *Claims in Conflict: Renewing and Retrieving the Human Rights Tradition* (New York: Paulist Press, 1979).

and self-centered individualism, recent approaches to natural law have helpfully been dedicated to the recovery of the common good.[65] Natural-law ethics supports efforts to establish the general conditions of human flourishing of the sort identified by UNICEF and WHO, and its uses of the language of human rights and the common good offer a strong moral framework for promoting goods that knowledge of evolution alone does not supply.

Christian ethics can be committed to moral engagement on these most general terms because of the urgency of the needs it addresses. Christian ethics also draws upon a religious vision of human life that includes but goes beyond the basic conditions of general human flourishing. It seeks conversion and holiness, recommends that we imitate the lives of the saints, responds to God's constant offer of grace, and strives for spiritual union with God. These goods do not necessarily compete with more basic goods. In metaphysical terms, what is most urgent is not in conflict with what is most excellent. While union with God is a higher level of excellence than is satisfying basic needs for food and shelter, it is difficult to consider the former while systematically deprived of the latter. Despite differences regarding the nature of human flourishing that distinguish different communities, traditions, and religions, there are basic conditions that most human beings recognize ought to be promoted. Christian particularity, then, should not be taken to undermine the human commitment to cross-cultural agreement about the basic conditions of human flourishing and to building a wide consensus on our common human responsibility to work for justice for the poor. On this point all the natural-law theorists examined above can agree.

One might supplement Cahill's point by noting that evolutionary theory can play a constructive moral role if it helps us to understand why it is more difficult for people to learn to care about members of "out-groups" and people living in communities very remote from their own. This is a major challenge to the virtue of neighbor-love, and especially solidarity, that must be faced and understood if it is to be significantly corrected. Evolutionary knowledge underscores the importance of creating cultures that extend our capacity to care and that enhance our sense of justice and fairness. It can be destructive, however, if it is interpreted in a fatalistic way to suggest that people can never learn to take responsibility for those who are neither kin nor reciprocators. Evolutionary psychology and sociobiology give us no reason why we should be concerned for non-kin

[65] See Hollenbach, *The Common Good and Christian Ethics*.

and non-reciprocators, but natural-law ethics argues that our common human dignity grounds the virtues and duties of love, justice, and solidarity. We do indeed have proclivities and biased learning that lean toward egoism and nepotism, but these do not mean that we are fated to a world of nepotism, indifference to outsiders, and moral myopia. By being critically appropriated within the context of Christian natural law rather than "Hobbesian Darwinism," however, we can transcend the evolutionists' moral blind spots, fatalism, and reductionism and develop a more credible and morally appealing vision of humanity.

NATURAL LAW IN LIGHT OF HUMAN EVOLUTION

The four moralists just examined all make significant contributions to the critical appropriation of our growing knowledge of human evolution by the field of Christian ethics. There are disputed points between them, for example between Porter and Cahill on the universality of the natural law, or MacIntyre and Arnhart on whether Hume is compatible with Thomistic natural law, but their points of agreement are more important for the purposes of this book. They are generally critical of evolutionary ideology and reductionism, insistent on our recognition of the embodied and material nature of humanity, alert to the interdependence of culture and biology in human nature, and aware of the fact that human rationality is always tradition-constituted. They all accept some kind of teleological view of human nature – the fact that we are naturally inclined to various kinds of goods, including biological goods – and understand these natural inclinations as slowly tutored by communal practices that teach moral norms, ideals, and good habits.

The evolutionary view of what is "natural" to human beings is based on what occurs with some frequency under various ecological conditions. In this sense of the word it is entirely "natural" for some male animals to practice infanticide, to kill conspecifics from other groups, and to engage in forced copulation with fertile females.[66] Acts such as these were probably also placed by natural selection on our own evolved menu of behavioral options in the course of our evolutionary past because they were in the distant past "fitness enhancing" under certain conditions. At the same time, empathy, parental caring, familial cooperation, reciprocity, and other forms of pro-social behavior are now widespread because, in a multitude of ways and over the long run, they promoted rather than undermined

---

[66] See Daly and Wilson, *Homicide*.

reproductive fitness. As we have seen, evolutionists hold that, under certain circumstances, egocentrism, lying, self-deception, out-group bias, and aggression can promote "fitness," but that under other circumstances the opposite strategies are "fitness maximizing." These behaviors are "natural" in the sense that they are favored by selection under certain circumstances. Thus kin altruism is adaptive over a wide range of environmental contexts, yet conditions of scarcity may be so severe that natural parental solicitude is jettisoned in favor of infanticide.[67] In some conditions, similarly, polygamy rather than monogamy would be considered the most natural arrangement between the sexes, at least in the sense that it might be the most effective "mating strategy."[68]

In natural law within the Christian tradition, on the other hand, what is "natural" includes biology but also our more distinctively human capacities. A right way of acting is not ethically obligatory or legitimate simply because it is "natural," in the scientific sense, as "evolved" or "genetically based," but it is obligatory because it accords with what is good for human beings, considered comprehensively. The obligatory character of morality – the "law" – binds the person to moral standards that promote the well-being, or flourishing, of the person and his or her community. It is wrong to murder, to be sexually unfaithful, to steal, to lie, and to cheat, because doing so undermines the good of both self and others. General norms are not matters of arbitrary taste or idiosyncratic preferences, but reflect judgments about structures of living that promote human flourishing. Our basic orientation to the good is not extinguished by wrongdoing: even liars resent being lied to, and those who steal are morally outraged when stolen from.

The normative ideal of natural law identifies certain human capacities from within the larger conglomeration of traits that constitute our evolved human nature, but it is selected on theological and moral rather than on biological grounds. Natural law promotes the activation of our created potentialities for loving the good and knowing the truth. It takes its bearings from human nature at its best or it is seen as most in accord with the image of God. The claim that God is eternal love is based on the classical Trinitarian theology, according to which God the Father eternally generates the Word of understanding, the Logos, the Son, and in which God the Spirit eternally proceeds as the love between the Father and the Son. Natural law is what promotes human flourishing comprehensively considered. Since God is eternal love and pure wisdom, and human beings

[67] See ibid., chs. 3–4. [68] See Buss, *The Evolution of Desire*.

are created in the image of God, the natural law for us lies in our capacity to know what is true and to love what is good. This means first of all to know and love God, and secondly to know and love ourselves and our neighbors. This is the deepest meaning of Thomas' description of natural law as "the rational creature's participation in the eternal law."[69]

The core of natural law does not lie in a detailed moral code, or a defined set of moral rules, or specific "action-guides" but in the virtues that shape and perfect our capacities for knowing and loving. The desire to know and to love are not merely two tendencies among others, but our central and most definitively human inclinations. In Christian ethics, knowing and loving are brought to their highest activity in the exercise of wisdom and charity, and charity as the "form of the virtues" shapes and inspires the cardinal virtues of prudence, justice, temperance, and fortitude. Virtues of knowing and loving provide the proper basis for ordering all other acts, habits, attitudes, and desires. They help to make human sense of the elementary inclinations that we share with other animals and allow us to move beyond the narrow orientation to fitness that pervades the rest of the animal world.

The natural-law tradition has historically tended to downplay the existence of conflict between goods and to attribute it to human sinfulness (either original or actual), but, as suggested already, some important forms of conflict are built into our nature. This tendency toward conflict is seen throughout the animal world. We find internal conflict over various goods (e.g. the need for security and the need for food), as well as external conflicts between different individuals over the same goods (e.g. two males and one female). In the human context, sin no doubt significantly intensifies the factor of conflict in human life, and it introduces new kinds of conflicts (e.g. over prestige, status, possessions) whose ferocity and ill effects extend very significantly beyond what is found in nature. Yet it is still the case that human beings have been "set up" by evolution to experience multiple inclinations that sometimes reinforce and at other times conflict with one another, and it is the case that communities are the scene of competition and conflict among and between both individuals and groups.

A view of natural law more attuned to the possibility of conflict of goods and responsibilities might be more willing to concede that at times the pursuit of one good, judged to be higher or otherwise more important, might require the direct negation of another good. This also implies that on occasion the exercise of one virtue (say, mercy) might not only be

---

[69] See Craig A. Boyd, "Participation Metaphysics in Aquinas's Theory of the Natural Law," *American Catholic Philosophical Quarterly* 79.3 (2005): 169–185.

neutral with regard to another (e.g. justice) but might actually be working against it. When a state takes up arms, even when it does so within the criteria of the just war, its pursuit of justified self-defense also entails the negation of mercy to its enemy.

This point raises the specter of consequentialism, the moral logic according to which the ends justify the means, and the "slippery slope" of situation ethics. One hedge against moral deterioration is absolutism, unwavering and rigid adherence to the letter of the law that allows for no exceptions, especially when it comes to cases such as the use of artificial birth control or remarriage after divorce. Absolutism sometimes leads to its own absurdities that do not need to be illustrated here.

In natural-law ethics, true moral objectivity is achieved in concrete acts through the exercise of the virtue of prudence. What is objectively binding in a particular situation, in other words, is what is most in keeping with the first principle of practical reason: do good and avoid evil. General moral knowledge includes various beliefs about which aspects of our inherited behavioral repertoire ought to be approved of, acted upon, and promoted – and which ought to be inhibited, sublimated, or closely monitored. It also includes general knowledge of which kinds of acts tend to undermine the human good and which kinds of acts promote it, but moral decision-making only succeeds when attention is focused on concrete goods and evils at stake in particular situations.

The virtue of prudence allows the moral agent to perceive the morally salient factors at stake in concrete human experiences. It is in and through concrete experiences that people discover, appropriate, and deepen their understanding of what constitutes true human flourishing. Interpretations of these experiences are influenced by membership in particular communities shaped by particular stories. Discovery of the natural law, Pamela Hall notes, "takes place within a life, within the narrative context of experiences that engage a person's intellect and will in the making of concrete choices."[70] The virtue of prudence avoids applying moral rules that, in concrete situations, damage human lives. Rather than prescribing a universal and exhaustive moral code proper for all times and all people, our understanding of natural law must be dynamic, flexible, and open to new developments as a result of changing human circumstances.[71]

---

[70] Pamela M. Hall, *Narrative and the Natural Law: An Interpretation of Thomistic Ethics* (Notre Dame, IN and London: University of Notre Dame Press, 1994), p. 94.

[71] See James F. Keenan, SJ, "The Virtue of Prudence (IIa IIae, qq. 47–56)," in Stephen J. Pope, ed., *The Ethics of Aquinas* (Washington, DC: Georgetown University Press, 2002), pp. 259–271.

Some Christian ethicists will complain that "flourishing" is more Aristotelian than Christian. The central image of Christianity is not that of a flourishing human being, but of a crucified man who called his followers to give up everything, follow him, and carry their own crosses. Instead of seeking enlightenment, or wisdom, or well-being, the argument runs, Christians are told to "strive first for the Kingdom of God" (Matt. 6:33, Luke 12:31). We should not begin to think ethically as Christians by starting with a generic viewpoint based on nature and then examining how the Christian tradition might fit into it. Christian ethics, on the contrary, pertains to the way of life of a particular community and particular tradition. Scientific theories of nature and evolution, and scientifically informed accounts of human flourishing, are irrelevant to this enterprise.[72]

In response to this criticism, I would argue that, while Christian ethics professes Jesus Christ as the normative human being, it nonetheless accepts that the basic standards for the moral life can be recognized by most human beings whether or not they profess the Christian faith. The golden rule, the ideal of peace, and the virtue of justice are not the exclusive possession of the Christian community either in theory or (even less) in practice. The locus of moral obligation in natural-law ethics comes from the human good, and ethics is based on the fundamental human desire for flourishing, happiness, or, as Thomas put it, "beatitude." I use the term "flourishing" because "beatitude" is archaic and because "happiness" in English tends to connote something like a subjectively pleasant emotional state. Morality provides a path to true and perfect happiness, experienced partially in this life and completely in the next. "Flourishing" functions as the justification for virtues and moral standards rather than as the intentional and direct goal of every act. The agent asks herself, "what is the right thing to do?" or "what would a good person do in this situation?" rather than "which of these actions will most contribute to my flourishing?" (let alone, "which of these options will make me 'happier'?").

We understand that vandalism is wrong not because it is condemned biblically (and actually there are lots of cases in the Hebrew Bible, such as Joshua 6, where vandalism is expressly commanded) but because it unjustifiably harms other people, their property, and their communities. Thomas Aquinas expresses the central importance of the good for Christian ethics when he writes that "we do not offend God, except by doing something contrary to our own good."[73] Sociobiologists and their

---

[72] This is the brunt of John Howard Yoder, *The Politics of Jesus* (Grand Rapids, MI: Eerdmans, 1972).
[73] *SCG* 3,122.

philosophical allies miss this central dimension of natural-law theory when they confuse it with "divine command" ethics.

Natural law is necessary but not sufficient for Christian ethics. The good calls for the transformation of our motivations, attitudes, and intentions, and in a way and to a degree that surpasses our natural human capacities. Christian ethics (and, really, grace) strives to shape the evolved "matter" of our inherited human nature so that it is more in keeping with the "form" of the good life revealed in Jesus. Thus Christian ethics attempts to inform our natural sense of fairness by shaping the Christian imagination according to the key narratives, parables, and injunctions present in Scripture. Jesus concludes the parable of the good Samaritan with "go and do likewise" (Luke 10:37). Christian ethics expands justice beyond our natural satisfaction with family and friends to include strangers and enemies. Indeed, Christian ethics inclines us to regard the good of others as in some sense also our own good; this of course is where a vision of the good that is more than biological offers a radically more humane ethic than anything envisioned by sociobiology or evolutionary psychology.

The moral ideal by which the strains of human nature are judged is not arrived at independently of faith and then supplemented by secondary religious concerns of Christianity, yet neither does it somehow "begin" with faith as if nature and culture did not provide the pervasive setting for religious and moral development. Christian narrative and participation within the Christian community supply a context for interpreting human evolution, yet the latter gave rise to capacities and inclinations developed within the Christian life. There is a kind of circularity, but not a vicious one, in the relation between faith and nature in this regard.

Christian natural-law theory identifies which actions comport with moral goodness and which do not. They are concerned not only with *selecting* which capacities, desires, and needs are part of the good life, but also what *shape* they are going to take in the moral life as well as how they are to be *ordered* to one another and to the common good. When it comes to the standard for Christian ethics, then, the central focus is not "naturalness" *per se* but transformation "in Christ," not in biological evolution but in spiritual and moral growth. Saints continue to be human beings with the same natural constitution as other human beings, but they face the challenge of habituation in the good and learn to order their affections so that they accord with the Christian vision. Christian ethics, then, regards human nature as properly developed in the virtues, and above all in faith, hope, and charity.

CHAPTER 12

Sex, marriage, and family

This chapter explores the practical significance of the approach to natural law presented in the previous chapter. I examine various types of scientifically based arguments about what evolution implies, or does not imply, concerning sex, marriage, and family, and investigate whether and how evolved capacities and needs are captured in institutions of marriage and the family and in moral norms governing sexual behavior. This cluster of topics provides a particularly interesting context for investigating whether there is an evolutionary basis for the moral norms advanced by the natural-law tradition of Christian ethics, and whether there might be evolutionary support for modifying these norms in some ways. In evolutionary terms, we can ask whether certain moral beliefs have supported acts or practices that are more adaptive than their alternatives. This involves descriptive as well as properly normative concerns. Identifying certain acts or practices as more in keeping with our "inclusive fitness" does not in and of itself provide sufficient moral justification for acts and practices under consideration.

The chapter has three sections: the first concerns evolutionary theory, the second offers a natural-law evaluation of evolutionary views of sex, and the third considers some aspects of normative ethics in the domain of sex, marriage, and family. My thesis is that contemporary natural-law ethics ought to attempt a critical and selective appropriation of evolutionary views of sex, marriage, and the family in light of its inclusive vision of the human good.

EVOLUTION AND SEX, MARRIAGE, AND FAMILY

Darwin's *The Descent of Man and Selection in Relation to Sex* (1871)[1] proposed that some traits would be produced by sexual selection if the

---

[1] See *Darwin*, ed. Appleman.

benefits of such traits were to outweigh their costs. Darwin proposed two kinds of sexual selection: (1) intra-sexual selection, which typically involves male–male competition, and (2) inter-sexual selection, which is usually seen in female choice of mate. For the last twenty years or so feminist evolutionary writers have criticized the inadequacy of the Victorian view of women as passive and have shown in great detail the significance of female choice of mate and female–female competition in this regard.[2] Women are described as more discriminating than men but no longer as simply "coy." Evolutionary biology here is seen as potentially liberating rather than as irredeemably sexist.

Sociobiologists investigate sexual selection in terms of differential "biological investments" made by males and females. As a general rule, males in most mammal species copulate and move on without giving any more attention to the resulting offspring. As Jared Diamond puts it, "Most male mammals have no involvement with either their offspring or their offspring's mother after inseminating her; they are too busy seeking other females to inseminate. Male animals in general, not just male mammals, provide much less parental care (if any) than do females."[3]

Males from our own species depart from the animal norm in this regard. A man produces 150 million sperm every day of his life and has, as one biology textbook puts it, "enough genetic information in a single ejaculation to repopulate half of the USA."[4] A woman, on the other hand, releases only 450 eggs in her entire reproductive lifetime. A man is theoretically capable of reproducing after every copulation and is often fertile even in old age. A woman, on the other hand, can normally bear one child a year and even then only until the onset of menopause.

Evolutionary psychologists argue that this asymmetry leads to two competing sexual "strategies." Men can afford to engage in casual sexual encounters because they are not forced by the facts of biology to face the consequences of impregnation. Women do not have this luxury. Because fertilization can follow a single sexual encounter and entails high costs, women must be much more selective and discriminating than men. This may account for the widespread recurrence across cultures of male sexual

---

[2] See Helena Cronin, *The Ant and the Peacock* (New York: Cambridge University Press, 1991), and Patricia Gowaty, "Evolutionary Biology and Feminism," *Human Nature* 3 (1992): 217–249. See also essays of a more critical nature by A. Fausto-Sterling, V. L. Sork, and Z. Tang-Martinz, in Patricia Gowaty, ed., *Feminism and Evolutionary Biology* (New York: Chapman and Hall, 1997).

[3] Jared Diamond, *Why Is Sex Fun? The Evolution of Human Sexuality* (New York: Basic Books, 1997), p. 15.

[4] Malcolm Potts and Roger Scott, *Ever Since Eve: The Evolution of Human Sexuality* (New York: Cambridge University Press, 1999), p. 30.

jealousy and sexual "double standard."[5] Sociobiologist Matt Ridley argues that male fear of cuckoldry plays a significant role in practices such as wearing veils and chastity belts, female circumcision, harems, and foot binding.[6] What Don Browning calls the "male problematic" – "the tendency of men to drift away from their families"[7] – is a problem for women attempting to raise children.

Pair-bonding can be a significant corrective to the "male problematic." Concealed ovulation and constant sexual receptivity – two other traits of human sexuality that depart from the "mammalian norm" – allow for nonreproductive sex that can create an affective bond between mates. The human sexual bond contrasts with the pattern even among most other primates, including the highly sexual bonobo chimpanzees, whose sexual activity does not seem to build individualized male–female bonds.[8] The capacity for orgasm may play a central role here as well. It is unclear, Potts and Scott observe, "whether any species other than humans, chimpanzees and some monkeys actually have orgasms."[9] One hypothesis, proposed by Frans de Waal, suggests that this bonding process might have originated when males began to stay with their female mates in order to guard their vulnerable newborns against infanticide by other males.[10] The symmetrical balance of male–female orgasms, some evolutionists suggest, functions as a reciprocal bonding mechanism.[11] The human sexual bond may provide a strong motivation for men to stay with their mates and to give them the kinds of aid, especially in the forms of protection and provisioning, that are critically important for the demanding task of rearing vulnerable and dependent children.

Some evolutionary psychologists argue that males have an innate proclivity for moderate polygamy but also a potential for emotionally bonded monogamy. Male sexuality is deeply conflicted, but men are not the only ones who struggle. Women are attracted to men other than their mates. Cuckoldry (female "extra-pair copulation") detracts from the man's

---

[5] David M. Buss, *The Dangerous Passion: Why Jealousy Is as Necessary as Love and Sex* (New York: Free Press, 2000).
[6] See Matt Ridley, *The Red Queen: Sex and the Evolution of Human Nature* (New York: Harper Collins, 1993).
[7] Don Browning et al., *From Culture Wars to Common Ground: Religion and the American Family Debate* (Louisville, KY: Westminster John Knox Press, 1997), p. 265.
[8] See Wrangham and Peterson, *Demonic Males*, ch. 10.
[9] See Potts and Scott, *Ever since Eve*, p. 108. For an earlier discussion (1981) of this issue see Sarah Hrdy, *The Woman that Never Evolved* (Cambridge, MA: Harvard University Press, 1981), ch. 8.
[10] See de Waal, *The Tree of Origin*.
[11] Potts and Scott, *Ever since Eve*, p. 102. See the arguments for and against the "sex as glue" theory by Jared Diamond in *Why Is Sex Fun?*, ch. 4.

interests much more than the woman's, but if discovered it can lead to desertion and significant reduction of resources.[12] The ultimate male deterrent to female infidelity is the threat and use of physical violence. Male on male attack is the most common and the most lethal form of violence; the least common is female on female violence.[13]

Because the reproductive interests of parents and their offspring do not always coincide, family life is also marked by competition and conflict. Applying the evolutionary logic, according to which the degree of concern is proportionate to the degree of genetic relatedness, sociobiologists argue that a child is inclined to care for his or her own interests twice as much as for the interests of a sibling (experience teaches that this preference can be a lot stronger). The biological roots of sibling rivalry lie in competition for parental resources. For their part, parents can be expected, other things being equal, to treat their children equally since they share the same degree of genetic relatedness. But parents are inevitably faced with conflict with individual children, each of whom has genetic reasons for preferring his own interests to those of either his or her parents or siblings.[14] Parents try to temper these tensions and channel the energy that they produce, but potential conflict cannot be completely eradicated.

The usual criticisms of sociobiology discussed above apply to its treatment of sex, marriage, and family. The most notorious recent example of offensive reductionism in this sphere of behavior comes from Randy Thornhill and Craig Palmer's book *The Natural History of Rape*,[15] where they argue that rape, although a crime and morally reprehensible, is an evolved adaptive behavior when its benefits outweigh its costs, and even an evolved predisposition in men. Most evolutionary psychologists distance themselves from this radically simplistic view of rape as something like the "forced copulation" displayed in the behavior of some other animals. Human action has to be understood on multiple levels, including the psychological and sociocultural.

Evolutionary psychologists generally attempt to avoid "panadaptationism" (or what Gould called "panglossianism").[16] Recognizing that the human mind has adaptive features does not imply that all behavior is

---

[12] See Martin Daly and Margo Wilson, *Sex, Evolution, and Behavior* (North Scituate, MA: Duxbury Press, 1978).
[13] See Daly and Wilson, *Homicide*, ch. 8.
[14] Robert Trivers, "Parent–offspring Conflict," *American Zoologist* 14 (1994): 249–264.
[15] Randy Thornhill and Craig Palmer, *The Natural History of Rape: The Biological Basis of Sexual Coercion* (Cambridge, MA: MIT Press, 2001). For a critical review, see Cheryl Brown Travis, ed., *Changing Evolutionary Theories of Sex Differences and Rape: An Interdisciplinary Response to Thornhill and Palmer* (Cambridge, MA: MIT Press, 2003).
[16] Gould and Lewontin, "The Spandrels of San Marcos and the Panglossian Paradigm": 581–598.

adaptive. Behavior as such did not evolve – "what evolved was the mind,"[17] Pinker writes, a complex functional system composed of "psychological faculties or mental modules" adapted to a vast array of specialized activities.[18] Evolutionary psychologists regard the mind as a highly complicated network of specialized functions suited to solving countless kinds of challenges, such as locating objects in motion, moving over difficult terrain and locating edible food. Moreover, human minds operate within cultures and cannot be interpreted apart from them. Even if we have specialized mental traits that at one time were adaptive for members of the species, the pervasive influence of culture and the widespread fact of cultural variability means that we have to be very careful about ascribing any ongoing adaptive value to particular biologically based inclinations that we experience in ordinary life or that evolutionary writers describe as "natural."

Thus even if there are evolved male–female differences beyond physiology, the fact that they are "natural" or "evolved" means neither that they necessarily continue to be adaptive in our present social context (they may or may not be) nor, even if they are adaptive, that they are morally normative. "Reproductively optimal acts" can be morally either right or wrong, and generalizations about human differences do not carry clear and direct implications for our normative moral judgments about sex, marriage, or family.

Evolutionary psychologist David Buss maintains that men and women are characterized by inherent psychological differences regarding "mate selection."[19] Men have evolved to be relatively more attracted to youth and physical beauty in women, and women to be relatively more attracted to social status and economic success in men. Men, he holds, have a proclivity for casual sex, desire sexual variety, and are more subject to visual arousal.[20] Buss argues that his cross-cultural studies suggest that these tendencies are innate, part of the species' "biogram," and not the product of culture alone. He moves from this account of an innate structure to argue that the best sexual ethic is one that practices wide toleration toward different sexual practices, that respects the choices of consenting adults, and that avoids Puritanism and self-righteousness, but there is little reason to think that these values are any more consistent with what we know about human sexuality than most other sexual ethics, including those taught by Christian traditions. Buss' account of male and female preferences, in and of themselves, does not compel us to adopt any particular ethic in relation to them.

---

[17] Pinker, *How the Mind Works*, p. 42. [18] Ibid., p. 27.
[19] See Buss, *The Evolution of Desire*. [20] Ibid., pp. 215–216.

His own normative perspective is adopted from modern culture, not science, and in fact he never gives it any conceptual justification.

Buss tends to think that behavior that appears repeatedly in different cultures is probably rooted in our evolved nature, but feminist anthropologist Sarah Hrdy rightly reminds us that the fact that a particular trait is pervasive across many cultures does not necessarily imply that it is innate or rooted in biology.[21] She is especially attuned to the ways in which appeals to biology have been used to "naturalize" class, racial, and ethnic inequalities by declaring them to be the results of natural selection and therefore best for the species overall (the latter of course is a social Darwinist rationale that gene-centered sociobiologists, as well as Hrdy, would reject). Hrdy finds more variation in mating-related behaviors within sexes than between them and maintains that women's control of resources influenced the mating strategies of both sexes.[22] She laments the fact that so much energy has been spent on "preference" studies because they do not capture the complex reality of sexuality and sexual behavior, particularly the variation within each sex and the environmentally sensitive and historically variable character of sexual strategies. Female preference for men with money might make sense in most contemporary societies because of the significant limits that societies impose on female economic opportunity,[23] but one might wonder whether a significant change in our present economic system might modify female preference.

What does an evolutionary approach to sex imply for marriage? Evolutionists often speak of marriage as a kind of "social contract," and as a form of mutually beneficial "reciprocal altruism" based on balanced "tit-for-tat" exchanges that benefit both parties and provide the social context for "parental investment" and "kin altruism." Our culture presumes that romantic love proceeds to marriage and that it yields benefits to both parties. Yet reciprocity also characterizes mating arrangements in cultures that practice arranged marriages or polygamy, and in those where one finds cohabiting couples with no formal legal commitment to one another.

---

[21] Hrdy, *The Woman that Never Evolved*, and "Introduction: Female Reproductive Strategies," in M. Small, ed., *Female Primates: Studies by Women Primatologists* (New York: Alan Liss, 1984), pp. 13–16.
[22] See Hrdy, "Raising Darwin's Consciousness: Female Sexuality and the Prehominid Origins of Patriarchy," *Human Nature* 8 (1997): 1–49.
[23] See John Horgan, "The New Social Darwinists," *Scientific American* 273 (October 1995): 176. *Contra* Buss, see A. H. Eagley, R. D. Ashmore, M. G. Makhijani, and I. C. Longo, "What Is Beautiful Is Good, But ... : A Meta-Analytic Review of Research on the Physical Attractiveness Stereotype," *Psychological Bulletin* 110 (1991): 109–128; A. H. Eagley, "The Science and Politics of Comparing Men and Women," *American Psychologist* 50 (1995): 145–158. More generally, see Ruth Bleier, ed., *Feminist Approaches to Science* (New York and London: Teachers College Press, 1986).

Marriage is often regarded as a social institution most appropriate for couples who wish for children, but marriage can be based on romantic love and interpersonal commitment rather than on child-bearing. Childless but loving marriages are regarded as successful, whereas procreative but emotionally cold marriages are not. Recent cultural trends have tended to diminish the place of procreation within marriage and seem to undercut what would seem to be (from a biological point of view) the primary significance of marriage. The current cultural ideal recognizes the "contractual" and "reciprocal" aspects of marriage but these are in the service of the interpersonal (rather than procreative) core of the relationship.[24] Sociobiologists claim that this focus on romantic emotions reflects the fact that our brains, and especially our emotions, have been "wired" to love and care for others. Marriage works as a reciprocal social contract made for the sake of reproduction when emotionally developed adults fall in love, bear children, and then selflessly care for them. The fitness interests of couples works best, in other words, when they do not calculate costs and benefits of behavior in relation to one another.

The term "reciprocity" should not disguise the fact that significant tension is built into the structure of the "mating relation." Competing sexual strategies lead to frequent conflict among married couples over sexual accessibility, the allocation of money and other material resources, the investment of time and effort, degrees of emotional commitment, and so forth. As Diamond puts it, behavior that is in a male's genetic interests "may not necessarily be in the interests of his female co-parent, and vice versa. That cruel fact is one of the fundamental causes of human misery."[25] Competing sexual strategies can generate various forms of mutual deception, sexual infidelity, and abuse. It certainly is not "a pretty picture," Buss says, but at least greater knowledge of these "psychological mechanisms" and of the "crucial contexts that activate those mechanisms" brings both understanding and a basis for intervention, modification, and perhaps even improvement of behavior.[26]

One brief if oversimplified example suffices to illustrate this approach. Buss argues: "A man's ability and willingness to provide a woman with resources are central to his mating value, central to her selection of him as a marriage partner, central to the tactics that men use in general to attract mates, and central to the tactics that men use to retain mates."[27] Marital dissolution worldwide is connected to male "provisioning failure" but the

---

[24] See Cahill, *Sex, Gender, and Christian Ethics*, ch. 1.   [25] Diamond, *Why Is Sex Fun?*, p. 19.
[26] Buss, *The Evolution of Desire*, p. 159.   [27] Ibid., p. 177.

converse, female "provisioning failure," does not constitute legitimate grounds for divorce in a single known culture.[28] Buss' perfectly reasonable point is that if society wants lasting and stable marriages, it must allow men to be good providers (and, from his point of view, this is by no means to suggest that we should not be equally committed to institutions and practices that enable women to be good providers as well).

What does this imply about ethics? Buss, attempting to avoid any hint of social Darwinism, insists that evolutionary psychology does not provide *moral* teachings but only "data" to be used by various moral perspectives in private moral decision-making and in public moral debate. Instead of deriving the "ought" from the "is," he seeks to use information about nature to clarify options for ethics, including means of empowering women. Science assists ethics as instrumental reason assists the end-determining activity of the human conscience.

Buss might be correct to argue, technically, that the practice of identifying differences between men and women – even differences rooted in one way or another in biological nature – need not be an inherently sexist enterprise. But Cahill rightly worries that evolutionary accounts of human sexuality, including "male promiscuity and female fidelity to mate and young ... seem to confer on these roles a normative status encouraging social acceptance if not exactly moral defense."[29] Scientific descriptions of human evolution are not in any way definitive for morality, unless what is "natural" is assumed *ipso facto* to be either inevitable or morally good. Even if it is the case that men "naturally" seek more partners and that women are naturally more sexually discriminating, there is no need to regard these predispositions as giving moral legitimacy to the "double standard." The same is true of "male proprietariness," sexual jealousy, and suchlike. Buss holds that understanding the conditions that tend statistically to favor various "mating strategies" (e.g. when it "pays" a female to "defect" from one mate and become available for another) provides "the possibility of choosing which to activate and which to leave dormant."[30] Buss assumes, without advancing the slightest argument, that we have free choice and are therefore not morally fated or otherwise bereft of moral resources, but he does not explain how he can justify such an assumption in evolutionary terms (or whether he thinks such a justification can be forthcoming from an evolutionary perspective).

The Darwinian theory of "parental investment" accords with "kin altruism" in predicting that, other things being equal, men and women

---

[28] Ibid., p. 178.   [29] Cahill, *Sex, Gender, and Christian Ethics*, p. 93.   [30] Ibid., p. 209.

will be more protective of their own children than other children. Men are more prone to allocate their resources to children known as their own, and to lose in evolutionary terms when they give resources to other men's children, so "certainty of paternity" plays an important role in paternal concern. This theory maintains that, among their own children, parents are inclined by nature to invest their limited resources in children who are most likely to be reproductively successful in the future. This means that parents have evolved to grow in their attachment to children as time passes (in rough proportion to their "investment" in them). It also implies that there is an evolutionary logic lying behind the practice of infanticide and child abandonment found in some cultures.[31]

Evolutionary psychologists Martin Daly and Margo Wilson argue that infanticide is found most often where paternity is unknown or inappropriate, or a newborn is deformed or otherwise compromised (rejection of an unfit newborn could be "fitness-promoting"), or there are insufficient familial resources (e.g. the mother is already overburdened by older siblings). The "parental investment" model is supported by the fact that individualized maternal love for an infant grows and deepens over the course of years, that younger children are at greater risk than older children and the latter tend to be more valued than the former, that mentally handicapped children are much more likely to be abused by their parents than other children, that women past their reproductive prime are less likely to harm a healthy, first-born child, and that stepchildren are at much greater risk of abuse and homicide than biological children.

Perhaps the earliest forms of "kin altruism" were expressed in child-care practices. Human parents are unique among primates in the length of time they spend rearing children. It would stand to reason that child-care practices such as maintaining physical proximity, holding, monitoring, feeding, clothing, would be highly adaptive for early humans. Instead of focusing on types of behavior, evolutionary psychologists are concerned with identifying a "kin recognition module" that contributes to the "discriminative parental solicitude"[32] – the differential care of genetic offspring as opposed to stepchildren and others. Altruistic behaviors were probably established in small kin groups before they were extended to others in the development of larger, multifamily groups. "Kin altruism" might also have contributed to the extension of altruistic behavior toward non-kin who were members of communities within which kin were the primary beneficiaries, especially if one of the markers of kinship was long-term association. In small, tight-knit

---

[31] See Daly and Wilson, *Homicide*, ch. 3. [32] Ibid., pp. 42ff.

bands or villages, caring for any member of the group would have a strong likelihood of benefiting kin.

The converse of the bias *for* kin and for strong parental investment is a tendency to be biased *against* non-kin in whom little or no investment has taken place. Evolutionists argue that human beings in every society tend to distinguish members of their own groups from those who are strangers or members of other groups and tend more easily to learn to trust members of their own group and to be suspicious of others. Sociobiologists hold that aggressive behavior today is made possible by natural capacities that were adaptive in the past. The closer the connection of one individual to another, the less willing one ought to be to inflict damage on the other; the more distant the relation, the less reluctant one ought to be to inflict damage. Nepotism and reciprocity give a bias against aggression and a bias in favor of toleration when it comes to kin. Daly and Martin have found that non-kin are killed much more often than kin and reciprocators, and, within the family, spouses are killed more often than children.[33]

Daly and Wilson connect Darwin's theory of sexual selection to aggressive behavior. The most common form of homicide is between unrelated young men lacking in three goods: status, resources, and mates. Sexual dimorphism among men and women is generally accompanied by male–male competition over differential status and respect. In this context, men are more prone to engage in risky confrontational interactions than are women. Evolutionists hold that in most tribal societies the capacity for violence is often regarded as a marker of virility, and retaliatory violence is regarded as justified or at least as an excusing condition that mitigates punishment. A willingness to extract vengeance for offenses can provide a deterrent to would-be aggressors. Though the social context of violence has changed enormously over the course of human history, the evolutionary theory of male sexual selection may be reflected in the fact that the most common context for urban homicide in the United States is the exchange of insults or an affront to an individual's status, reputation, or honor (all of which can function as social resources that signal "fitness").

"Kin altruism" is not belied by the recurrence of violence among intimates and within households. Evolutionists hold that sexual selection has shaped men to put their reproductive effort into "mating competition" and "status competition." Unlike most other mammals, men also cooperate in the upbringing of their children (minimally in providing resources). Because of the demands of paternal investment of one kind or another,

---

[33] Ibid., esp. ch. 2: "Killing Kinfolks."

however, men tend to place a high value on knowledge of paternity. "Male sexual proprietariness" is connected to the "double standard" that defines adultery as a crime of the wife but not of the husband. The evolution of male sexual jealousy is a way in which nature hedges against cuckoldry. The inclinations of "male sexual proprietariness" may play a role in why the most common form of homicide within families is that of husbands killing their wives.[34] By virtue of their size, men who want to control their women can employ coercion, intimidation, and beatings as well as the threat of death. Sociobiologists speculate that the capacity for uncontrollable rage following a sexual betrayal may also have evolved as an emotional deterrent to adultery. Again, evolutionary theorists mentioned here do not regard these tendencies as morally good, but rather as "data" for ethical reflection.

MORAL STANDARDS REGARDING SEX, MARRIAGE, AND FAMILY

I would next like to consider the implications of this view of natural-law ethics for normative standards regarding sex, marriage, and the family. The complexity of these domains, of course, cannot be examined in detail here, so I will have to restrict my comments to only the most basic points. My general position is that natural-law ethics responds in different ways to various aspects of these evolutionary accounts of some of the evolved dynamics underlying behavior in sex, marriage, and family. It must accept whatever factual claims are forthcoming from science. At the same time, it can distinguish realistic information from highly speculative and even egregiously attention-getting proposals. There is nothing in contemporary evolutionary theory that makes irrelevant the central Christian norms regarding sex, marriage, and family. It does offer interesting considerations for some of the reasons why nature, in addition to culture, makes monogamy and sexual fidelity difficult. It suggests that nature, and not only distinctive personalities, psychological difficulties, or moral weaknesses, might be responsible for some of the tensions within marriage and family. It also proposes that there are evolutionary bases for why people sometimes experience marriage and family as so meaningful – the scene of great happiness and not only one of tension, conflict, and trade-offs.

Evolutionary theory and natural-law ethics seem to be at opposite ends of the spectrum when it comes to describing what is going on in sex, marriage, and family. Thomas Aquinas, for example, regarded monogamous and heterosexual marriage as naturally human and other kinds of

[34] Ibid., chs. 6–7.

sexual activity as "unnatural." While adultery, seduction, fornication, and rape accord with the procreative end of sex, he thought, these acts were attacks on natural law as based in "right reason."[35] More grave for him were sins "against nature" such as bestiality, masturbation, and homosexual acts. "Natural" for Thomas meant "good," not morally neutral or pre-moral. He valued marriage and family as social institutions that allow for the expression of natural inclinations in a way that promotes the good of individuals and their communities. They are "natural" but not reducible to mere biology. Theologically, he understood our natural inclinations as part of the order of creation produced by God, the "Author of nature." Because we live in a "fallen state," we are constantly prone to deviate from the order of nature. Ideally, sacraments, prayer, and the moral life are shaped by grace to restore our behavior to its proper natural order. And of course for Thomas the richest level of human existence is not nature but grace, the realm of sanctity, asceticism, and spiritual transformation. As the rational creature's "participation" in the eternal law of God, the natural law provides the *basic* and *elementary* – if not most *elevated* or *excellent* – moral framework regulating sexual activity, spousal interaction, and family relations.

I believe we ought to appropriate selectively and critically insights about love and human nature from Thomas Aquinas. We no longer understand the world as ordered according to the plans of the "Author of nature" or human nature as originally created with perfect integrity and then corrupted by a single fateful human decision in paradise. Cosmic, biological, and human evolution to this point have proceeded through the interplay of "contingency" and "necessity," namely the "laws" of physics, chemistry, and biology. These "laws," or at least "law-like regularities," as Conway Morris, Howard Van Till, and others have argued, were constructed in such a way that, working on their own through the course of events, they eventually gave rise to creatures capable of consciously knowing and loving God and one another – and in this sense made in "the image of God." God constantly "interacts with" as well as "upholds" the creation, so this is not a remote and indifferent deistic Creator. Emergent complexity implies that this process is developmental, creative, and open to the future; it is anything but a closed material system.

In this context we can retain the notion that love "comes naturally" to us, at least in the sense that we are naturally primed to receive and give love ("natural" does not mean that it comes without training, effort, or choice). Our emergent cognitive, emotional, and social capacities have yielded a powerful need and desire to love. Our usual sense of the term "love"

---

[35] See *ST* II-II, 154–155.

emphasizes commitment, concern, and action, and in Christian ethics there is a tendency to emphasize compassion, service, and self-sacrifice, but I believe that Thomas Aquinas was right to hold that the deeper roots of love lie in our capacity for affective affirmation of the goodness of the beloved.[36] The person who loves experiences a moment of appreciation and quiet acceptance of what is good before he or she moves into a decision actively to promote what is good for the beloved.

Love is a human potentiality that must be shaped and guided by training and education. Because we love many people and things, and in different ways, our love must be given an appropriate ordering or be guided by a proper set of priorities. As social animals with the same human nature, we have the capacity to love one another. Important aspects of the elementary natural bases of love are seen in Damasio's analysis of the biological basis of the emotions and in sociobiological accounts of kin altruism and reciprocity. I emphasize that these account for "aspects" of love but that they do not explain the entirety of human nature.

We also ought to acknowledge that there is a certain kind of "naturalness" in the ancient axiom that we should love our friends and hate our enemies, advice that fits the "tit-for-tat" strategy of cooperating with cooperators and punishing cheats. Reciprocity and mutuality are deeply ingrained in human psychology: we tend to love those who love us and to be indifferent to those who do not. Yet Christian ethics recognizes every person as created in the image of God, capable of eternal fellowship with God, and therefore as a "neighbor." No one is excluded from the "universe of moral obligation"[37] on any grounds, so we are required to love both friends and enemies though not in the same way. The natural law holds that we are bound to have good will for every person, even enemies, but because of conditions and finitude and individual psychological and moral limits, we cannot be expected to develop deep bonds of affection with those who would deliberately harm us. The Christian ideal takes active steps to invite the enemy into a new relation of friendship, and it moves to mutuality as much as possible. But though we can and must have what Martin Luther King, Jr. called "redemptive good will" for every person,[38] we are not obligated to offer personal friendship, and the intimacy, trust,

---

[36] See Frederick E. Crowe, SJ, "Complacency and Concern in the Thought of St. Thomas," *Theological Studies* 20 (1959): 1–39, 198–230, 343–395; also Jules Toner, *Love and Friendship* (Milwaukee: Marquette University Press, 2003), ch. 7.

[37] See Helen Fein, *Imperial Crime and Punishment: The Massacre at Jallianwalla Bagh and British Judgment, 1919–1920* (Honolulu: University Press of Hawaii, 1977), ch. 1, especially pp. 18–19.

[38] Martin Luther King, Jr., *Strength to Love* (Cleveland: Collins World, 1963), p. 50.

and special concern it entails, equally to every human being regardless of their interest in or worthiness for such a relationship.

Christian ethics recognizes that there are limits to what can be expected of any one finite person. If one accepts the principle that grace perfects nature, then Christian love moves us not only to have "redemptive good will" for enemies and to meet the needs of strangers, but also to care for those people who are "the nearest and dearest" to us. This implies that Christian ethics gives a special place to the bonds and responsibilities within marriage and family. We are tied by affection and social communion to friends outside marriage and family, whose bonds should not wall us off from "non-kin" and "nonreciprocators." We naturally have greater affection for those nearest to us than for those who are more distant. It is morally "fitting" that we love our parents more than acquaintances, even if the latter are more interesting or attractive in other ways, and it is right that we love our own children more than those of others.[39] The subject of love cares for the needy object of love, both inside and outside marriage and family, but, as Thomas put it, parental care is "most necessary for the common good."[40] Because one's children are "part" of oneself, parents have greater love of care for their children than vice versa. This natural connection even grounds the moral legitimacy of using violence – at least as a last resort – to defend one's children from lethal harm because "one is bound to take more care of one's own life than of another's."[41]

Love for those who are needy is expressed in the virtue of mercy. Jules Toner captures the sense of mercy when he defines care as "the form love takes when the lover is attentive to the beloved's need."[42] We should not understand care-giving in a narrowly parochial way, as a love of "kith and kin" that is essentially unconcerned about others. Because care intends to meet the needs of others in a sufficient manner, care-giving should not be distributed indiscriminately.[43] Care-giving must be guided by the judgment of the virtue of prudence, which seeks to meet both primary and long-standing responsibilities as well as those of the most needy.

---

[39] Thomas used the notion of "right" in an objective sense rather than in the sense of moral power possessed by individual moral agents. See Odon Lottin, *Etudes de morale, histoire et doctrine* (Belgium: Gembloux, 1961), pp. 171–173, and *Le droit naturel chez Saint Thomas d'Aquin et ses prédécesseurs* (second edn., Bruges: Beyaert, 1931).
[40] *ST* II-II, q. 153, a. 3. By this Thomas meant not only physical care but also, and especially, spiritual guidance. See II-II, q. 10, a. 12.
[41] *ST* II-II, q. 64, a. 7.
[42] Jules Toner, *The Experience of Love* (Washington and Cleveland: Corpus Books, 1968), p. 80.
[43] *ST* II-II, q. 32, a. 10.

The ties of marriage and family, however, do not always take priority over other responsibilities. Multiple schemes of priority are relevant to different spheres of life, modes of connection, and shared goods. Because different spheres of social life require different priority systems, claims of kinship only precede other claims in the arena of basic well-being. Thus in cases of conflict between needs of a friend and needs of a stranger, the urgency of the latter can override the prima facie priority given to the former. Claims of kinship can be overridden by more urgent needs from other quarters.

Traditional Roman Catholic sexual ethics was based on a view of the natural law that presumed a normative and primarily philosophical view of nature based on Aristotelian finality: a person functioning according to his or her in-built inclinations is acting "according to nature"; departures from these natural inclinations amount to harmful contradictions to nature. This treatment of our obligations to those whom we love provides the larger framework for the norms that regulate sex, marriage, and family. This moral vision regards sexual intercourse as properly taking place between a man and a woman in a way that does not deliberately obstruct conception, the spontaneous outcome of sexual intercourse. We have an ethical obligation to conform to the natural purpose of sex because doing so is good for the agent and because it adheres to the will of God, the "Author of nature." This view assumes that human reproductive biology was deliberately designed by God for an end. For this reason, any act involving a deliberate violation of this end – homosexual sexual acts, masturbation, the use of artificial birth control, and sterilization – is inherently wrong, "intrinsically evil," and an affront to the divine will.

As mentioned above, this design-like understanding of God as the Author of nature can no longer be sustained by those who accept an evolutionary framework for the cosmos, life on our planet, and human nature. The theology assumed by older natural law might be in some ways acceptable to those who advocate the theory of "intelligent design" but it is simply incompatible with our knowledge of evolution.[44] (It is interesting to note that evangelical Christians who accept the "intelligent design" position as an alternative to Darwinism have not also been known to apply it to sexual ethics in a way that leads to the prohibition of artificial

---

[44] On intelligent design, see Michael Behe, *Darwin's Black Box: The Biochemical Challenge to Evolution* (New York: The Free Press, 1996) and William Dembski, ed., *Mere Creation: Science, Faith, and Intelligent Design* (Downers Grove, IL: InterVarsity Press, 1998). Phillip E. Johnson is one of the major opponents of theistic evolutionism. See his *Darwin on Trial*.

contraception.) If God creates in an improvisational way through the patterns and contingencies of the evolutionary process, then the natural structures of reproductive biology and the evolved basis of human sexuality have to be appropriated in a more circumspect and selective manner than one finds in classical natural law. This way of understanding God in relation to nature cannot help but weaken the kind of normative weight one puts on the biological structures involved in sex and reproduction.

Contemporary Catholic moral theologians do not disagree over whether sex has natural ends – at least in the "teleonomic" sense discussed earlier – but they do differ over whether these ends have *normative* status and, if so, what kind of normative status they possess. Moral theology in the last half-century has moved away from the physicalism of previous approaches to sexual morality. John Paul II continued to maintain that each and every act of intercourse ought to remain open to procreation, but he tended to make the personalist argument that we are morally bound to do so because it is the only way fully to give oneself in radical love. This kind of argument, however, displays a kind of phenomenological circularity that proceeds this way: phenomenon "X" has the following traits A, B, and C. Anyone who does not find these traits does not grasp the nature of X, and anyone who claims to understand phenomenon X in the absence of A, B, and C is simply mistaken about the meaning of X. True self-giving love cannot include artificial contraception, and anyone who claims that true self-giving love is compatible with the use of artificial contraception simply has misunderstood the nature of true self-giving love. This approach posits that a particular virtue, marital love, includes certain necessary conditions and then argues that any failure to include all the stipulated conditions by definition not only falls short of the virtue but also implicates the agent in moral evil. This kind of argument only leads one to surmise that whoever defines the starting premises controls the conclusion.

Revisionist natural-law moralists interpret sexual ethics more significantly on the basis of ordinary experience and contemporary science than do traditionalists, to stress the distinctively human and interpersonal aspects of marital love and commitment more than the physical aspects of sexual intercourse, and to attend carefully to the broad social and cultural context of sexuality rather than to emphasize its biological nature. They hold that marriages need to be open to procreation as a *general* good shared over the course of their life together but not that the couple has a moral duty to be open to procreation in *every* act of sexual intercourse. Indeed, as Cahill points out with reference to natural family planning, the magisterium's willingness in *Humanae Vitae* to accept the moral legitimacy of married couples engaging "in sex acts while intentionally avoiding procreation ... envisioned a meaning of sex that was non-procreative, that expressed the mutual commitment of the couple."[45]

Revisionists claim that complete self-giving love is not necessarily frustrated by the attempt to foreclose the possibility of further procreation. They insist that there is a massive moral difference between teenagers using a condom after the dance and the use of artificial contraception by an impoverished forty-year-old married couple who already have seven children. While traditionalists hold that artificial birth control is never licit, revisionists regard it as a general "pre-moral disvalue"[46] but judge its concrete moral status on a case-by-case basis and with sensitivity to how particularities shape cases. The revisionist argues that intention, relationship, circumstances, and consequences all have to be considered when making a moral decision with this degree of complexity and conflict. The former position carries the advantage of clarity and consistency, but it is attended by the disadvantage of confusing goods that *in general* and in the *abstract* contribute to human flourishing with the specific and practical needs of *concrete* individuals who are striving to live well under very difficult situations. The same kinds of concerns obtain in difficult cases involving, for example, divorce and remarriage, same sex marriage, and sterilization. I believe that knowledge of evolution supports this greater sensitivity to concrete particularities in sexual ethics because it does not draw such an intimate connection between the Creator's will and the natural reproductive end of sex.

NATURAL LAW AND THE EVOLUTION OF SEX

We can now turn to the relation between natural law and contemporary evolutionary theory in light of the evolution of sex. It would seem that their prima facie relation is one of uncompromising mutual opposition. Nature is a competitive if also a cooperative evolutionary experiment, not a smoothly integrated "order of nature," as Thomas thought. Evolution undercuts any assumption that we ought to strive to return to an original moral order. There is no reason to think that there was ever a time when we were not conflictual, manipulative, selfish, and prone to deceit and violence – as well as cooperative, generous, empathic, and altruistic. Christian ethics

---

[45] Cahill, *Sex, Gender, and Christian Ethics*, p. 200.
[46] See Louis Janssens, "Norms and Priorities in a Love Ethics," *Louvain Studies* 6 (1977): 207–238; and "Ontic Good and Evil – Premoral Values and Disvalues," *Louvain Studies* 12 (1987): 62–82.

regards the existence of creation in general, and human creatures in particular, as (ontologically) "good" or "valuable" or "worthwhile" but not necessarily as morally "virtuous." Religion and morality, writes Cahill, are charged with the "'humanization' of the species."[47] "Far from a simple loosening of social controls on the body, the moral question for a Christian ethics of sex and gender becomes how to socialize the body – as male and female, as sexual, as parental – in ways which enlarge our social capacities for compassion toward others and solidarity in the common good."[48] Christianity strives to "humanize" by raising us to higher and more inclusive ways of being and acting. It seeks to develop and extend valuable aspects of human nature given as the "raw materials" of our evolutionary heritage for the sake of the highest good, union with God and one another. Evolutionists, on the other hand, seek to "humanize" us by giving us knowledge of the levers and pulleys of human nature that we can manipulate for the sake of maximum damage control. They strive to use morality for the sake of physical survival. Natural-law ethics regards marriage as the normative context of sex and procreation, evolutionism views it as only one among many "mating strategies" to which it is bound to be morally neutral. The former regards the whole sphere of the "moral" as a way of describing the pursuit of true happiness through virtue and right action, the latter regards the moral as, at best, a means of resolving intra-group conflict and attaining a modicum of social stability and order.

Nature in evolutionary theory is neither purposive nor normative. This is for two important reasons: (1) purposes come from the *minds* of human agents and do not reside in physical structures, acts, or processes unless a mind puts them there (as in an automobile engine) and (2) even if there were natural purposes in organisms, one cannot infer, at least from a purely scientific standpoint, that such purposes are *morally* obligatory (biologists do not say that bats "ought" to eat bugs or that our lungs have a "moral obligation" to breathe).

In response to the first claim, we have already seen that evolution is replete with examples of functionality: organs evolve because they are adaptive – the eye to see, the heart to pump blood – this is their end, why they exist in the first place. Ends, unlike intentions, do not have to be produced by minds to exist in nature. Applied to sex, natural-law ethics is right to hold that the natural function of sex, at least in the human species, includes reproduction and affective bonding (or, in the language of Christian ethics, "procreative" and "unitive" ends). Organisms function

---

[47] Cahill, *Sex, Gender, and Christian Ethics*, p. 77.   [48] Ibid., p. 164.

in the way they do because of the interaction of contingency and natural selection. As Mayr explains, "It is illegitimate to describe evolutionary processes or trends as goal-directed (teleological). Selection rewards past phenomena (mutation, recombination, etc.) but does not plan for the future, at least not in any specific way."[49] Evolution is a wasteful process of trial and error; changes in the evolutionary process are not goal-directed. Species evolved because they had specific traits that gave them a fitness advantage under specific evolutionary circumstances. There is no fixed order of nature, no progress, no inherent natural end unrelated to reproduction.

"Natural" in evolutionary terms simply means predictable and occurring with some regularity under set conditions. Thus it is "natural" for people to form "in-groups," to discriminate against members of "out-groups," to help friends and to be suspicious of strangers. It is "natural" to deceive, threaten, manipulate, mutilate, and kill one's enemies under certain conditions. Some primatologists have argued that engaging in infanticide can be "natural."[50] As T. H. Huxley put it some time ago, "The thief and the murderer follow nature just as much as the philanthropist."[51]

Is there any way to harmonize natural law and evolutionary theory, or at least put them within shouting distance of one another? No, not if they are understood to be two competing theories of science, moral perspectives, or ideologies. Yet it seems to me that differentiating scientific and moral modes of discourse supplies a basis for thinking about rapprochement between the *science* provided by evolutionary theory and the *moral insight* offered by natural-law ethics. Science has to be distinguished from subtle, or not so subtle, forms of inappropriate moralizing, such as when sociobiologists endorse the value of tolerance, seem to be resigned not only to the dominance of the "selfish gene" but also to the selfish individual, encourage a sexual ethic of "consenting adults," and describe the universe as an "amoral" construction of a "Blind Watchmaker"[52] – a metaphor that is no less philosophical than the contrary theistic claim that the universe has a moral purpose. Natural lawyers, for their part, often make assumptions about empirical matters that are subject to scientific investigation, such as when they describe human beings as "social," make judgments about the status of embryonic life, and offer generalizations about the psychological

---

[49] Mayr, *Towards a New Philosophy of Biology*, p. 60.
[50] See G. Hausfater and S.B. Hrdy, eds., *Infanticide: Comparative and Evolutionary Perspectives* (New York: Aldine de Gruyter, 1984).
[51] T. H. Huxley, *Evolution and Ethics*, p. 77.
[52] See Richard Dawkins, *The Blind Watchmaker* (New York and London: W. W. Norton, 1986).

and social effects of certain kinds of sexual activity. In response to a traditionalist natural-law view of homosexuality, for example, one can ask: must all forms of gay sex always be inherently morally debilitating?[53] And how would one ever know enough to make this kind of sweeping empirical generalization? Evolutionists *qua* scientists can speak about biological functions, and not about moral purposes, which are by definition intended by moral agents. Conversely, advocates of natural law increase the plausibility of their positions to the extent that they show some empirical basis for them.

Applied to sex, marriage, and the family, this suggests that natural law can learn from scientific evolutionary claims about human behavior without also accepting ideological positions that are often joined to them. Evolutionary psychology is unhelpful, and even downright destructive, when it advances a moral ideology based on unsubstantiated speculation, metaphysical materialism, ontological reductionism, genetic determinism, and ethical egoism. Tolerance might be a value that is needed in sexual ethics, but sociobiology has to make the moral case for it and not simply assert that we ought to be tolerant. The moral defense for tolerance is provided by ethics, not science. Natural law challenges evolutionism when it paves over the complexities, richness, and diversity of human behavior. This having been said, there is no reason why natural-law ethics cannot accept the findings of evolution regarding the strictly *biological* dimensions of human sexuality, sexual behavior, marriage, and family. As in Thomas' day, natural law today needs to build upon available knowledge of nature rather than ignore it. "Truth cannot be against truth."[54] Natural-law ethics should acknowledge that there are some ultimate, remote evolutionary reasons for the patterns we find in nature. Nor is it odd, as I argued above, to think of the development of morality as a way of monitoring compliance and noncompliance in situations of reciprocity and cooperation and of inhibiting forms of conflict and promoting order in the midst of competing incentives.

Sexual desire is of interest to both evolutionists and natural lawyers. It presents an enormously powerful motivation devised by nature for spreading genes into future generations. Sociobiologists point out that sexual desire preceded by many millions of years the exponential growth of the human

---

[53] See Andrew Sullivan, *Virtually Normal* (New York: Vintage, 1996).

[54] See John Paul II, *Fides et Ratio: On the Relationship between Faith and Reason* (Boston: Daughters of St. Paul, 1998) and also "Message to the Pontifical Academy of Sciences on Evolution" (October 22, 1996).

brain between 600,000 and 100,000 years ago. In fact, these millions of years have "wired" us to be among the most sexually active of all species. And of course advertising agencies have become extremely clever at manipulating our evolved pleasure circuitry with artificial stimuli that lead to forms of pleasure detached from both interpersonal love and procreation.

Evolutionary thinkers tend to stress the dark side of human nature. This is in part because they have followed the modern reductionistic tendency to explain our higher capacities in terms of our lower drives and more elementary factors. Contemporary natural law reverses this direction by attempting to relate the lower impulses of our nature to our most noble aspirations. We need to be realistic about the various causes of human discontent and disorder while also appreciating the fact that men and women have significant natural capacities for love, care, and commitment within marriage and family.

Natural law relates moral standards to the human good, which includes sex, companionship, reproduction, and kinship. This affirmation is significant not only for communities in which the value of marriage, sex, and procreation is called into question (as it has been with periodic revivals of Manicheanism) but also those, like our own, in which erotic desires are overvalued, limited only by voluntary consent, and not adequately balanced with the pursuit of more important human goods.

This having been said, natural law does not regard the human good as exhausted by sex, companionship, reproduction, and kinship, which when taken as supreme goods imply, respectively, various forms of hedonism, sentimentality, familialism, nepotism, and tribalism. The essential implication of the notion of human emergent complexity is that the human good embraces *all aspects* of human nature: not only the physical, chemical, and biological but also the social, cultural, psychological and emotional, intellectual, moral, and religious dimensions of human existence. This is crucial in societies that place so much importance on the family as a "haven in a heartless world" that they fail to appreciate both goods higher than the family and the significance of wider responsibilities to the larger human community.

Human flourishing incorporates the full range of human goods and assimilates amenable components of human nature. Presuming they exist, evolved "in-built mechanisms" of the sort discussed by evolutionary psychologists provide "matter" to be related to the larger "form" of our movement to the human good. Christian ethics should not adopt a monolithic response to these "mechanisms." Some are to be resisted, some softened, and some developed, depending on the relation to the human good, including the goods involved in sex, marriage, and family. Sexual jealousy, for example, can be virtuous if it means that one is vigilant in protecting one's beloved or intolerant of unfaithfulness; indifference to a spouse's philandering is not a sign of a healthy marriage. But jealousy can obviously also be a terrible vice when it leads to disordered possessiveness, attempts unjustly to control the beloved, and irrational suspicion.

The same differentiation can be made regarding the purported "in-built mechanism" that identifies individuals as one of "us" or one of "them." This tendency can be used for good, as when it creates strong bonds in a business or small company, or the camaraderie of Henry V's "band of brothers." But it has also been liable to be manipulated by dictators and co-opted to support xenophobia, racism, mob violence, exploitation, colonialism, and genocide. Most evolved mechanisms can be developed in virtuous or vicious ways, and their behavioral expression depends on human choices, social context, and the moral ideals for which we strive.

The greatest human moral ideal is love – the affective affirmation of, and movement toward, what is recognized to be good. This includes not only romantic love, but benevolence, friendship, and solidarity. Natural law strives to shape every truly human act to be an expression of love, either as affirmation of what is good or as the rejection of what is evil. One of the most powerful expressions of romantic love is sex, which has the capacity to put in bodily form the content of one's affections, will, and ideals. Human sex is complex and multifunctional, and, for Christian ethics, sexual activity needs to be placed as much as possible at the service of love and commitment. The moral capacity to hold all three of these goods together – sex, love, and commitment – radically separates us from other primates. One of the greatest ways in which we care for one another is through sex. Conversely, some of the most painful ways in which people harm one another is through the misuse of sex, particularly when pursued without love and commitment. Whereas an evolutionist might regard Don Juan's pursuit of sexual conquests as nothing more than a flamboyant example of the male "mating strategy," natural law would recognize it as cruel exploitation.

The institutionalization of love is found in marriage and family. Marriage is a human way of ordering the sexual and reproductive inclinations we have in common with animals as well as the distinctive in-built mechanisms that compose our own distinctive human biology. It is, however, much more than biology. Animals go into heat, mate, and rear young according to patterns set by evolution. Human beings, on the other hand, undertake interpersonal and social commitments, forge strong affective bonds, engage in frequent sex, and deliberately take up a way of life that promotes the ongoing growth of the human purposes of sexuality. Marriage and family are thus "natural" in the same sense that Aristotle regarded forming friendships and living in political communities as "natural" to us – they are institutions that build upon and allow for the satisfaction of natural desires and in a way that contributes to human well-being in a comprehensive sense – individually, interpersonally, and communally.

Human moral development moves toward ideals higher than the purely reproductive functionality of animal behavior. Religiously speaking, this development can be included in a larger sacramental view of reality. Because of our unprecedented cognitive powers, it is sure to be the case that if we do not make sex, marriage, and family the scene of altruism, integrity, and other forms of moral goodness, they will lead to levels of harm and self-destruction far worse than anything ever seen in the animal kingdom. The natural-law tradition of Christian ethics thus seeks to build on what is best in human nature in all its emergent complexity while striving for an ultimate good that is bestowed on us by the gift of divine grace.

Bibliography

Alexander, Richard D., *Darwinism and Human Affairs*, Seattle and London: University of Washington Press, 1979.
The Biology of Moral Systems, New York: Aldine de Gruyter, 1987.
Andolsen, Barbara Hilkert, "Agape in Feminist Ethics," in Lois K. Daly, ed., *Feminist Theological Ethics*, Louisville, KY: Westminster John Knox, 1994, pp. 146–159.
Aquinas, Thomas, *Summa Theologiae*, 3 vols., trans. English Dominican Province, New York: Benziger Brothers, 1946; Rome: Marietti, 1948.
Summa contra Gentiles, On the Truth of the Catholic Faith: Summa contra Gentiles, Book One, trans. Anton C. Pegis., New York: Image, 1955.
Summa contra Gentiles, On the Truth of the Catholic Faith: Summa contra Gentiles, Book Two, trans. James F. Anderson, New York: Image, 1955.
Summa contra Gentiles, On the Truth of the Catholic Faith: Summa contra Gentiles, Book Three, trans. Vernon J. Bourke, New York: Image, 1955.
De Veritate, Disputed Questions on the Virtues, trans. E. M. Atkins, New York: Cambridge University Press, 2005.
Aristotle, "The Rhetoric," in Richard McKeon, ed., trans. ed. W. Rhys Roberts, *The Basic Works of Aristotle*, New York: Random House, 1941.
Nicomachean Ethics, Indianapolis, IN: Bobbs-Merrill, 1962.
Arnhart, Larry, *Darwinian Natural Right: The Biological Ethics of Human Nature*, Albany, NY: SUNY, 1998.
"Thomistic Natural Law as Darwinian Natural Right," *Social Philosophy and Policy* 18 (winter 2001), 1–33.
"The Darwinian Moral Sense and Biblical Religion," in Philip Clayton and Jeffrey Schloss, eds., *Evolution and Ethics: Human Morality in Biological and Religious Perspective*, Grand Rapids, MI: Eerdmans, 2004, pp. 205–220.
Artigas, Mariono, *Galileo in Rome: The Rise and Fall of a Troublesome Genius*, New York: Oxford University Press, 2003.
Augustine, *On Christian Doctrine*, trans. D. W. Robertson, Jr., Indianapolis: Bobbs-Merrill, 1958.
Concerning the City of God against the Pagans, trans. Henry Bettenson, New York: Penguin, 1984.
Confessions, trans. Henry Chadwick, New York: Penguin, 1992.
Axelrod, David, *The Evolution of Cooperation*, New York: Basic Books, 1984.

Ayala, Francisco J., "The Autonomy of Biology as a Natural Science," in A. D. Breck and W. Youngrau, eds., *Biology, History, and Natural Philosophy*, New York: Plenum Press, 1972, pp. 1–16.
  "Introduction," in Francisco J. Ayala and Theodosius Dobzhansky, eds., *Studies in the Philosophy of Biology: Reduction and Related Problems*, Berkeley: University of California Press, 1974.
  "Reduction in Biology: A Recent Challenge," in David Depew and Bruce Weber, eds., *Evolution at the Crossroads*, Cambridge, MA: MIT Press, 1985, pp. 67–78.
  "Biological Reductionism: The Problems and Some Answers," in F. E. Yates, ed., *Self-Organizing Systems: The Emergence of Order*, New York: Plenum Press, 1987, pp. 315–324.
  "The Biological Roots of Morality," in Paul Thompson, ed., *Issues in Evolutionary Ethics*, Albany: State University of New York, 1995, pp. 293–316.
  "The Difference of Being Human," in Holmes Rolston III, ed., *Biology, Ethics, and the Origins of Life*, Boston and London: Jones and Bartlett Publishers, 1995, pp. 117–135.
  "Biological Evolution: An Introduction," in James B. Miller, ed., *An Evolving Dialogue: Scientific, Historical, Philosophical, and Theological Perspectives on Evolution*, Washington, DC: American Association for the Advancement of Science, 1998, pp. 9–54.
  "Darwin's Devolution: Design without a Designer," in Robert John Russell, William R. Stoeger, SJ, and Francisco J. Ayala, eds., *Evolutionary and Molecular Biology: Scientific Perspectives on Divine Action*, Vatican City State: Vatican Observatory Publications, and Berkeley, CA: Center for Theology and the Natural Sciences, 1998, pp. 101–116.
Bacon, Francis, *The Advancement of Learning, Book I*, ed. Hugh G. Dick, New York: Random House, 1955.
Barash, David, *The Whisperings Within: Evolution and the Origin of Human Nature*, New York: Harper and Row, 1979.
Barbour, Ian, *Religion in an Age of Science*, San Francisco: Harper, 1990.
  Religion and Science: Historical and Contemporary Issues, San Francisco: HarperCollins, 1998.
  "Neuroscience, Artificial Intelligence, Human Nature," in Robert John Russell, Nancey Murphy, Theo C. Meyering, and Michael A. Arbib, eds., *Neuroscience and the Person: Scientific Perspectives on Divine Action*, Vatican City State: Vatican Observatory Publications, and Berkeley, CA: Center for Theology and the Natural Sciences, 2002, pp. 249–280.
Barker, Eileen, "Apes and Angels: Reductionism, Selection, and Emergence in the Study of Man," *Inquiry* 19 (1976), 367–399.
Barkow, Jerome H., Cosmides, Leda, and Tooby, John, *The Adapted Mind: Evolutionary Psychology and the Generation of Culture*, New York: Oxford University Press, 1992.
Baron-Cohen, Simon, "The Evolution of a Theory of Mind," in Michael C. Corballis and Stephen E. G. Lea, eds., *The Descent of Mind: Psychological Perspectives on Hominid Evolution*, New York: Oxford University Press, 1999, pp. 261–277.

Barr, James, *Biblical Faith and Natural Theology*, Oxford: Clarendon Press, 1993.
Barth, Karl, *Church Dogmatics* III/4, trans. A. T. Mackay *et al.*, Edinburgh: T. & T. Clark, 1961.
Batson, C. Daniel, *The Altruism Question: Toward a Social Psychological Answer*, Hillsdale, NJ: Lawrence Erlbaum Associates, 1991.
Beckstrom, E. G., *Darwinism Applied: Evolutionary Paths to Social Goals*, Westport, CT: Praeger, 1993.
Behe, Michael, *Darwin's Black Box: The Biochemical Challenge to Evolution*, New York: The Free Press, 1996.
Beilby, James, ed., *Naturalism Defeated? Essays on Plantinga's Evolutionary Argument against Naturalism*, Ithaca, NY: Cornell University Press, 2002.
Bellah, Robert, Sullivan, William M., Tipton, Steven M., Madsen, Richard, and Swidler, Ann, eds., *Habits of the Heart: Individualism and Commitment in American Life*, San Francisco: Harper and Row, 1985.
Berlin, Isaiah, *Four Essays on Liberty*, London and Oxford: Oxford University Press, 1970.
Berra, Tim M., *Evolution and the Myth of Creationism: A Basic Guide to the Facts in the Evolution Debate*, Stanford, CA: Stanford University Press, 1990.
Blackburn, Simon, *The Oxford Dictionary of Philosophy*, New York: Oxford University Press, 1994.
Blanshard, Brand, "The Case for Determinism," in Sidney Hook, ed., *Determinism and Freedom in the Age of Modern Science*, New York: Collier Books, 1958, pp. 3–15.
Bleier, Ruth, ed., *Feminist Approaches to Science*, New York and London: Teachers College Press, 1986.
Boehm, Christopher, *Hierarchy in the Forest: The Evolution of Egalitarian Behavior*, Cambridge, MA: Harvard University Press, 1999.
Bohm, David, *Wholeness and the Implicate Order*, Boston: Routledge, 1980.
Bourke, Vernon, *Will in Western Thought: An Historico-Critical Survey*, New York: Sheed and Ward, 1964.
Bowler, Peter J., *Evolution: The History of an Idea*, revised edn., Berkeley: University of California Press, 1989.
Boyd, Craig A., "Was Thomas Aquinas a Sociobiologist? Thomistic Natural Law, Rational Goods, and Sociobiology," *Zygon* 39.3 (September 2004), 659–680.
  "Thomistic Natural Law and the Limits of Evolutionary Psychology," in Philip Clayton and Jeffrey Schloss, eds., *Evolution and Ethics: Human Morality in Biological and Religious Perspective*, Grand Rapids, MI: Eerdmans, 2004, pp. 221–237.
  "Participation Metaphysics in Aquinas's Theory of the Natural Law," *American Catholic Philosophical Quarterly* 79.3 (2005), 169–185.
Boyd, Robert and Richerson, Peter J., *Culture and the Evolutionary Process*, Chicago: University of Chicago Press, 1985.
Boyer, Pascal, *Religion Explained: The Evolutionary Origins of Religious Thought*, New York: Basic Books, 2001.

Bradie, Michael, "The Moral Status of Animals in Eighteenth Century British Philosophy," in Jane Maienschein and Michael Ruse, eds., *Biology and the Foundations of Ethics*, New York: Cambridge University Press, 1999, pp. 32–51.

Brooke, John H., "The Relations between Darwin's Science and His Religion," in John Durant, ed., *Darwinism and Divinity*, New York and Oxford: Basil Blackwell, 1985, pp. 40–75.

Science and Religion: Some Historical Perspectives, Cambridge: Cambridge University Press, 1991.

Brooke, John H., Lindberg, David C., and Numbers, Ronald L., eds., *God and Nature: Historical Essays on the Encounter between Christianity and Science*, Berkeley: University of California Press, 1986.

Brosnan, Sarah F. and de Waal, Frans B. M., "Monkeys Reject Unequal Pay," *Nature* 425 (September 2003), 297–299.

Brosnan, Sarah F., Schiff, Hillary C., and de Waal, Frans B. M., "Tolerance for Social Inequity may Increase with Social Closeness in Chimpanzees," *Proceedings of the Royal Society B: Biological Sciences* 272, no. 1560 (February 2005), 253–258.

Brown, Raymond E., *The Community of the Beloved Disciple*, Ramsey, NY: Paulist Press, 1979.

Brown, Warren, Murphy, Nancey, and Maloney, H. Newton, eds., *Whatever Happened to the Soul?*, Minneapolis: Fortress Press, 1998.

Browne, E. Janet, *Charles Darwin: Voyaging*, Princeton: Princeton University Press, 1996.

Charles Darwin: The Power of Place, New York: Knopf, 2002.

Browning, Don, "Science and Religion on the Nature of Love," in Stephen G. Post, Lynn G. Underwood, Jeffrey P. Schloss, and William B. Hurlbut, eds., *Altruism and Altruistic Love: Science, Philosophy, and Religion in Dialogue*, New York: Oxford University Press, 2002, pp. 335–345.

Christian Ethics and the Moral Psychologies, Grand Rapids, MI: Eerdmans, 2007.

Browning, Don, Miller-McLemore, Bonnie J., Coture, Pamela D., Lyon, K. Brinolf, and Franklin, Robert, eds., *From Culture Wars to Common Ground: Religion and the American Family Debate*, Louisville, KY: Westminster John Knox Press, 1997.

Bube, Richard H., "The Failure of the God-of-the-Gaps," in James E. Huchingson, ed., *Religion and the Natural Sciences: The Range of Engagement*, Fort Worth: Harcourt Brace, 1993, pp. 131–140.

Burnaby, John, *Amor Dei*, London: Hodder and Stoughton, 1938.

Burrell, David B., CSC, *Freedom and Creation in Three Traditions*, Notre Dame, IN: University of Notre Dame Press, 1993.

Burrell, David B., CSC, and Malits, Elena, CSC, *Original Peace: Restoring God's Creation*, Mahwah, NY: Paulist Press, 1997.

Buss, David M., *The Evolution of Desire*, New York: Basic Books, 1994.

The Dangerous Passion: Why Jealousy Is as Necessary as Love and Sex, New York: Free Press, 2000.

Buss, David M., Haselton, M. G., Shackelford, T. K., Bleske, A. L., and Wakefield, J. C., "Adaptations, Exaptations, and Spandrels," *American Psychologist* 53 (1998), 533–548.

Byrne, Richard W. and Whiten, Andrew, eds., *Machiavellian Intelligence: Social Expertise and the Evolution of Intellect in Monkeys, Apes, and Humans*, Oxford: Clarendon Press, 1988.

Cahill, Lisa S., "Women, Marriage, Parenthood: What Are Their 'Natures'?" *Logos: Philosophical Issues in Christian Perspective* 9 (1988), 11–35.

Sex, Gender, and Christian Ethics, New York: Cambridge University Press, 1996.

"Toward Global Ethics," *Theological Studies* 63.2 (June 2002), 324–344.

Calvin, John, *Institutes of the Christian Religion*, 2 vols., trans. Lewis Ford Battles, Philadelphia: Westminster Press, 1960.

Campbell, Donald T., "Downward Causation in Hierarchically Organized Biological Systems," in Francisco J. Ayala and Theodosius Dobzhansky, eds., *Studies in the Philosophy of Biology: Reduction and Related Problems*, Berkeley: University of California Press, 1974, pp. 179–186.

"On the Conflict between Biological and Social Evolution and the Concept of Original Sin," *Zygon* 10 (1975), 234–249.

"On the Conflicts between Biological and Social Evolution and between Psychology and Moral Tradition," *American Psychologist* 30 (December 1975), 1103–1126.

Carnegie, Andrew, *The Autobiography of Andrew Carnegie*, Lebanon, NH: Northeastern University Press, 1986.

Carroll, Robert L., *Patterns and Processes of Vertebrate Evolution*, Cambridge and New York: Cambridge University Press, 1997.

Catechism of the Catholic Church, second edn., Vatican City: Libreria Editrice Vaticana, Ligouri Publications, 2000.

Cates, Diana Fritz, "The Virtue of Temperance," in Stephen J. Pope, ed., *The Ethics of Aquinas*, Washington, DC: Georgetown University Press, 2002, pp. 321–339.

"The Religious Dimension of Ordinary Human Emotions: Working with Gustafson and Nussbaum," *Journal of the Society of Christian Ethics* 25.1 (2005), 35–53.

Cavalli-Sforza, Luigi L. and Feldman, Marcus W., *Cultural Transmission and Evolution*, Princeton, NJ: Princeton University Press, 1981.

Clark, Stephen R. L., *The Political Animal: Biology, Ethics and Politics*, London and New York: Routledge, 1999.

Biology and Christian Ethics, Cambridge: Cambridge University Press, 2000.

Clark, William R. and Grunstein, Michael, *Are We Hardwired? The Role of Genes in Human Behavior*, New York: Oxford University Press, 2000.

Clayton, Philip, "Neuroscience, the Person, and God: An Emergentist Account," in Robert John Russell, Nancey Murphy, Theo C. Meyering, and Michael A. Arbib, eds., *Neuroscience and the Person: Scientific Perspectives on Divine Action*, Vatican City State: Vatican Observatory Publications, and Berkeley, CA: Center for Theology and Natural Sciences, 2002, pp. 181–214.

"Biology and Purpose," in Philip Clayton and Jeffrey Schloss, eds., *Evolution and Ethics: Human Morality in Biological and Religious Perspective*, Grand Rapids, MI: Eerdmans, 2004, pp. 332–333.
Clifford, Richard J., "The Hebrew Scriptures and the Theology of Creation," *Theological Studies* 46 (1985), 507–523.
"Creation in the Hebrew Bible," in Robert J. Russell, William R. Stoeger, SJ, and George V. Coyne, SJ, eds., *Physics, Philosophy, and Theology: A Common Quest for Understanding*, Notre Dame, IN: University of Notre Dame Press, and Vatican City State: Vatican Observatory Publications, 1988, pp. 160–164.
Cole, R. David, "Genetic Predestination?," *Dialog* 33:1 (winter 1994), 17–22.
Conn, Walter, *The Desiring Self*, New York: Paulist Press, 1998.
Conors, Russell B., Jr. and McCormick, Patrick T., *Character, Choices, and Community: Three Faces of Christian Ethics*, Mahwah, NY: Paulist Press, 1998.
Conway Morris, Simon, *The Crucible of Creation: The Burgess Shale and the Rise of Animals*, New York: Oxford University Press, 1998.
Life's Solution: Inevitable Humans in a Lonely Universe, Cambridge: Cambridge University Press, 2003.
Crick, Francis C. H., *Of Molecules and Man*, Seattle: University of Washington Press, 1966.
Crippen, T. and Machalak, R., "The Evolutionary Foundations of Religious Life," *International Review of Sociology* 3 (1989), 61–84.
Cronin, Helena, *The Ant and the Peacock*, New York: Cambridge University Press, 1991.
Crowe, Frederick E., SJ, "Complacency and Concern in the Thought of St. Thomas," *Theological Studies* 20 (1959), 1–39, 198–230, 343–395.
Crowley, Paul G., SJ, *Unwanted Wisdom: Suffering, the Cross, and Hope*, New York and London: Continuum, 2005.
Curran, Charles E., *Catholic Higher Education, Theology, and Academic Freedom*, Notre Dame, IN: University of Notre Dame Press, 1990.
"Natural Law in Moral Theology," in Charles E. Curran and Richard McCormick, SJ, eds., *Readings in Moral Theology No. 7: Natural Law and Theology*, Mahwah, NY: Paulist Press, 1991, pp. 247–295.
Curtiss, Susan, *Genie: A Psycholinguistic Study of a Modern-Day "Wild Child,"* New York: Academic Press, 1977.
Daly, Martin and Wilson, Margo, *Sex, Evolution, and Behavior*, North Scituate, MA: Duxbury Press, 1978.
Homicide, New York: Aldine de Gruyter, 1988.
Damasio, Antonio, *Descartes' Error: Emotion, Reason, and the Human Brain*, New York: Avon Books, 1994.
The Feeling of What Happens: Body and Emotion in the Making of Consciousness, New York: Harcourt Brace and Co., 1999.
"A Note on the Neurobiology of Emotions," in Stephen G. Post, Lynn G. Underwood, Jeffrey P. Schloss, and William B. Hurlbut, eds., *Altruism and Altruistic Love: Science, Philosophy, and Religion in Dialogue*, New York: Oxford University Press, 2002, pp. 264–271.

D'Arcy, Martin C., *The Mind and Heart of Love, Lion and Unicorn: A Study in Eros and Agape*, Cleveland: World, 1964.
Dartmouth Medical School, the YMCA of the USA, and the Institute for American Values, *Hardwired to Connect: The New Scientific Case for Authoritative Communities*, New York: Institute for American Values, 2003.
Darwin, Charles, *The Autobiography of Charles Darwin*, ed. Nora Barlow, London: Collins, 1958.
　Autobiography, in *Charles Darwin and Thomas Henry Huxley: Autobiographies*, ed. Gavin de Beer, London: Oxford University Press, 1974.
　Darwin: A Norton Critical Edition, second edn., ed. Philip Appleman, New York: W. W. Norton, 1979.
　The Origin of Species, New York: Penguin, 1986.
Dawkins, Richard, *The Selfish Gene*, New York: Oxford, 1976.
　The Extended Phenotype, New York: Oxford University Press, 1982.
　The Blind Watchmaker, New York and London: W. W. Norton, 1986.
　"Viruses of the Mind," in B. Dahlbom, ed., *Dennett and His Critics*, New York: Blackwell, 1993, pp. 13–27.
　River out of Eden, New York: HarperCollins, 1995.
　"A Reply to Poole," *Science and Christian Belief* 7.1 (April 1995), 45–50.
　The God Delusion, Boston: Houghton Mifflin, 2006.
Dawkins, Richard and Krebs, J. R., "Arms Races within and between Species," *Proceedings of the Royal Society*, series B, 205 (1979), 489–511.
Dembski, William, ed., *Mere Creation: Science, Faith, and Intelligent Design*, Downers Grove, IL: InterVarsity Press, 1998.
Dennett, Daniel, *Consciousness Explained*, Boston: Back Bay Books, 1992.
　Darwin's Dangerous Idea: Evolution and the Meanings of Life, New York: Simon and Schuster, 1995.
Descartes, René, *Discourse on Method and Meditations on First Philosophy*, trans. Donald A. Cress, third edn., Indianapolis and Cambridge: Hackett Publishing Co., 1993.
Desmond, Adrian and Moore, James, *Darwin*, New York: Penguin, 1992.
　Darwin: The Life of a Tormented Evolutionist, New York: W. W. Norton, 1994.
De Waal, Frans B. M., *Good Natured: The Origins of Right and Wrong in Humans and Other Animals*, Cambridge, MA and London: Harvard University Press, 1996.
　Tree of Origin: What Primate Behavior Can Tell Us about Human Social Evolution, Cambridge, MA: Harvard University Press, 2001.
Dewan, Lawrence, OP, "Jean Porter on Natural Law," *The Thomist* 66.2 (April 2002), 275–309.
Dewey, John, *The Impact of Darwin on Philosophy: And Other Essays in Contemporary Thought*, Bloomington, IN: Indiana University Press, 1965 [1910].
Diamond, Jared, *The Third Chimpanzee: The Evolution and Future of the Human Animal*, New York: HarperCollins, 1992.
　Why Is Sex Fun? The Evolution of Human Sexuality, New York: Basic Books, 1997.

Dobzhansky, Theodosius, "Nothing in Biology Makes Sense Except in the Light of Evolution," *The American Biology Teacher* 35 (March 1973), 125–129.

Dodds, Michael J., *The Unchanging God of Love: A Study of the Teaching of St. Thomas Aquinas on Divine Immutability in View of Certain Contemporary Criticisms of this Doctrine*, Fribourg, Switzerland: Editions Universitaires, 1986.

"Thomas Aquinas, Human Suffering, and the Unchanging God of Love," *Theological Studies* 52 (1991), 330–344.

Dorey, Emma, Dove, Alan, Fox, Jeff, Hodgson, John, and Niiler, Eric, "GMO Roundup," *Nature Biotechnology* 18 (January 2000), 7.

Dostoevsky, Fyodor, *The Brothers Karamazov*, trans. Richard Pevear and Larissa Volokhonsky, New York: Vintage, 1990.

Drummond, Henry, *The Ascent of Man*, New York: James Pott and Company, 1894.

Duffy, Stephen, "Our Hearts of Darkness: Original Sin Revisited," *Theological Studies* 49 (1988), 597–622.

Dupré, John, *Darwin's Legacy: What Evolution Means Today*, New York: Oxford University Press, 2003.

Eagley, A. H., "The Science and Politics of Comparing Men and Women," *American Psychologist* 50 (1995), 145–158.

Eagley, A. H., Ashmore, R. D., Makhijani, M. G., and Longo, I. C., "What Is Beautiful Is Good, But . . .: A Meta-Analytic Review of Research on the Physical Attractiveness Stereotype," *Psychological Bulletin* 110 (1991), 109–128.

Edelman, Gerald M., *Bright Air, Brilliant Fire: On the Matter of the Mind*, New York: Penguin Press, 1992.

Edwards, Jonathan, *Charity and Its Fruits*, London: Banner of Truth Press, 1978.

Edwards, Paul, ed., *Encyclopedia of Philosophy*, New York: MacMillan Publishing Co. Inc., and London: The Free Press, 1967.

Ehrlich, Paul, *Human Natures: Genes, Cultures, and the Human Prospect*, New York: Penguin Books, 2002.

Elders, Leo, *The Philosophy of Nature of St. Thomas Aquinas: Nature, the Universe and Man*, Frankfurt am Main: Peter Lang, 1997.

Ellis, George F. R., *Before the Beginning: Cosmology Explained*, London and New York: Boyars/Bowerdean, 1993.

"The Thinking Underlying the New 'Scientific' World-Views," in Robert John Russell, William R. Stoeger, SJ, and Francisco J. Ayala, eds., *Evolutionary and Molecular Biology: Scientific Perspectives on Divine Action*, Vatican City State: Vatican Observatory Publications, and Berkeley, CA: Center for Theology and the Natural Sciences, 1998, pp. 251–280.

Engelhardt, H. Tristram, *The Foundations of Ethics*, New York: Oxford University Press, 1996.

Farrer, Austin, *A Science of God?*, London: Geoffrey Bles, 1966.

Faith and Speculation, New York: New York University Press, 1967.

Fein, Helen, *Imperial Crime and Punishment: The Massacre at Jallianwalla Bagh and British Judgment, 1919–1920*, Honolulu: University Press of Hawaii, 1977.

Finnis, John, *Natural Law and Natural Rights*, New York: Oxford University Press, 1980.
   "Is Natural Law Theory Compatible with Limited Government?" in Robert P. George, ed., *Natural Law, Liberalism, and Morality*, New York: Oxford University Press, 1996, pp. 1–26.
   Aquinas: Moral, Political, and Legal Theory, New York: Oxford University Press, 1998.
Fitzmyer, Joseph A., SJ, "The Letter to the Romans," in Raymond E. Brown, SS, Joseph A. Fitzmyer, SJ, and Roland A. Murphy, OCarm., eds., *The Jerome Biblical Commentary*, 2 vols., Englewood Cliffs, NJ: Prentice-Hall, Inc., 1968, pp. 830–868.
Flanagan, Owen, *The Science of the Mind*, second edn., London and Cambridge, MA: MIT Press, 1991.
   Varieties of Moral Personality, Cambridge, MA: Harvard University Press, 1991.
   Self Expressions: Mind, Morals, and the Meaning of Life, New York: Oxford University Press, 1996.
Flannery, Austin, OP, ed., *Vatican Council II*, in rev. trans., Northport, NY and Dublin: Costello Publishing Company, 1996.
   "Dei Verbum: Dogmatic Constitution on Divine Revelation," *Vatican Council II*, in rev. trans., Northport, NY and Dublin: Costello Publishing Company, 1996, pp. 97–115.
   "Gaudium et spes: Pastoral Constitution on the Church in the Modern World," *Vatican Council II*, in rev. trans., Northport, NY and Dublin: Costello Publishing Company, 1996, pp. 163–282.
Frank, Robert H., *Passions within Reason: The Strategic Role of the Emotions*, London and New York: W. W. Norton, 1988.
Frank, Robert H., Golivich, Thomas, and Regan, Dennis T., "Does Studying Economics Inhibit Cooperation?," *Journal of Economic Perspectives* 7.2 (1993), 168–170.
Frankfurt, Harry, "Freedom of the Will and the Concept of the Person," *Journal of Philosophy* 68 (January 1971), 5–20.
Fuchs, Josef, SJ, *Personal Responsibility and Christian Morality*, trans. William Cleves et al., Washington, DC: Georgetown University Press, and Dublin: Gill and MacMillan, 1983.
Fuentes, Augustin, "Revisiting Conflict Resolution: Is There a Role for Emphasizing Negotiation and Cooperation instead of Conflict and Reconciliation?," in Robert W. Sussman and Audrey R. Chapman, eds., *The Origins and Nature of Sociality*, Hawthorne, NY: Aldine de Gruyter, 2004, pp. 215–234.
Gadamer, Hans Georg, *Truth and Method*, trans. J. Weinsheimer and D. Marshall, second rev. edn., New York: Continuum, 1989; originally published 1972.
Gallagher, David, "The Will," in Stephen J. Pope, ed., *The Ethics of Aquinas*, Washington, DC: Georgetown University Press, 2002, pp. 69–89.
Gallup, G., Anderson, J., and Shillito, D., "The Mirror Test," in M. Berkoff, C. Allen, and G. Burghardt, eds., *The Cognitive Animal: Empirical and*

Theoretical Perspectives on Animal Cognition, Cambridge, MA: MIT Press, 2002, pp. 325–333.
George, Robert P., and Bradley, Gerald V., "Marriage and the Liberal Imagination," *Georgetown Law Journal* 84 (1995), 301–320.
Ghiseln, Michael, *The Economy of Nature and the Evolution of Sex*, Berkeley: University of California Press, 1974.
Giberson, Karl W. and Yerxa, Donald, *Species of Origins: America's Search for a Creation Story*, Lanham, MD: Rowman and Littlefield, 2002.
Gilkey, Langdon, *Creationism on Trial*, San Francisco: Harper and Row, 1985.
Nature, Reality, and the Sacred, Minneapolis: Augsburg Fortress Press, 1993.
Gillespie, Neil, *Charles Darwin and the Problem of Creation*, Chicago and London: University of Chicago Press, 1979.
Gilson, Etienne, *History of Christian Philosophy in the Middle Ages*, New York: Random House, 1955.
The Christian Philosophy of St. Thomas Aquinas, New York: Random House, 1956.
"The Corporeal World and the Efficacy of Secondary Causes," in Owen Thomas, ed., *God's Activity in the World: The Contemporary Problem*, Chicago: Scholars Press, 1983, pp. 213–230.
Gingerich, Owen, "Truth in Science: Proof, Persuasion, and the Galileo Affair," *Perspectives in Science and Christian Faith* 55.2 (June 2003), 80–87.
Goldberg, Steven, *Seduced by Science*, New York: New York University Press, 1999.
Goodall, Jane, *In the Shadow of Man*, New York: Random House, 1974.
Gould, Stephen Jay, *Ever since Darwin: Reflections in Natural History*, New York: W. W. Norton, 1977.
"Biological Potential vs. Biological Determinism," in Arthur L. Caplan, ed., *The Sociobiology Debate*, New York: Harper and Row, 1978, pp. 343–351; originally printed in *Natural History Magazine*, May 1976.
The Mismeasure of Man, New York: W. W. Norton, 1981.
Hens' Teeth and Horses' Toes: Reflections in Natural History, New York: Norton, 1983.
Wonderful Life: The Burgess Shale and the Nature of History, New York and London: W. W. Norton, 1989.
"Tires to Sandals," *Natural History* 98 (April 1989), 8–15.
"Exaptation: A Crucial Tool for Evolutionary Psychology," *Journal of Social Issues* 47 (1991), 43–65.
Eight Little Piggies, New York: Penguin, 1993.
Flamingo's Smile: Reflections in Natural History, New York: Norton, 1997.
Life's Grandeur: The Spread of Excellence from Plato to Darwin, New York: Random House, 1997.
"On Embryos and Ancestors," *Natural History* (July/August 1998), 20–22, 58–65.
Rock of Ages: Science and Religion and the Fullness of Life, New York: Ballantine, 1999.

Gould, Stephen Jay and Eldredge, Niles, "Punctuated Equilibria: The Tempo and Mode of Evolution Reconsidered," *Paleobiology* 3 (1977), 115–151.

Gould, Stephen Jay and Lewontin, S., "The Spandrels of San Marcos and the Panglossian Paradigm: A Critique of the Adaptationist Programme," *Proceedings of the Royal Society*, series B, 205 (1979), 581–598.

Gowaty, Patricia, "Evolutionary Biology and Feminism," *Human Nature* 3 (1992), 217–249.

Gowaty, Patricia, ed., *Feminism and Evolutionary Biology*, New York: Chapman and Hall, 1997.

Gregerson, Niels, Drees, Williem B., and Gorman, Ulf, eds., *The Human Person in Science and Theology*, Edinburgh: T. & T. Clark, 2000.

Grisez, Germaine, "The First Principle of Practical Reason: A Commentary on Summa Theologica Ia IIae, q.94 a. 2," *Natural Law Forum* 10 (1965), 162–201.

Gula, Richard M., SS, *Reason Informed by Faith: Foundations of Catholic Morality*, Mahwah, NY: Paulist Press, 1989.

Gustafson, James M., *Christian Ethics and the Community*, New York: Pilgrim Press, 1971; second edn. 1979.

Theology and Christian Ethics, Philadelphia: Pilgrim Press, 1974.

Protestant and Roman Catholic Ethics: Prospects for Rapprochement, Chicago: University of Chicago Press, 1978.

"Sociobiology: A Secular Theology," *The Hastings Center Report* 9.1 (February 1979), 44–45.

Ethics from a Theocentric Perspective, 2 vols., Chicago: University of Chicago Press, 1981; second edn., 1983.

"The Sectarian Temptation: Reflections on Theology, the Church, and the University," *Proceedings of the Catholic Theological Society of America* 40 (1985), 83–94.

Varieties of Moral Discourse: Prophetic, Narrative, Ethical, and Policy, Grand Rapids, MI: Calvin College and Seminary, 1988.

A Sense of the Divine: The Natural Environment from a Theocentric Perspective, Cleveland: Pilgrim Press, 1994.

Intersections: Science, Theology, and Ethics, Cleveland, Ohio: Pilgrim Press, 1996.

An Examined Faith: The Grace of Self-Doubt, Minneapolis: Fortress Press, 2004.

Gutierrez, Gustavo, *A Theology of Liberation*, trans. Sister Caidad Inda and John Eagleson, revised edn., Maryknoll, NY: Orbis Press, 1988.

Hall, Pamela M., *Narrative and the Natural Law: An Interpretation of Thomistic Ethics*, Notre Dame, IN and London: University of Notre Dame Press, 1994.

Hallie, Philip, *Lest Innocent Blood Be Shed*, New York: Harper and Row, 1979.

Hamilton, W. D., "The Genetic Evolution of Social Behavior," *Journal of Theoretical Biology* 7:1–2 (1964), 1–52.

"Innate Social Aptitudes of Man: An Approach from Evolution Genetics," in Robin Fox, ed., *Biosocial Anthropology*, New York: John Wiley and Sons, 1975, pp. 133–135.

Hare, John, "Is There an Evolutionary Foundation for Human Morality?," in Philip Clayton and Jeffrey Schloss, eds., *Evolution and Ethics: Human*

Morality in Biological and Religious Perspective, Grand Rapids, MI: Eerdmans, 2004, pp. 187–203.

Harrington, Daniel J., SJ, *The Church according to the New Testament*, Franklin, WI and Chicago: Sheed and Ward, 2001.

Harrington, Daniel J., SJ, and Keenan, James F., SJ, *Jesus and Virtue Ethics: Building Bridges between New Testament Studies and Moral Theology*, Lanham: Sheed and Ward, 2002.

Hauerwas, Stanley, *A Community of Character: Toward a Constructive Christian Social Ethic*, Notre Dame, IN: University of Notre Dame Press, 1981.

Against the Nations, Minneapolis: Winston Press, 1985.

"Why the Truth Demands Truthfulness," in Stanley Hauerwas and L. Gregory Jones, eds., trans. ed. David Smith, *Why Narrative? Readings in Narrative Theology*, Eugene: Wipe Stick Publications, 1997, pp. 303–310.

Hauerwas, Stanley and Burrell, David, "From System to Story: An Alternative Pattern for Rationality in Ethics," in Stanley Hauerwas and L. Gregory Jones, eds., trans. ed. David Smith, *Why Narrative? Readings in Narrative Theology*, Eugene: Wipe Stick Publications, 1997, pp. 158–190.

Haught, John, *Science and Religion: From Conflict to Conversation*, Mahwah, NY: Paulist Press, 1995.

Hausfater, Glenn and Hrdy, Sarah B., eds., *Infanticide: Comparative and Evolutionary Perspectives*, New York: Aldine de Gruyter, 1984.

Hawkes, Nigel, "Science into Religion Won't Go – Or Will It? Rock of the Ages: Science and Religion in the Fullness of Life," *The Tablet*, March 3, 2001, pp. 309–10.

Hawking, Stephen, *A Brief History of Time*, New York: Bantam Books, 1988.

Hefner, Philip, *The Human Factor: Evolution, Culture, and Religion*, Minneapolis: Fortress Press, 1993.

"The Evolution of the Created Co-Creator," in James B. Miller, ed., *An Evolving Dialogue: Scientific, Historical, Philosophical, and Theological Perspectives on Evolution*, Washington, DC: American Association for the Advancement of Science, 1998, pp. 403–420.

"Imago Dei: The Possibility and Necessity of the Human Person," in Niels Gregerson, Willem B. Drees, and Ulf Gorman, eds., *The Human Person in Science and Theology*, Edinburgh: T. & T. Clark, 2000, pp. 73–94.

Technology and Human Becoming, Minneapolis: Fortress Press, 2003.

Hittinger, Russell, *A Critique of the New Natural Law Theory*, Notre Dame, IN: University of Notre Dame Press, 1989.

Hobbes, Thomas, *Leviathan*, New York: Collier, 1962.

Leviathan: Or the Matter, Forms and Power of a Commonwealth Ecclesiastical and Civil, ed. Michael Oakeshott, New York: Collier, 1977.

Hoffmann, Bruce, *Inside Terrorism*, New York: Columbia University Press, 1999.

Hollenbach, David, SJ, *Claims in Conflict: Renewing and Retrieving the Human Rights Tradition*, New York: Paulist Press, 1979.

The Common Good and Christian Ethics, New York: Cambridge University Press, 2002.

The Global Face of Public Faith: Politics, Human Rights, and Christian Ethics, Washington, DC: Georgetown University Press, 2003.

The Holy Bible, New Revised Standard Version, New York and Oxford: Oxford University Press, 1989.

Horgan, John, "The New Social Darwinists," *Scientific American* 273 (October 1995), 174–182.

Howard, David, *Conquistador in Chains: Cabeza de Vaca and the Indians of the Americas*, Tuscaloosa, AL: University of Alabama, 1997.

Hrdy, Sarah, *The Woman that Never Evolved*, Cambridge, MA: Harvard University Press, 1981.

    "Introduction: Female Reproductive Strategies," in M. Small, ed., *Female Primates: Studies by Women Primatologists*, New York: Alan Liss, 1984, pp. 13–16.

    "Raising Darwin's Consciousness: Female Sexuality and the Prehominid Origins of Patriarchy," *Human Nature* 8 (1997), 1–49.

Hütter, Reinhard and Dieter, Theodor, eds., *Ecumenical Ventures in Ethics: Protestants Engage Pope John Paul's Moral Encyclicals*, Grand Rapids, MI: Eerdmans, 1998.

Huxley, Aldous, *The Olive Tree*, New York: Harper and Brothers, 1937.

Huxley, Thomas H., *Evolution and Ethics and Other Essays*, New York: D. Appleton, 1894; reprinted in *Issues in Evolutionary Ethics*, ed. Paul Thompson, Albany, NY: SUNY Press, 1995, pp. 111–150.

Irenaeus, "Irenaeus against Heresies," in Alexander Roberts and James Donaldson, eds., *The Ante-Nicene Fathers*, Grand Rapids, MI: Eerdmans, and Edinburgh: T. and T. Clark, 1989, pp. 309–567.

Irons, William, "How Did Morality Evolve?," *Zygon* 26 (March 1991), 49–89.

Jaki, Stanley, *The Road of Science and the Ways to God*, Chicago: University of Chicago Press, 1978.

Janssens, Louis, "Norms and Priorities in a Love Ethics," *Louvain Studies* 6 (1977), 207–238.

    "Artificial Insemination: Ethical Considerations," *Louvain Studies* 5 (1980), 3–29.

    "Ontic Good and Evil – Premoral Values and Disvalues," *Louvain Studies* 12 (1987), 62–82.

John Paul II, *Familiaris consortio: Apostolic Exhortation on the Family*, (November 22, 1981), United States Catholic Conference, Washington, D.C., 1982.

    "Mulieris dignitatem: 'On the Dignity and Vocation of Woman,'" *Origins* 18/17 (October 6, 1988), 261–83.

    "Letter to the Rev. George V. Coyne, S. J.," *Origins* 18 (November 17, 1988), 375–378.

    Centesimus Annus: On the Hundredth Anniversary of Rerum Novarum, Washington, DC: United States Catholic Conference, 1991.

    "Lessons of the Galileo Case: Address to the Pontifical Academy of Sciences, October 31, 1992," *Origins* 22 (1992), 371–373.

    The Splendor of Truth: Veritatis Splendor, Washington, DC: United States Catholic Conference, 1993.

"*Evangelium Vitae*," *The Gospel of Life: On the Value and Inviolability of Human Life*, Washington, DC: United States Catholic Conference, 1995.

"Address to the United Nations General Assembly," *Origins* 25 (October 19, 1995), 270.

"Message to the Pontifical Academy of Sciences on Evolution," *Origins* 26.2 (November 14, 1996), 349–352.

Fides et Ratio: On the Relationship between Faith and Reason, Boston: Daughters of St. Paul, 1998.

"God Made Man the Steward of Creation," Papal Audience January 24, 2001, *L'Osservatore Romano* (January 2001), 11.

"Communion and Stewardship: Human Persons Created in the Image of God," *Origins* 34.15 (September 23, 2004), 233–248.

Johnson, Elizabeth, CSJ, "Does God Play Dice? Divine Providence and Chance," *Theological Studies* 56 (1996), 3–18; republished in James B. Miller, ed., *An Evolving Dialogue: Scientific, Historical, Philosophical, and Theological Perspectives on Evolution*, Washington, DC: American Association for the Advancement of Science, 1998, pp. 355–373.

Johnson, Phillip E., *Darwin on Trial*, second edn., Downers Grove, IL: InterVarsity Press [1991] 1993.

Jonas, Hans, *The Imperative of Responsibility: In Search of an Ethics for the Technological Age*, Chicago and London: University of Chicago Press, 1984.

Kaufman, Stuart, *At Home in the Universe*, New York: Oxford University Press, 1995.

Kayzer, William, *A Glorious Accident: Understanding Our Place in the Cosmic Puzzle*, New York: Freeman and Co., 1999.

Keenan, James F., SJ, "The Virtue of Prudence (IIa IIae, qq. 47–56)," in Stephen J. Pope, ed., *The Ethics of Aquinas*, Washington, DC: Georgetown University Press, 2002, pp. 259–271.

Moral Wisdom: Lessons and Texts from the Catholic Tradition, New York: Sheed and Ward, 2003.

The Works of Mercy: The Heart of Catholicism, Lanham: Sheed and Ward, 2005.

Kierkegaard, Søren, *Works of Love*, San Francisco: Harper, 1962.

King, Martin Luther, Jr., *Strength to Love*, Cleveland: Collins World, 1963.

Kirkpatrick, Lee A., *Attachment, Evolution and the Psychology of Religion*, New York: Guilford Press, 2004.

Kitcher, Philip, *Abusing Science: The Case against Creationism*, Cambridge, MA: MIT Press, 1982.

Vaulting Ambition: Sociobiology and the Quest for Human Nature, Cambridge, MA and London: MIT Press, 1985.

"Darwin's Achievements," in N. Rescher, ed., *Reason and Rationality in Natural Science*, New York: University Press of America, 1985, pp. 127–189.

Koltermann, Rainer, "Evolution, Creation, and Church Documents," *Theology Digest* 48.2 (2001), 124–132.

Kopelman, P. G., "Obesity as a Medical Problem," *Nature* 404 (2000), 635–643.

Kopfensteiner, Thomas R., "The Metaphorical Structure of Normativity," *Theological Studies* 58 (1997), 331–346.
Kramer, Peter D., *Listening to Prozac*, New York: Penguin, 1997.
Kuhn, Thomas S., *The Structure of Scientific Revolutions*, Chicago: University of Chicago Press, 1970.
Larson, Edward J., *Summer for the Gods: The Scopes Trial and America's Continuing Debate over Science and Religion*, San Francisco: Basic Books, 1997.
Lash, Nicholas, *The Beginning and the End of "Religion,"* New York: Cambridge University Press, 1996.
  "Ideology, Metaphor, and Analogy," in Stanley Hauerwas and L. Gregory Jones, eds., trans. ed. David Smith, *Why Narrative? Readings in Narrative Theology*, Eugene: Wipe Stick Publications, 1997, pp. 113–157.
Lauritzen, Paul, "Emotion and Religious Ethics," *Journal of Religious Ethics* 16.2 (fall 1988), 307–324.
Lennox, James, "Aristotle on the Biological Roots of the Virtues: The Natural History of Natural Virtues," in Jane Maienschein and Michael Ruse, eds., *Biology and the Foundations of Ethics*, New York: Cambridge University Press, 1999, pp. 10–31.
Lewis, C. S., *The Abolition of Man*, San Francisco: HarperSanFrancisco, 1974 [1944].
  The Four Loves, New York: Harcourt Brace, 1960.
Lewontin, Richard C., "Billions and Billions of Demons," *New York Review of Books* 44.1 (January 9, 1997), 28–32.
Lewontin, Richard C., Rose, Steven, and Kamin, Leon J., eds., *Not in Our Genes: Biology, Ideology, and Human Nature*, New York: Pantheon, 1984.
Lindbeck, George, *The Nature of Doctrine: Religion and Theology in a Postliberal Age*, Philadelphia: Westminster Press, 1984.
Lorenz, Konrad, *On Aggression*, trans. Marjorie Kerr Wilson, New York: Harcourt, Brace and World, 1966.
Lottin, Odon, *Le droit naturel chez Saint Thomas d'Aquin et ses prédécesseurs*, second edn., Bruges: Beyaert, 1931.
  Etudes de morale histoire et doctrine, Belgium: Gembloux, 1961.
Lowe, E. J., "Ontology," in Ted Honderich, ed., *The Oxford Companion to Philosophy*, New York: Oxford University Press, 1995, pp. 634.
Lukes, Steven, *Emile Durkheim*, Stanford: Stanford University Press, 1985.
Lumsden, Charles J. and Wilson, Edward O., *Genes, Mind, and Culture*, Cambridge, MA: Harvard University Press, 1981.
  Promethean Fire, Cambridge, MA: Harvard University Press, 1983.
MacIntyre, Alasdair, *After Virtue: A Study in Moral Theory*, second edn., Notre Dame, IN: University of Notre Dame Press, 1984 [1980].
  Whose Justice? Which Rationality?, Notre Dame, IN: University of Notre Dame Press, 1988.
  Dependent Rational Animals: Why Human Beings Need the Virtues, Chicago and La Salle, IL: Open Court Press, 1999.
Maurer, Armand, "Darwin, Thomists, and Secondary Causality," *The Review of Metaphysics* 57 (March 2004), 491–514.

Mayr, Ernst, *The Growth of Biological Thought*, Cambridge, MA: Harvard University Press, 1982.
Towards a New Philosophy of Biology, Cambridge, MA: Harvard University Press, 1988.
One Long Argument: Charles Darwin and the Genesis of Modern Evolutionary Thought, Cambridge, MA: Harvard University Press, 1991.
What Evolution Is, New York: Basic Books, 2001.
McCabe, Herbert, *God Matters*, London: Cassell Publishers Limited, 1987.
The Good Life: Ethics and the Pursuit of Happiness, ed. Brian Davies, OP, London: Continuum International Publishing Group, 2005.
McCormick, Patrick, *Sin as Addiction*, Mahwah, NY: Paulist Press, 1989.
McCormick, Richard, SJ, *The Critical Calling: Reflections on Moral Dilemmas since Vatican II*, Washington, DC: Georgetown University Press, 1989.
McGill, Arthur, *Suffering: A Test of Theological Method*, Philadelphia: Westminster Press, 1982.
McMullin, Ernan, "Values in Science," in Peter D. Asquith and Thomas Nikles, eds., *PSA 1982: Proceedings of the 1982 Biennial Meeting of the Philosophy of Science Association*, vol. II, East Lansing, MI: Philosophy of Science Association, 1982, pp. 1–25.
"Evolution and Special Creation," *Zygon* 28.1 (September 1993), 299–335.
"Evolutionary Contingency and Cosmic Purpose," in Michael Himes and Stephen J. Pope, eds., *Finding God in All Things: Essays in Honor of Michael J. Buckley, S. J.*, New York: Crossroad, 1996, pp. 140–161.
Metz, Johann Baptist, *Faith in History and Society*, trans. David Smith, New York: Crossroad, 1980.
"A Short Apology of Narrative," in Stanley Hauerwas and L. Gregory Jones, eds., trans. ed. David Smith, *Why Narrative? Readings in Narrative Theology*, Eugene: Wipe Stick Publications, 1997, pp. 251–62.
Midgley, Mary, *Beast and Man: The Roots of Human Nature*, Ithaca, NY: Cornell University Press, 1978.
Wickedness: A Philosophical Essay, London and New York: ARK, 1984.
Evolution as a Religion: Strange Hopes and Stranger Fears, London and New York: Methuen, 1985.
The Ethical Primate: Humans, Freedom and Morality, London and New York: Routledge, 1994.
Science as Salvation: A Modern Myth and Its Meaning, Florence, KY: Routledge, 1994.
"Consciousness, Fatalism and Science," in Niels Gregerson, William B. Drees, and Ulf Gorman, eds., *The Human Person in Science and Theology*, Edinburgh: T. & T. Clark, 2000, pp. 21–40.
Miller, Kenneth, *Finding Darwin's God: A Scientist's Search for Common Ground between God and Evolution*, New York: Cliff Street Books/HarperCollins, 1999.
Mivart, St. George, *On the Genesis of Species*, second edn., London: Macmillan, 1871.

Monod, Jacques, *Chance and Necessity*, trans. Austryn Wainhouse, New York: Vintage, 1972.
Montagu, Ashley, *Science and Creationism*, New York: Oxford University Press, 1983.
Moore, James R., *The Post-Darwinian Controversies*, Cambridge: Cambridge University Press, 1979.
  "Herbert Spencer's Henchmen: The Evolution of Protestant Liberals in Late Nineteenth-Century America," in John Durant, ed., *Darwinism and Divinity: Essays on Evolution and Religious Belief*, New York and Oxford: Basil Blackwell, 1985, pp. 76–100.
Murphy, Jeffrie, *Evolution, Morality, and the Meaning of Life*, Totowa: Roman and Littlefield, 1982.
Nash, James A., *Loving Nature: Ecological Integrity and Christian Responsibility*, Nashville: Abingdon Press, 1991.
National Academy of Science, *Teaching about Evolution and the Nature of Science*, Washington, DC: National Academy Press, 1998.
Neuhaus, Richard John, "A Continuing Survey of Religion and Public Life," *First Things* 69 (1997), 56–70.
Nichols, Terence L., *The Sacred Cosmos: Christian Faith and the Challenge of Naturalism*, Grand Rapids, MI: Brazos, 2003.
Niditch, Susan, *War in the Hebrew Bible: A Study in the Ethics of Violence*, New York: Oxford University Press, 1993.
Niebuhr, H. Richard, *The Responsible Self*, San Francisco: Harper and Row, 1963.
Niebuhr, Reinhold, *An Interpretation of Christian Ethics*, New York: Crossroad, 1935; second edn. 1979.
  Beyond Tragedy: Essays on the Christian Interpretation of History, New York: Charles Scribner's Sons, 1937.
  The Nature and Destiny of Man: A Christian Interpretation, 2 vols., New York: Charles Scribner's Sons, 1941.
Noonan, John T., Jr., *The Lustre of Our Country: The American Experience of Religious Freedom*, Berkeley: University of California Press, 1998.
Numbers, Ronald R., *The Creationists: The Evolution of Scientific Creationism*, Berkeley: University of California Press, 1993.
Nussbaum, Martha, *The Therapy of Desire: Theory and Practice in Hellenistic Ethics*, Princeton, NJ: Princeton University Press, 1994.
O'Hear, Anthony, *Beyond Evolution: Human Nature and the Limits of Evolutionary Explanation*, Oxford: Clarendon Press, 1997.
Outka, Gene, *Agape: An Ethical Analysis*, New Haven and London: Yale University Press, 1972.
  "Universal Love and Impartiality," in Edmund N. Santurri and William Werpehowski, eds., *The Love Commandments: Christian Ethics and Moral Philosophy*, Washington, DC: Georgetown University Press, 1992, pp. 1–103.
Pannenberg, Wolfhart, "Human Life: Creation versus Evolution?," in Ted Peters, ed., *Science and Theology: The New Consonance*, Cambridge, MA: Westview, 1998, pp. 137–148.

Paul VI, *Humanae Vitae: On the Regulation of Birth (July 25, 1968)*, in *The Gospel of Peace and Justice: Catholic Social Teaching since Pope John*, Maryknoll, NY: Orbis Books, 1975, pp. 427–444.

Peacocke, Arthur R., *Creation and the World of Science*, Oxford: Clarendon Press, 1979.

God and the New Biology, London: Dent, 1986.

Theology for a Scientific Age: Being and Becoming – Natural, Divine and Human, London: SCM Press, 1993.

"A Map of Scientific Knowledge: Genetics, Evolution, and Theology," in Ted Peters, ed., *Science and Theology: The New Consonance*, Cambridge, MA: Westview, 1998, pp. 189–210.

Penizias, Arno, "The Elegant Universe," in Mark W. Richardson and Gordy Slack, eds., *Faith in Science: Scientists Search for Truth*, London: Routledge, 2001, pp. 18–34.

Penrose, Roger, *The Emperor's New Mind: Concerning Computers, Minds, and the Laws of Physics*, New York: Oxford University Press, 2002.

Shadows of the Mind: A Search for the Missing Science of Consciousness, New York: Oxford, reprint, 1996.

Percy, Walker, *Lost in the Cosmos: The Last Self-Help Book*, New York: Macmillan, 2000.

Peters, Ted, *Playing God?: Genetic Determinism and Human Freedom*, New York and London: Routledge, 1997.

Peters, Ted, ed., *Science and Theology: The New Consonance*, Cambridge, MA: Westview, 1998.

Peterson, Gregory, "Falling Up: Evolution and Original Sin," in Philip Clayton and Jeffrey Schloss, eds., *Evolution and Ethics: Human Morality in Biological and Religious Perspective*, Grand Rapids, MI: Eerdmans, 2004, pp. 273–286.

Phipps, William E., *Darwin's Religious Odyssey*, Harrisburg, PA: Trinity Press International, 2002.

Pinckaers, Servais, *La vie selon l'Esprit: essai de théologie spirituelle selon saint Paul et saint Thomas d'Aquin*, Luxembourg: Editions Saint Paul, 1966.

The Sources of Christian Ethics, Washington, DC: Catholic University of America, 1995.

Pinker, Steven, *How the Mind Works*, New York and London: W. W. Norton, 1997.

Placher, William, *Unapologetic Theology: A Christian Voice in a Pluralistic Conversation*, Louisville, KY: Westminster John Knox, 1989.

Narratives of a Vulnerable God: Christ, Theology, and Scripture, Louisville, KY: Westminster John Knox, 1994.

"Being Postliberal: A Response to James Gustafson," *Christian Century* (April 7, 1999), 390–392.

Plantinga, Alvin, *Warrant and Proper Function*, New York: Oxford University Press, 1993.

"When Faith and Reason Clash: Evolution and the Bible," in David Hull and Michael Ruse, eds., *The Philosophy of Biology*, New York: Oxford University

Press, 1998, pp. 674–697. Reprinted from *Christian Scholars' Review* 21 (1991), 8–32.

Plato, "Lysis," in C. D. C. Reed, ed., *Plato on Love: Symposium, Phaedrus, Alcibiades, Lysis, with Selections from Republic and Laws*, Indianapolis, IN: Hackett Publishing Co., 2006, pp. 1–25.

Polkinghorne, John, *Science and Providence*, London: SPCK, 1989.

Science and Christian Belief, London: SPCK, 1994.

"The Metaphysics of Divine Action," in Robert Russell, Nancey Murphy, and Arthur R. Peacocke, eds., *Chaos and Complexity: Scientific Perspectives on Divine Action*, vol. II., Vatican City State: Vatican Observatory Publications, 1996, pp. 147–156.

Poole, Michael, "A Critique of Aspects of the Philosophy and Theology of Richard Dawkins," *Science and Christian Belief* 6.1 (April 1994), 41–59.

"A Reply to Dawkins," *Science and Christian Belief* 7.1 (April 1995), 51–58.

Pope, Stephen J., "Expressive Individualism and True Self-Love: A Thomistic Perspective," *The Journal of Religion* 71 (1991), 384–398.

The Evolution of Altruism and the Ordering of Love, Washington, DC: Georgetown University Press, 1994.

"'Equal Regard' versus 'Special Relations'? Reaffirming the Inclusiveness of Agape," *Journal of Religion* 25.1 (1997), 353–379.

"Neither Enemy Nor Friend: Nature as Creation in the Theology of Saint Thomas Aquinas," *Zygon* 32 (1997), 219–230.

"Natural Law and Christian Ethics," in Robin Gill, ed., *The Cambridge Companion to Christian Ethics*, Cambridge: Cambridge University Press, 2001, pp. 77–95.

"Relating Self, Others, and Sacrifice in the Ordering of Love," in Stephen G. Post, Lynn G. Underwood, Jeffrey P. Schloss, and William B. Hurlbut, eds., *Altruism and Altruistic Love: Science, Philosophy, and Religion in Dialogue*, New York: Oxford University Press, 2002, pp. 168–181.

Porter, Jean, *Natural and Divine Law: Reclaiming the Tradition for Christian Ethics*, Grand Rapids, MI: Eerdmans, and Cambridge: Novalis, 1999.

Nature as Reason: A Thomistic Theory of the Natural Law, Grand Rapids, MI: Eerdmans, 2005.

Post, Stephen G., *The Theory of Agape: On the Meaning of Christian Love*, Lewisburg, PA: Bucknell University Press, 1990.

The Moral Challenge of Alzheimer's Disease: Ethical Issues from Diagnosis to Dying, Baltimore: Johns Hopkins University Press, 2000.

Post, Stephen G., Underwood, Lynn G., Schloss, Jeffrey P., and Hurlbut, William B., eds., *Altruism and Altruistic Love: Science, Philosophy, and Religion in Dialogue*, New York: Oxford University Press, 2002.

Potts, Malcom and Scott, Roger, *Ever since Eve: The Evolution of Human Sexuality*, New York: Cambridge University Press, 1999.

Premack, David, and Woodruff, Guy, "Does the Chimpanzee Have a Theory of Mind?" *Behavioral and Brain Sciences* 4 (1978), 515–526.

Provine, William, "Evolution and the Foundation of Ethics," *Marine Biological Laboratory Science* 3 (1988), 27–28.
Rachels, James, *Created from Animals: The Moral Implications of Darwinism*, New York: Oxford University Press, 1990.
Rahner, Karl, SJ, *On the Theology of Death*, New York: Herder and Herder, 1961.
  Hominization: The Evolutionary Origin of Man as a Theological Problem, New York: Herder and Herder, 1965.
  "Christology within an Evolutionary View of the World," trans. Karl H. Kruger, in *Theological Investigations*, vol. V: *Later Writings*, New York: Crossroad, 1966, pp. 157–192.
  "The Secret of Life," trans. Karl H. Kruger and Boniface Kruger, in *Theological Investigations*, vol. VI: *Concerning Vatican Council II*, New York: Seabury Press, and London: Darton, Longman and Todd, 1974, pp. 141–152.
  "The Theology of Freedom," trans. Karl H. Kruger and Boniface Kruger, in *Theological Investigations*, vol. VI: *Concerning Vatican Council II*, New York: Seabury Press, and London: Darton, Longman and Todd, 1974, pp. 178–196.
  "The Unity of Spirit and Matter in the Christian Understanding of Faith," trans. Karl H. Kruger and Boniface Kruger, in *Theological Investigations*, vol. VI: *Concerning Vatican Council II*, New York: Seabury Press, and London: Darton, Longman and Todd, 1974, pp. 153–177.
  "The Sin of Adam," trans. David Bourke, in *Theological Investigations*, vol. XI: *Confrontations One*, London: Darton, Longman and Todd, and New York: Crossroad, 1982, pp. 247–262.
  "Why Does God Allow Us to Suffer?," trans. Edward Quinn, in *Theological Investigations*, vol. XIX: *Faith and Ministry*, New York: Crossroad, 1983, pp. 194–208.
  Foundations of the Christian Faith: An Introduction to the Idea of Christianity, trans. William V. Dych, New York: Crossroad, 1985.
  "Natural Science and Reasonable Faith," trans. Hugh M. Riley, in *Theological Investigations*, vol. XXI: *Science and Christian Faith*, New York: Crossroad, 1988, pp. 16–55.
  "Nature as Creation," in Karl Lehmann and Albert Raffelt, eds., trans. ed. Harvey D. Egan, SJ, *The Content of Faith*, New York: Crossroad, 1999, pp. 82–88.
  "The Mystery of the Human Person," in Karl Lehmann and Albert Raffelt, eds., trans. ed. Harvey D. Egan, SJ, *The Content of Faith*, New York: Crossroad, 1999, pp. 73–81.
Reiss, D. and Marino, L., "Mirror Self-Recognition in Bottlenose Dolphin: A Case of Cognitive Convergence," *Proceedings of the National Academy of Sciences of the United States of America* 98.10 (May 2, 2001), 5937–5942.
Richards, Robert J., *Darwin and the Emergence of Evolutionary Theories of Mind and Behavior*, Chicago and London: University of Chicago Press, 1987.
Richerson, Peter J. and Boyd, Robert, "Complex Societies: The Evolutionary Origins of a Crude Superorganism," *Human Nature* 10 (1999), 253–290.

"Darwinian Evolutionary Ethics: between Patriotism and Sympathy," in Philip Clayton and Jeffrey Schloss, eds., *Evolutionary Ethics: Human Morality in Biological and Religious Perspectives*, Grand Rapids, MI: Eerdmans, 2004, pp. 50–77.

Ridley, Mark and Dawkins, Richard, "The Natural Selection of Altruism," in J. Philippe Rushton and Richard M. Sorrentino, eds., *Altruism and Helping Behavior: Social, Personality, and Developmental Perspectives*, Hillsdale, NJ: Lawrence Erlbaum Associates, 1981, pp. 19–39.

Ridley, Matt, *The Red Queen: Sex and the Evolution of Human Nature*, New York: HarperCollins, 1993.

Rolston, Holmes, III, *Genes, Genesis, and God: Values and Their Origins in Natural and Human History*, New York: Cambridge University Press, 1999.

Romero, Oscar, *Voice of the Voiceless: The Four Pastoral Letters and Other Statements*, trans. Michael J. Walsh, Maryknoll, NY: Orbis Press, 1999.

Rose, Steven, *Lifelines: Biology beyond Determinism*, New York: Oxford University Press, 1998.

Rousseau, Jean-Jacques, *Emile or On Education*, trans. Alan Bloom, New York: Basic Books, 1979 [1762].

Rudwick, Martin J.S., *The Meaning of Fossils*, Chicago: University of Chicago Press, 1985.

Rue, Loyal, *By the Grace of Guile: The Role of Deception in Natural History and Human Affairs*, New York: Oxford University, 1994.

Ruse, Michael, *Darwinism Defended: A Guide to the Evolution Controversies*, Reading, MA: Addison-Wesley, 1982.

Taking Darwin Seriously: A Naturalistic Approach to Philosophy, New York: Basil Blackwell, 1986.

"Evolutionary Ethics: Healthy Prospect or Last Infirmity?," *Canadian Journal of Philosophy*, Supplementary Volume 14 (1988), 27–73.

"Evolutionary Ethics and the Search for Predecessors: Kant, Hume, and All the Way Back to Aristotle?," *Social Philosophy and Policy* 8 (1990), 59–85.

"Evolutionary Theory and Christian Ethics," *Zygon* 29 (1994), 5–24.

Can a Darwinian Be a Christian?, New York: Cambridge University Press, 2000.

Ruse, Michael, ed., *But Is It Science?*, Amhurst, NY: Prometheus Press, 1996.

Ruse, Michael and Wilson, Edward O., "The Evolution of Ethics," *New Scientist* 108 (October 17, 1985), 50–52.

"The Evolution of Morals," in Michael Ruse, ed., *The Philosophy of Biology*, New York: Macmillan, 1989, pp. 313–317.

Russel, P. R., "Darwinism Examined," *Advent Review and Sabbath Herald* 47 (1876), 153.

Russell, Sharman Apt, *Hunger: An Unnatural History*, New York: Basic Books, 2005.

Schloss, Jeffrey P., "Emerging Accounts of Altruism: 'Love Creation's Final Law?'" in Stephen G. Post, Lynn G. Underwood, Jeffrey P. Schloss, and William B. Hurlbut, eds., *Altruism and Altruistic Love: Science, Philosophy, and Religion in Dialogue*, New York: Oxford University Press, 2002, pp. 212–242.

Schneewind, Jerome B., *The Invention of Autonomy: A History of Modern Moral Philosophy*, New York: Cambridge University Press, 1998.

Schultz, Janice, "Is–Ought: Prescribing and a Present Controversy," *The Thomist* 49 (January, 1985), 1–23.

Seabright, Paul, "The Evolution of Fairness Norms," *Politics, Philosophy & Economics* 5.1 (2006), 33–50.

Shea, William R. and Artigas, Mariono, *Galileo in Rome: The Rise and Fall of a Troublesome Genius*, New York: Oxford University Press, 2003.

Shklar, Judith N., *The Faces of Injustice*, New Haven and London: Yale University Press, 1990.

Simon, Yves, *Freedom of Choice*, ed. Peter Wolff, New York: Fordham University Press, 1969.

Simpson, George G., *This View of Life: The World of an Evolutionist*, New York: Harcourt, Brace and World, 1963.

Singer, Peter, *Practical Ethics*, second edn., New York: Cambridge University Press, 1993.

A Darwinian Left: Politics, Evolution, and Cooperation, New Haven: Yale University Press, 1999.

Animal Liberation, San Francisco: Harper, 2001.

Smith, David L., *Why We Lie: The Evolutionary Roots of Deception and the Unconscious Mind*, New York: St. Martin's Press, 2003.

Sober, Elliott, "When Natural Selection and Culture Conflict," in Holmes Rolston III, ed., *Biology, Ethics and the Origins of Life*, Boston: Jones and Bartlett, 1995, pp. 137–161.

Sober, Elliot and Wilson, David Sloan, *Unto Others: The Evolution and Psychology of Unselfish Behavior*, Cambridge, MA and London: Harvard University Press, 1998.

Sobrino, Jon, *Witnesses to the Kingdom: The Martyrs of El Salvador and the Crucified Peoples*, Maryknoll, NY: Orbis Press, 2003.

Solomon, Robert, *The Passions: Emotions and the Meaning of Life*, Indianapolis, IN: Hackett Publishing Company, 1976; second edn. 1993.

Spencer, Herbert, *The Data of Ethics*, New York: D. Appleton and Co., 1895.

Sperry, Roger, "The New Mentalist Paradigm and Ultimate Concern [part I]," *Perspectives in Biology and Medicine* 29.3 (spring 1986), 413–422.

Spohn, William C., *Go and Do Likewise: Jesus and Ethics*, New York: Continuum, 1999.

Stanford, Craig, *Significant Others: The Ape–Human Continuum and the Quest for Human Nature*, New York: Basic Books, 2001.

Stark, Rodney, *The Rise of Christianity: A Sociologist Reconsiders History*, Princeton, NJ: Princeton University Press, 1996.

Steiner, Henry J. and Alston, Philip, *International Human Rights in Context: Law, Politics, Morals*, second edn., New York: Oxford University Press, 2000.

Stocker, Michael, *Valuing Emotions*, New York: Cambridge University Press, 1996.
Stoeger, William R., SJ, "Contemporary Cosmology and Its Implications for the Science–Religion Dialogue," in Robert John Russell, William R. Stoeger, SJ, and George V. Coyne, SJ, eds., *Physics, Philosophy, and Theology: A Common Quest for Understanding*, Notre Dame, IN: University of Notre Dame Press and Vatican City State: Vatican Observatory Publications, 1988, pp. 219–247.
  "Contemporary Physics and the Ontological Status of the Laws of Nature," in Robert John Russell, William R. Stoeger, SJ, and George V. Coyne, SJ, eds., *Quantum Cosmology and the Laws of Nature – Scientific Perspectives on Divine Action*, Vatican City State: Vatican Observatory Publications, and Berkeley, CA: Center for Theology and the Natural Sciences, 1993, pp. 209–234.
  "Faith Reflects on the Evolving Universe," in Michael Himes and Stephen J. Pope, eds., *Finding God in All Things: Essays in Honor of Michael J. Buckley, S. J.*, New York: Crossroad, 1996, pp. 162–182.
  "The Immanent Directionality of the Evolutionary Process, and Its Relationship to Teleology," in Robert John Russell, William R. Stoeger, SJ, and Francisco J. Ayala, eds., *Evolutionary and Molecular Biology: Scientific Perspectives on Divine Action*, Vatican City State: Vatican Observatory Publications, and Berkeley, CA: Center for Theology and Natural Sciences, 1998, pp. 163–190.
Strawson, Peter F., *Skepticism and Naturalism: Some Varieties*, New York: Columbia University Press, 1985.
Sullivan, Andrew, *Virtually Normal*, New York: Vintage, 1996.
Sussman, Robert W. and Chapman, Audrey R., eds., *The Origins and Nature of Sociality*, Hawthorne, NY: Aldine de Gruyter, 2004.
Sussman, Robert W. and Garber, Paul A., "Rethinking Sociality: Cooperation and Aggression among Primates," in Robert W. Sussman and Audrey R. Chapman, eds., *The Origins and Nature of Sociality*, Hawthorne, NY: Aldine de Gruyter, 2004, pp. 161–190.
Tannehill, Robert C., *Luke*, Nashville: Abingdon Press, 1996.
Tattersall, Ian, *The Fossil Trail*, New York: Oxford University Press, 1995.
  "Human Evolution," in James B. Miller, ed., *An Evolving Dialogue: Scientific, Historical, Philosophical, and Theological Perspectives on Evolution*, Washington, DC: American Association for the Advancement of Science, 1998, pp. 99–210.
  Becoming Human: Evolution and Human Uniqueness, San Diego, New York, and London: Harcourt Brace and Co., 1998.
  The Monkey in the Mirror: Essays on the Science of What Makes Us Human, Orlando, FL: Harvest Books, 2002.
Taylor, Charles, *Human Agency and Language: Philosophical Papers*, vol. I, New York: Cambridge University Press, 1985.
  "Atomism," *Philosophy and the Human Sciences: Philosophical Papers*, vol. II, New York: Cambridge University Press, 1985, pp. 187–210.

Tertullian, *De Carne Christi, Treatise on the Incarnation*, trans. Ernest Evans, London: SPCK, 1956.
Theissen, Gerd, *Biblical Faith: An Evolutionary Approach*, Philadelphia: Fortress Press, 1985.
Thornhill, Randy and Palmer, Craig, *The Natural History of Rape: The Biological Basis of Sexual Coercion*, Cambridge, MA: MIT Press, 2001.
Toner, Jules, *The Experience of Love*, Washington and Cleveland: Corpus Books, 1968.
  Love and Friendship, Milwaukee: Marquette University Press, 2003.
Tooby, J. and Cosmides, L., "The Psychological Foundations of Culture," in J. Barkow, J. Tooby, and L. Cosmides, eds., *The Adapted Mind*, New York: Oxford University Press, 1992, pp. 19–136.
Torrance, Thomas, *Space, Time, and Resurrection*, Edinburgh: Handsel, 1976.
  Reality and Scientific Theology, Edinburgh: Scottish Academic Press, 1985.
  Theological Science, New York: Oxford, 1969; reissued by T. & T. Clark, 1996.
Toulmin, Stephen, *Cosmopolis: The Hidden Agenda of Modernity*, Chicago: University of Chicago Press, 1990.
Travis, Cheryl Brown, ed., *Changing Evolutionary Theories of Sex Differences and Rape: An Interdisciplinary Response to Thornhill and Palmer*, Cambridge, MA: MIT Press, 2003.
Trivers, Robert, "The Evolution of Reciprocal Altruism," *Quarterly Review of Biology* 46 (1971), 35–57.
  Social Evolution, Menlo Park, CA: Benjamin/Cummings, 1985.
  "Parent–Offspring Conflict," *American Zoologist* 14 (1994), 249–264.
Vacek, Edward Collins, *Love Human and Divine: The Heart of Christian Ethics*, Washington, DC: Georgetown University Press, 1998.
Vanier, Jean, *Becoming Human*, Mahwah, NY: Paulist Press, 1998.
Vann, Gerald, OP, *The Pain of Christ and the Sorrow of God*, Oxford: Blackfriars, 1947.
Van Till, Howard J., "The Character of Contemporary Natural Science," in Howard J. Van Till, Robert E. Snow, John H. Steck, and Davis A. Young, *Portraits of Creation: Biblical and Scientific Perspectives on the World's Creation*, Grand Rapids, MI: Eerdmans, 1990, pp. 141–145.
  "God and Evolution: An Exchange," *First Things* (June/July 1993), pp. 32–41.
Van Till, Howard J., Snow, Robert E., Steck, John H., and Young, Davis A., *Portraits of Creation: Biblical and Scientific Perspectives on the World's Formation*, Grand Rapids, MI: Eerdmans, 1990.
Vatican Congregation for the Doctrine of the Faith, *Donum Vitae: The Gift of Life, Vatican Instruction of Respect for Human Life in Its Origin and the Dignity of Procreation*, February 22, 1987.
Vine, Ian, "Inclusive Fitness and the Self-System: The Roles of Human Nature and Sociocultural Processes in Intergroup Discrimination," in V. Reynolds, V. Falger, and I. Vine, eds., *The Sociobiology of Ethnocentrism*, Athens, GA: The University of Georgia Press, 1987, pp. 60–80.
Vogel, Gretchen, "The Evolution of the Golden Rule," *Science* 303 (February 2004), 1128–1131.

Ward, Keith, *God, Chance, and Necessity*, Oxford: Oneworld Publications, 1996.
Watts, Fraser, "Are We Really Nothing More than Neurons?" *Journal of Consciousness Studies* 1 (1994), 275–279.
Westermann, Claus, *Genesis 1–11: A Commentary*, trans. John J. Scullion, second edn., Minneapolis: Augsburg Publishing House, 1984.
Whitehead, Alfred, *Process and Reality*, New York: The Free Press edn., 1969.
Wiener, Philip P., *Evolution and the Founders of Pragmatism*, Cambridge, MA: Harvard University Press, 1949.
Wilberforce, Samuel, "On the Origin of Species," *Quarterly Review* (1860), 225–264.
Wilkinson, Gerald S., "Reciprocal Food Sharing in the Vampire Bat," *Nature* 308 (1984), 181–184.
  "Food Sharing in Vampire Bats," *Scientific American* (February 1990), 76–82.
William, Daniel Day, *The Spirit and the Forms of Love*, New York: Harper and Row, 1968.
Williams, Bernard, *Ethics and the Limits of Philosophy*, Cambridge, MA: Harvard University Press, 1985.
Williams, George C., "Huxley's Evolution and Ethics in Sociobiological Perspective," *Zygon* 23 (1988), 383–438.
  The Pony Fish's Glow, New York: HarperCollins, 1997.
Williams, Patricia A., *Doing Without Adam and Eve: Sociobiology and Original Sin*, Minneapolis: Fortress Press, 2001.
Williams, Rowan, *Lost Icons: Reflections on Cultural Bereavement*, Edinburgh: T. & T. Clark, 2000.
  On Christian Theology, Malden, MA: Blackwell, 2000.
Wilson, David Sloan, *Darwin's Cathedral*, Chicago: University of Chicago Press, 2001.
  "Evolution, Morality, and Human Potential," in Steven Scher and Frederick Rauscher, eds., *Evolutionary Psychology: Alternative Approaches*, Boston: Kluwer, 2003.
Wilson, Edward O., *Sociobiology: The New Synthesis*, Cambridge, MA: Harvard University Press, 1975.
  Biophilia: The Human Bond with Other Species, Cambridge, MA and London: Harvard University Press, 1984.
  The Diversity of Life, Cambridge, MA: Harvard University Press, 1992.
  Consilience: The Unity of Knowledge, New York: Alfred A. Knopf, 1998.
  On Human Nature, Cambridge, MA: Harvard University Press, 2004.
Wrangham, Richard and Peterson, Dale, *Demonic Males: Apes and the Origins of Human Violence*, Boston and New York: Houghton Mifflin, 1996.
Wright, John H., SJ, "The Eternal Plan of Divine Providence," *Theological Studies* 27 (1966), 27–57.
  "Divine Knowledge and Human Freedom: The God Who Dialogues," *Theological Studies* 38 (1977), 450–77.
Wright, Robert, *The Moral Animal*, New York: Vintage, 1995.
  "God," in Joseph A. Komonchak *et al.*, eds., *The New Dictionary of Theology*, Wilmington, DE: Michael Glazier, Inc., 1987, pp. 423–436.

Yoder, John Howard, *The Politics of Jesus*, Grand Rapids, MI: Eerdmans, 1972.
Xenophon, *Memorabilia*, trans. Amy L. Bonnette, Ithaca and London: Cornell University Press, 1994.
Zirkle, Conway, "Species before Darwin," *Proceedings of the American Philosophical Society* 103.5 (October 15, 1959), 636–644.

ELECTRONIC RESOURCES

NOVA, "Secret of the Wild Child" (March 4, 1997). Available at http://www.pbs.org/wgbh/nova/transcripts/2112gchild.html (accessed August 19, 2005).
The United Nations Children's Fund, "State of the World's Children 2005." Available at http://www.unicef.org/media/media_10958.html (accessed December 20, 2004).
The United Nations Educational, Scientific, and Cultural Organization, Gen. Conf. Res. 29 C/ Res. 16, "Universal Declaration on the Human Genome and Human Rights" (1997, 1999). Available at www1.umn.edu/humanrts/instree/Udhrhg.htm (accessed August 24, 2005).
World Health Organization and the United Nations Children's Fund, "Malaria Is Alive, Well & Killing more than 3000 Children a Day in Africa" (April 25, 2003). Available at http://www.unicef.org/media/media_7701.html (accessed December 20, 2004).

Index of scriptural citations

Genesis
   1:1–2:3, 205
   1:26–27, 153, 198, 205, 206,
      235, 236
   1:27, 197
   1:28, 206
   1:31, 15
   2:4–25, 205
   2:18, 197
   2:24, 230
   4:1–16, 135
   49, 244

Exodus
   12:31–36, 96
   14:10–31, 96
   19–20, 96
   19:4–6, 96
   20:12, 230
   34:10, 96

Numbers 16:30, 96

Deuteronomy 20:16–18,
   244

Joshua
   6, 295
   11:20, 244

1 Samuel 15:3, 244

2 Samuel 10–12, 135

1 Chronicles 244

2 Chronicles 244

Job
   10:8–11, 96
   38:1–42:6, 110

Psalms
   1:1–6, 237
   1:6
   8:3–4, 90
   19:2, 10
   21:9–10, 244
   42:2, 235
   104:29–30, 104
   104:30, 96
   107, 96
   120:1, 96
   121:58, 96
   139, 96
   145:9, 234
   148:7, 96

Isaiah
   7:17–19, 96
   42:5, 96
   45:7, 96
   55:8–9, 110

Jeremiah
   1:5, 173
   25:9–14, 96

Amos
   1–2, 96, 244
   5:8–9, 89
   9:7, 89

Malachi 1:2–3, 230

Wisdom
   4:10, 234
   7:28, 234
   9:3, 199

Matthew
   5:3–12, 237
   5:45, 12

6:33, 295
7:12, 239
10:37, 230
10:37–39, 237, 247
12:50, 236
13:24–30, 233
19:17–19, 230
19:30, 239
25:31–46, 235, 236

Mark
  8:34–35, 237
  10:1–12, 230

Luke
  6:31, 239
  6:33, 243
  10:29–37, 240
  10:37, 296
  12:31, 295
  14:26, 230, 237
  17:33, 237

John
  13:1, 243
  15:15, 243
  15:17, 232

Acts 14:15–17, 88
  20:35, 239

Romans
  2:14, 88
  2:15, 49
  3:8, 52n63
  5, 157
  5:5, 235

5:8, 235
5:12, 155
6:23, 156
8:9–22, 96
11:36, 94

1 Corinthians
  7:15–16, 52n63
  12:12–27, 243
  13:12, 110
  15:21–22, 157

2 Corinthians 5:17, 156, 232

Ephesians
  1:3–14, 96
  1:5, 235
  6:1–3, 230

Philippians
  2:5, 237
  2:5–11, 232

1 Timothy
  5:8, 247
  6:20, 87

2 Timothy 1:12, 14, 87

Hebrews 1:2–3, 96

1 John 4:10, 235
  4:16, 202
  4:20, 235

Jude 3, 87

Index of names and subjects

abortion and *image Dei* doctrine, 196
absolutism, 294
accountability. *See* freedom and responsibility
adaptive view of morality, 259–263
agape. See love and altruism
AI (Artificial Intelligence), 209
Albert the Great, 104
Alexander, Richard D., 146–147, 180n80, 218–219, 237, 240
altruism, *See* love and altruism
ambiguity, moral, 250–252
Andolsen, Barbara Hilkert, 237n64
anthropocentrism of human dignity, Gustafson's deflation of, 203–206
anthropology. *See* human nature and human flourishing
anthropomorphism, moral, 11, 12
Aquinas. *See* Thomas Aquinas and Thomism
Aristotelian Darwinism, 273–278
Aristotle
   on freedom and responsibility, 180, 181, 185n94
   on human dignity, 195
   on human nature and human flourishing, 142
   natural-law tradition and, 271, 272, 273–278, 279, 283, 319
   origins of morality and, 260
Arnhart, Larry, 45, 49, 136, 268, 273–278, 279, 284, 285–291
Artificial Intelligence (AI), 209
atheistic naturalism, 74
Augustine, 5, 17, 73, 102, 160, 181, 233
australopithecines, 77, 132
Axelrod, David, 217n13, 217n15
Ayala, Francisco, 60, 66, 77, 91, 115, 176, 255–257

Bacon, Francis, 183n87
Barash, David, 134
Barbour, Ian, 33, 67, 74, 83n32, 127, 131
Barker, Eileen, 66–67
Baron-Cohen, Simon, 133n13, 137
Barth, Karl, 45n42

basic goods in new natural-law theory, 51–54
Batson, Daniel, 222
beatitude, Thomistic notion of, 295
Beckstrom, E. G., 282
Bellah, Robert, 214, 262
Bible. *See* Index of scriptural citations; Scripture
Big Bang theory, 20, 95, 120
birth control, 198, 205, 311–313
Blanshard, Brand, 160n5
Blind Watchmaker theory, 315
Boehm, Christopher, 219n24
Bourke, Vernon, 181
Bowler, Peter J., 73
Boyd, Craig A., 278
Boyd, Robert, 62, 218n16, 219n24
Brooke, John, 4, 189
Browning, Don S., 72, 247, 299
Bryan, William Jennings, 79, 192
Bube, Richard H., 175
Burgess Shale, 112
Burrell, David, 35
Buss, David, 252, 261, 262, 301–302, 303–304
by-product of natural selection
   morality as, 255–257
   religion as, 23–24
Byrne, Richard W., 133n14

Cahill, Lisa Sowle, 215, 268, 281, 285–291, 304, 312–313, 314
Calvin, Jean, 22
Camus, Albert, 71
caritas. See love and altruism
Carnegie, Andrew, 191
causality
   downward, 67
   in free will vs. determinism debate, 180–181
   primary and secondary, 103–107
chance and purpose in evolution. *See* contingency problem in evolution; freedom and responsibility
charity. *See* love and altruism

348

Index of names and subjects 349

choice, 179–180. *See also* freedom and responsibility
Christian ethics and evolution. *See* evolution and Christian ethics
Christian faith, nature of, 86–90
Christ's death and resurrection as climax of Christian anthropology, 157
Clark, Steven R. L., 130–131, 194
Clark, William R., 166
Clayton, Philip, 172, 178–179
cognitive capacity. *See* intelligence, evolution of
Cole, R. David, 165
common descent and human dignity. *See* human dignity and common descent
compassion
  freedom and responsibility, 166
  human dignity and, 202
  human nature and human flourishing, role in, 142, 144–145, 146, 147, 152
  love and altruism, 214–215, 230, 236–237, 242
  in natural-law tradition, 269, 273
  origins of morality and, 254
  sex, marriage, and family issues, 307, 311
complexity
  adaptive and non-adaptive, 116
  emergent (*See* emergent complexity)
  relationships between disciplines and, 65–69
conceptual frameworks or maps, 68–69
Conn, Walter, 269
Conors, Russell, 177
Conquistadors, 227
consciousness and self-consciousness, 133–134, 142
consequentialism, 294
contingency problem in evolution, 111–128
  defined, 111
  emergent complexity and, 112, 116, 127–128
  Gould's argument for contingency, 111, 112–115
  meanings of chance, 123–124
  Morris on evolutionary convergence, 111, 116–119
  progressivism
    Gould vs. Dawkins on, 113–115
    human moral development and, 116
  Stoeger's theory of immanent directionality, 111, 119–123
  theological interpretation of, 123–127
contraception, 198, 205, 311–313
convergence, evolutionary, 116–119
Conway Morris, Simon, 7, 57, 79, 111, 116–119, 207, 308
Cosmides, Leda, 2, 166
creation, as theological doctrine, 93–95
creationism, scientific, 1n2, 93n59
Crick, Francis, 61
critical-realist alternative to scientism, 65

Cro-Magnons, 132
Cross and resurrection as climax of Christian anthropology, 157
Crusaders, 227
culture
  Christian faith and, 90–91
  evolution in non-human species, 117
  human nature and, 129, 130
    emotions, emergence of, 140
    intelligence, evolution of, 132–133, 135–136
    morality as product of, 266–267
  sociobiology
    acceptance of impact of culture by, 163
    explanations of culture by, 63–65
Curran, Charles, 177

Daly, Martin, 305, 306
Damasio, Antonio, 133, 140–142, 144, 222–223, 256–257, 258, 309
Darwin, Anne, 9
Darwin, Charles
  on animal and human cognitive capacity, 135
  fixed essences, end of concept of, 148
  free will vs. determinism debate, 159–160
  human dignity, Darwinian challenge to, 189–194
  on love and altruism, 239, 241–242
  natural-law tradition and, 273–276
  reductionism and, 56–57, 70
  religion, evolutionary accounts of, 8–10, 11
  science and theology, dialogue between, 4
  on sex, marriage, and family, 297
  social Darwinism rejected by, 82
  theological interpretation of evolution and, 91, 96, 103–104
  theory of evolution produced by, 77–79
Darwinian self-interest, 214–215
Darwinism, Aristotelian, 273–276
Darwinism, Hobbesian, 273, 291
Dawkins, Richard
  evolutionary accounts of religion and, 18, 20, 21, 23, 24
  on free will vs. determinism debate, 163, 166, 181
  on human dignity and common descent, 193
  on love and altruism, 216, 242
  on meme theory, 63, 254
  natural law critique of, 49
  natural-law tradition and, 269, 280, 282
  Placher's critique of scientific narrative and, 43
  on problem of evil, 14–15, 19
  progressivism of, 114–115
  reductionism
    blurring of forms of, 70
    meme theory, epistemological reductionism of, 63

Dawkins, Richard (cont.)
    methodological, 58
    ontological, 70, 71, 74
    as sociobiologist, 1, 4
    truth, human desire for, 138
De Waal, Frans, 179, 231, 259, 275, 282, 299
death and resurrection of Jesus as climax of
    Christian anthropology, 157
death and sin, 157
definitions. See terminology
Dennett, Daniel, 60, 84, 136–137
Descartes, René, 182, 190
descent, common. See human dignity and
    common descent
descent with modification, 77
design, argument from, 4, 8, 10
design, intelligent, 44, 73, 83n32, 308
desires vs. inclinations, 135
determinism. See freedom and responsibility
Dewey, John, 159
Diamond, Jared, 298, 303
Dieter, Theodor, 45n42
dignity. See human dignity and common descent
directionality in evolution. See contingency
    problem in evolution: progressivism
Dobzhansky, Theodosius, 81
Dostoevsky, Fyodor, 245, 255
double agency, 107
downward causality, 67
dual inheritance theory, 62
dualism
    of Christian ethics (See indifference of
        Christian ethics to evolution)
    of evolutionists (See evolutionary accounts of
        religion)
    in free will vs. determinism debate
        free will, as term, 179
        John Paul II, 158, 170, 171–173
        sociobiological and evolutionist theories,
            168–169
Durkheim, Emile, 1, 22, 33

ecology and human dignity, 197–198, 204–205
economics and self-interest, 204
Edelman, Gerald, 64
Edwards, Jonathan, 5, 237
efficient causality, 180–181
egoism
    genotypical altruism and egoism, 224–226, 252
    phenotypical altruism and egoism, 221, 225,
        226, 246
    psychological, 224
Eldredge, Niles, 79
Ellis, George, 80
emanation, Neoplatonic doctrine of, 94

emergence, 66–67
emergent complexity
    contingency problem and, 112, 115, 127
    in dialogue between theologians and scientists,
        3, 6
    essential implication of, 317
    free will vs. determinism debate and, 158,
        176–177, 187
    human nature and human flourishing, 135,
        139–140, 147–153
emergent materialism, 117, 172
emotion
    emergence of capacity for, 140–143
    motivation and, 222–223
    value of training and controlling, 262
emotivism, 251
environmentalism and human dignity, 196–197,
    202–203
Epicurus, 286
epistemological naturalism, 34
epistemological reductionism, 61–69
    complexity and relationships between
        disciplines, 67–69
    critical-realist alternative to, 65
    culture, sociobiological explanations of, 63–64
    maps or conceptual frameworks, 68–69
    scientism, 64–65
    social context of scientific inquiry, 65
eternity, divine, 101–102
ethics, Christian. See evolution and Christian
    ethics
evil, problem of. See problem of evil
evolution and Christian ethics, 1–7
    common descent, repercussions of (See human
        dignity and common descent)
    emotions and moral life, 142
    indifference of Christianity regarding
        evolution (See indifference of Christian
        ethics to evolution)
    meaning, scientific vs. theological
        presuppositions regarding, 6
    origins of morality (See origins of morality)
    separation theory of, 24–27
    theological interpretations (See theological
        interpretation of evolution)
    use and meaning of evolution as term, 74–75,
        76–85
        erymology of term, 76–77
        fact, evolution as, 77
        ontology, evolution as, 82–85
        theory, evolution as, 77–82
evolutionary accounts of religion, 8–31
    critique of, 28–31
    Gould, separation thesis of, 24–27
    Pinker, by-product theory of, 23–24

rejection of religion and problem of evil
(*See* problem of evil)
Wilson, D. S., group selection theory of, 21–23
Wilson, E. O., replacement theory of, 18–21, 29, 87
evolutionary convergence, 116–119
evolutionary ethics, 1
evolutionary ethics/evolution of morality. *See* origins of morality
evolutionary psychology, 2
  on freedom and responsibility, 160, 161, 163–164, 166, 167, 185, 186–187
  on human nature and human flourishing, 129–130, 135, 138–139
  indifference of Christian ethics to, 49
  on love and altruism, 221, 245
  natural-law tradition and, 269–270, 282, 291, 296
  on origins of morality, 250–251, 255, 256
  reductionism of, 57–58
  religion, accounting for, 23–24, 28–29
  on sex, marriage, and family issues, 298–301, 306–307, 315, 316–317
existentialism, 251

fact
  evolution viewed as, 77
  theory, relationship to, 79–81
faith, Christian, nature of, 86–90
family. *See* sex, marriage, and family
fatalism, 160
feminism, 237, 298, 302
final causality, 180–181
Finnis, John, 45, 50–54, 206, 280, 283
first principles and new natural-law theory, 50–54
Flanagan, Owen, 174
Frank, Robert H., 214
Frankfurt, Harry, 164, 181
freedom and responsibility, 158–170
  in Christian ethics, 158, 170–176
    John Paul II, theological dualism of, 158, 170–171, 172
    Nichols, Terence, 170, 171–173
    theological approaches to free will vs. determinism, 176–178
  definition of terms
    choice, 179–180
    determinism, 159–160
    free choice, 182–183
    free will, 177–178
    freedom, 183–185
    responsibility, 185–186
    will, 181–182
  dualism regarding
    free will, as term, 176–177

  John Paul II, 158, 170–171, 172
  sociobiological and evolutionist theories, 168–169
  emergent complexity and, 159, 176, 187
  indeterminacy vs., 167
  problem of evil and, 14–15
  reductionism applied to, 158, 164, 165, 186
  science and theology, dialogue between, 5–6
  sociobiological and evolutionist views on, 159–170
    attempts to protect free will concept, 167–170
    choice, evolutionary explanations of, 164–165
    dualistic theories, 168–169
    fatalism, 160–161
    genetic components of behavior, 162–163
    natural proclivities or tendencies, 163–164
    passivity, 162
Frei, Hans, 38
friendship, 243, 248. *See also* love and altruism
Fuchs, Josef, 177
fundamentalism
  creationism, 1n2
  problem of evil and, 11

Gandhi, Mahatma, 151
gene-culture theory, 62, 168, 254
genetic components of behavior, 162–163
genetic discrimination, 188
"Genie" the "wild child," 147
German Constitution on human dignity, 211
Gewirth, Alan, 284
Gilkey, Langdon, 26
Gingerich, Owen, 73
God, love of, 31, 48, 86, 207, 234–237
"God of the gaps" arguments, 175–176
golden rule, 239–241
good Samaritanism, 145, 146, 225, 240–241, 243, 246, 259, 296
Goodall, Jane, 133n12, 136
good(s). *See also* human nature and human flourishing; problem of evil
  basic goods in new natural-law theory, 50–54
  causality and, 105–106
  Christian ethics and, 6
  freedom and responsibility, 158, 163–164, 176, 179, 180–181, 183
  human dignity and pursuit of, 201–202, 210
  love as affirmative response to, 227
  motivation towards, 222
  natural-law tradition and, 53–54, 268–269, 287–289, 317–318
  neighbor, love of, 242
  providence, doctrine of, 95–103

good(s) (cont.)
　of species, 58
　theological interpretations of evolution and, 85–86, 90–91
Gould, Steven Jay
　on contingency in evolution, 111, 112–116, 118, 119, 127, 138
　on free will vs. determinism debate, 167
　on human nature, 130n4
　intelligence, evolution of, 138–139
　panadaptationism or panglossianism, 300
　reductionism and, 70–71
　separation thesis for evolution of religion, 24–27
　theological interpretation of evolution and, 85–86, 103
　theory, evolution viewed as, 79
grace
　evolutionary accounts of religion and, 30
　freedom and responsibility, 175
　human nature and human flourishing, role in, 151, 153–154, 157
　love and altruism, 229, 233, 235, 237, 242, 248, 249
　natural-law tradition and, 50, 270, 290, 296
　origins of morality and, 265
　sex, marriage, and family issues, 308, 310, 319
　theological interpretation of evolution and, 86, 90–91, 97, 98
gradualism, 79
Grant, Colin, 229–232, 248
Gray, Asa, 131
Grisez, Germaine, 45, 50–54, 206, 280, 283
group selection, 21–23, 79, 219–221
Grunstein, Michael, 166
Gula, Richard, 177
Gustafson, James M.
　adaptive view of morality and, 250
　human dignity, deflationary hermeneutics of, 203–209, 213
　narrative theology, critique of, 35–40, 41–42
　on original sin, 155–156
　on theological interpretation of evolution, 84, 85, 89
Gutierrez, Gustavo, 177

Haeckel, Ernst, 191
Hall, Pamela, 294
Hamilton, William D., 215
hard vs. soft naturalism, 83
Hauerwas, Stanley, 33–35, 39, 41
Haught, John, 12, 32
Hawking, Stephen, 95
Hefner, Philip, 30, 241
Henry V, 318

Hobbes, Thomas, 28, 29, 145, 147, 273, 291
holistic view as alternative to reductionism, 58–59
Hollenbach, David, 177, 287
Holy Spirit, 157
Homo habilis, 132
Hrdy, Sarah, 302
human dignity and common descent, 188–200
　Church, internal challenge to, 199
　Darwinian challenge to, 189–194
　dialogue between theologians and scientists on 5–6
　Gustafson's deflationary hermeneutics of, 203–209
　human rights and, 188–189
　imago Dei (*See imago Dei* doctrine)
　intelligence and rationality thesis, 195, 209
　John Paul II on, 196–200, 206–209
　Peacocke's reconsideration of, 200–203, 206–209
　Rachels' neo-Darwinian philosophical naturalism, 194–196, 206–209
　sociobiologists on, 193–194
　Thomistic theology and, 196, 197–200
human freedom. *See* freedom and responsibility
human nature and human flourishing
　basic goods in new natural-law theory, 51–54
　culture, role of, 129, 135
　emotions, emergence of, 140
　intelligence, evolution of, 135–136, 140
　desires vs. inclinations, 131
　emergent complexity, 131, 139, 140–147
　emotional capacities, emergence of intelligence (*See* intelligence, evolution of)
　in natural-law tradition, 280, 281, 283, 287–291, 292, 293, 294
　scientific views of, 129–130
　sex, marriage, and family, 313, 316
　social capacities, emergence of, 147–150
　theological interpretation of evolution and Christian ethicists' views, 129–131
　Christian theological anthropology, 153–157
　emergent human complexity, relevance of, 147–153
human responsibility. *See* freedom and responsibility
human rights and human dignity, 188–189
Humanae Vitae, 312–313
Humani Generis, 170
Hume, David, 275, 291
Hütter, Reinhard, 45n42
Huxley, Aldous, 244
Huxley, T. H., 8, 10, 16, 191, 315

imago Dei doctrine, 68, 197
　essential congruity with evolutionary theory, 210–213

in Gustafson's thought, 208–209
in John Paul II's thought, 197, 198–200
in Peacocke's thought, 201–203
immanent directionality, 119–120
inclinations. *See* proclivities, predispositions, or tendencies
indifference of Christian ethics to evolution, 32–44
disciplinary specialization as reason for, 32–33
narrative theology, 33–44
Gustafson's critique of, 35–38
Hauerwas on, 33–35
Placher on, 38–44
natural-law theory, 44–55
new natural-law theory of Grisez and Finnis, 50–55, 206
personalist view of John Paul II, 45–47
Thomist view of Pinckaers, 47–50
indirect reciprocity, 218, 240, 256
individualism
in modern culture, 217
moral individualism of Rachels, 194
infanticide, 165, 305
intelligence, evolution of, 132–140
animal behavior and, 132, 133–134, 135
brain size, 132
choice and cognition, 229–230
consciousness and self-consciousness, 133–134
contingency problem and, 118
culture, role of, 135–136, 138–139
fitness requirements, behavior transcending, 140–143
human dignity and, 195, 209
social interaction and, 133–134
symbolic representation, 132, 135, 136
theory of mind, 133, 137
truth, human desire for, 138
intelligent design, 44, 73, 83n32, 308
intention, 222, 223–224
interdisciplinary relationships complexity theory of, 65–69
indifference of Christian ethics to evolution, disciplinary specialization as reason for, 32–33
maps or conceptual frameworks, 68–69
Irenaeus of Lyons, 86
Irons, William, 256

Jaki, Stanley, 60–61
James, William, 159
Jesus' death and resurrection as climax of Christian anthropology, 157
John Damascene, 98
John Paul II
on basic human needs, 289
on freedom and responsibility, 158, 170, 171–173

on human dignity and common descent, 196–200, 201, 202, 205, 206–209, 210
natural-law tradition and, 45–47, 49
on sexuality, 312
theological interpretations of evolution, 81–82, 81n29
Johnson, Phillip E., 26, 74–75
Jonas, Hans, 185n95

Kamin, Leon J., 161
Kant, Immanuel, 194, 253, 273–274
Kaufman, Stuart, 79
Kierkegaard, Søren, 238–239
kin altruism
Christian ethics and, 230–231, 246–247
as evolutionary theory, 215–216
familial violence and, 304–305
human freedom and, 165
human nature and, 146–147
moral ambiguity of, 261
non-kin, extension towards, 304–305
parental investment and, 302, 305
social context form, 301
King, Martin Luther, Jr., 151, 309
Kirkpatrick, Lee A., 239
Kitcher, Philip, 70, 183
Kramer, Peter, 138
Kuhn, Thomas, 41

Lamarck, Jean-Baptiste, 78
language and symbolic representation, 135–136
Lash, Nicholas, 40, 157
"left wall" phenomenon, 113–114
Lennox, James, 273
Lewis, C. S., 212
Lewontin, Richard, 63–64, 71–72, 161–162
liberation theology, 152–153, 177
Lindbeck, George, 38
Linnean hierarchy, 189
Locke, John, 28
Lorenz, Konrad, 129
love and altruism, 214–239
Christian love, 227–237
familial love, 24, 230, 231, 232, 238, 246–249
friendship, 242, 249
God, love of, 31, 48, 86, 207, 234–237
golden rule, 239–241
neighbor, love of, 31, 48, 86, 207, 230–231, 232–233, 239–248
recent analyses of, 229–234
self-love, 237–239
types of, 227–229
unilateralism and mutualism of, 239n75, 243, 244, 248
universality of, 249

love and altruism (cont.)
    Darwinian self-interest and, 214–215
    defining altruism, 221–223
    dialogue between scientists and theologians regarding, 5–6
    evolutionary theories of altruism, 215–227
        critique of, 221–227
        genotypical altruism and egoism, 225–227, 252
        group selection, 21–23, 79, 219–221
        indirect reciprocity, 215, 218, 240
        kin altruism or nepotism (*See* kin altruism)
        manipulation, altruism as product of, 215, 219
        phenotypical altruism and egoism, 221, 226, 246
        reciprocal altruism, 147, 215, 216–218
    friendship, 242, 249
    good Samaritanism, 145, 146, 225, 240–241, 243, 246, 259, 296
    social capacities, emergence of, 147
Lukes, Steven, 22
Lumsden, Charles, 62, 254
Lyell, Charles, 8

Machiavelli, Niccolo, 28
Machiavellian intelligence, 133, 252
Machiavellian module, 219n22
MacIntyre, Alasdair, 182, 268, 271–273, 284, 285, 286, 291
Malthus, Thomas, 8
manipulation, altruism as product of, 215, 219
maps or conceptual frameworks, 68–69
marriage. *See* sex, marriage, and family
Marx, Karl, 1, 41
materialism
    emergent, 171, 172
    scientific, 20, 71–72, 74, 83
Mayr, Ernst
    on freedom and responsibility, 161–167
    on human dignity, 189, 193
    on human nature and human flourishing, 138
    on meanings of evolution, 77
    on natural-law tradition, 284, 315
    on reductionism, 58–59
McCabe, Herbert, 94, 222n28
McCormick, Patrick, 177
McCormick, Richard, 46
McMullin, Ernan, 80, 102–103
meaning and reality
    chance and purpose (*See* contingency problem in evolution)
    Gould's separation thesis and, 25
    natural-law tradition vs. evolutionary theory, 269
    ontological reductionism, 70–74
    problem of evil and, 14
    providence, doctrine of, 103

scientific vs. theological presuppositions regarding, 5–6
    terminological meaning (*See* terminology)
meme theory, 63, 254
Mendel, Gregor, 148
metaphysics, evolutionists' rejection of, 253
methodological naturalism, 73
methodological reductionism, 56–61, 72–74
Midgley, Mary, 20, 68–69, 85, 160, 174, 222n29
Mill, John Stuart, 280
Millennium Declaration, United Nations General Assembly, 288
Miller, Kenneth R., 127
mind, theory of, 133, 137
modulary theory
    free will vs. determinism debate and, 168
    Machiavellian module, 219n22
    panadaptationism or panglossianism, 300
    religion, evolutionary accounts of, 23–24
Monod, Jacques, 282
monogamy, 299
moral ambiguity, 250–252
moral anthropomorphism, 11, 12
moral individualism, 194
moral realism, 6
morality generally. *See* evolution and Christian ethics
Mother Theresa, 226, 254
motivation, 221–223
Murphy, Nancey, 172
Murray, John Courtney, 287
mutuality in Christian love, 231–233, 236
mystery, evil as, 14

narrative theology, 33–44
    defined, 33
    Gustafson's critique of, 35–38
    Hauerwas on, 33–34
    Placher on, 38–44
natural-law tradition, 268–296
    Arnhart on, 273–278, 291
    available knowledge, need to build on, 316
    Cahill on, 285–291
    evolutionary theory and, 149, 268–270, 291–296
    human nature and human flourishing in, 280, 281, 283, 287–91, 292, 293, 294
    indifference to evolution in, 47–55
    new natural-law theory of Grisez and Finnis, 50–54, 280
    personalist view of John Paul II, 45–47, 49
    Thomist view of Pinckaers, 47–50
    MacIntyre on, 271–273
    moral claims in, 92
    objectivity in, 294
    origins of morality and, 250, 254

natural-law tradition (cont.)
  Porter on, 278–285, 291
  science and theology, dialogue between, 3, 6
  on sex, marriage, and family
    coordination with evolutionary theory on, 313–319
    normative moral standards of, 307–313
  Thomist
    Arnhart and, 275–278
    available knowledge, need to build on, 316
    beatitude, notion of, 295
    Cahill and, 285, 287
    evolutionary theory and, 268, 291, 293, 295–296
    John Paul II's personalist view, 45, 46–47
    MacIntyre and, 271
    Pinckaers, 47–50
    Porter and, 279, 282, 283–284
    on sex, marriage, and family, 307–309, 310
natural revelation, 88–89
natural selection, 78. *See also* evolution and Christian ethics
naturalism
  atheistic, 93
  defined, 82–83
  epistemological, 43
  holistic approach to evolution and, 59
  methodological, 73
  ontological, 6, 32, 34, 70, 72, 75, 83–85
  origins of morality, accounts of, 250
  philosophical naturalism of Rachels regarding human dignity, 194–196
  Pinckaers' account of natural law and, 49
  reductive, 43, 59, 70–74, 83
  soft vs. hard, 83
  theological, 34
  of Wilson, E. O., 20
naturalistic fallacy, 50, 52, 54, 280
nature
  Christian faith and, 89–90
  defined, 5n11
  human nature (*See* human nature and human flourishing)
Neanderthals, 136
neighbor, love of, 31, 48, 86, 207, 230–231, 232–235, 239–248
neo-Darwinism
  alternative modern positions to, 79
  anti-evolutionary objections to methodological reductionism and, 74
  contingency in evolution and, 111
  free will vs. determinism debate, 149
  individual vs. group selection, 220
  narrative theology's critique of evolution and, 43
  philosophical naturalism of Rachels regarding human dignity, 194–196

  problem of evil and, 10, 15
  selective appropriation of, 263
  social capacities, emergence of, 145–147
Neoplatonism, 94
nepotism. *See* kin altruism
Neuhaus, Richard John, 80n21
new natural-law theory, 50–54, 204
Newton, Isaac, 91
Nichols, Terence, 170–176
Niditch, Susan, 244n91
Niebuhr, Reinhold, 45n42, 154–155, 185n95, 230
Nietzsche, Friedrich, 25, 280
Non-Overlapping Magisteria (NOMA), 25, 26, 27
Noonan, John, 177
Nuer tribe, Sudan, 220–221

objectivity
  in natural-law tradition, 294
  in science, 43, 64, 178
O'Hear, Anthony, 137–138
ontological naturalism, 6, 32, 34, 70, 72, 75, 83–85
ontological reductionism, 69–75
ontology, evolution as, 82–85
open programming, 257
origin of species, 78. *See also* evolution and Christian ethics
original sin, 154–157, 265
origins of morality, 1, 250–267
  Christian ethical appropriation of, 259–267
    implications of, 265–267
    natural-law tradition, 261–264
    objections to, 263–265
    role of evolutionary theory in, 261–263
    transcendence of evolutionary past, 259, 260, 263
  evolutionary accounts of, 250–259
    adaptive, morality as, 250–255
    by-product of natural selection, morality as, 255–257
    culture, morality as product of, 257–259
    proclivities or tendencies, 257, 260, 261
    reductionism in views of, 259, 265
Outka, Gene, 239n75, 243n86

Paley, William, 4, 8, 283
Palmer, Craig, 300
panadaptationism or panglossianism, 300
pantheism, 101
paradigms, Kuhnian, 41
parental investment, 302, 304
Parliament of Instincts, 129
part-whole relation and methodological reductionism, 58–59

Partially Overlapping Magisteria (POMA), 27
passivity, biological determinism as
　　encouraging, 162
Paul VI, 46
Pauline Principle, 52
Peacocke, Arthur
　　on freedom and responsibility, 183, 187
　　on human dignity and common descent,
　　　　200–203, 204, 205, 206–209, 213
　　on narrative theology, 33
　　on reductionism, 57, 65, 67–69
　　theological interpretations of evolution, 84,
　　　　108–109
Peirce, C. S., 159
Penzias, Arno, 64
personalist natural law of John Paul II, 45–47
personhood, unified center of, 181–182
Peterson, Gregory, 130n5
phenotypical altruism and egoism, 221,
　　226, 246
phenotypical plasticity, 149, 167
philosophical naturalism of Rachels regarding
　　human dignity, 194–196
Pinckaers, Servais, 45, 47–50, 55
Pinker, Steven, 2, 23–24, 169, 181, 182, 301
pity. *See* compassion
Pius XI, 46
Pius XII, 81, 170, 197
Placher, William, 38–44
Plantinga, Alvin, 72–74
Plato, 251
Polkinghorne, John, 95, 106, 108–109, 123n44
polygamy, 299
POMA (Partially Overlapping Magisteria), 27
population thinking, 78
Porter, Jean, 45, 268, 278–285, 291
Post, Steven, 243
Potts, Malcolm, 299
predispositions. *See* proclivities, predispositions,
　　or tendencies
preference studies, 302
prepared learning, 257
primary causation, 103–107
prisoner's dilemma game, 217, 236
problem of evil, 8–18
　　Christian response to, 14–18
　　Darwinism and, 8–10
　　moral vs. natural evil, 15
　　providence, Christian doctrine of, 98
　　sociobiology and, 9–12
proclivities, predispositions, or tendencies
　　desires vs. inclinations, 135
　　freedom and responsibility, 163–164
　　moral ambiguity of, 250–252
　　origins of morality, 257, 260, 261

progressivism
　　Dawkins on, 114–115
　　Gould's rejection of, 113–115
　　human moral development and, 116
　　of social Darwinists, 119
proto-morality, 275
providence, doctrine of, 95–103
Provine, William, 159
prudence, 294
psychological egoism, 224
punctuated equilibrium, 79
purpose and chance in evolution. *See* contingency
　　problem in evolution

Rachels, James, 194–196, 201, 205, 206–209
Rahner, Karl
　　dialogue between theologians and scientists
　　　　and, 5
　　evolutionary accounts of religion and, 16
　　on freedom and responsibility, 172
　　on human nature and human flourishing, 153,
　　　　155–156
　　on narrative theology, 35, 42
　　natural-law tradition and, 269, 270
　　theological interpretations of evolution in
　　　　writings of, 88, 91
rape, evolutionary explanation of, 300
rationality thesis of Aristotle, 195
realism, moral, 6
reality. *See* meaning and reality
reciprocity
　　adaptive view of morality and, 250–251
　　alrruism, reciprocal, 146, 214, 216–218
　　golden rule differentiated from, 238
　　indirect, 218, 240, 256
　　origins of morality and, 250–251, 255
　　sex, marriage, and family, 302–303
reductionism, 7, 56–75
　　blurring of forms of, 70
　　epistemological (*See* epistemological
　　　　reductionism)
　　in free will vs. determinism debate, 158–159,
　　　　165–166, 186
　　love and altruism, 242
　　methodological, 56–61, 72–74
　　ontological, 69–75
　　origins of morality, in accounts of, 259, 265
　　problem of evil and, 14
　　in sex, marriage, and family issues, 300, 316
reductive naturalism, 43, 59, 70–74, 83
religion
　　evolution and (*See* evolution and Christian
　　　　ethics; evolutionary accounts of religion)
　　modern concept of, 28–31
responsibility. *See* freedom and responsibility

resurrection as climax of Christian
    anthropology, 157
revelation and Christian faith, 87–89
Richerson, Peter, 62, 218n16, 219n24
Ridley, Mark, 216
Ridley, Matt, 299
rights and human dignity, 188–189
Riley, William Bell, 79
Rolston, Holmes, 225, 239, 257–258
Romero, Oscar, 152–153
Rose, Steven, 57, 59–60, 65, 68, 79, 161, 163, 182, 226
Rousseau, Jean-Jacques, 145, 147
Ruse, Michael
    on freedom and responsibility, 164–165, 168
    on human dignity, 193n23
    on human nature and human flourishing, 138
    on love and altruism, 230–233, 242, 246
    on origins of morality, 250–251, 252, 253, 256, 264
    on Porter, 282
Russel, P. R., 190
Ryle, Gilbert, 182

Sartre, Jean-Paul, 251
Schloss, Jeffrey, 224
science
    defined, 1n1
    objectivity of, 43, 64, 178
    theology, dialogue with, 1–7
scientific creationism, 1n2, 93n59
scientific materialism, 20, 71–72, 74, 83
scientism, 63–64
Scopes Monkey Trial, 192
Scott, Roger, 299
Scripture
    methodological reductionism and, 60
    personalist natural law of John Paul II, 46
secondary causation, 103–107
self, concept of, 180–181, 218
self-consciousness, 132–134, 142
self-interest, 214–215, 218, 252–253
self-love, Christian, 237–239
self-organization, 79
self-transcendence, 269
selfish gene theory. *See* Dawkins, Richard; sociobiology
separation theories. *See* dualism
sex, marriage, and family, 279–319
    abortion and *imago Dei* doctrine, 198
    birth control, 198, 205, 311–313
    Christian ambivalence regarding sex, 152
    Christian concepts of familial love, 24, 230, 231, 232, 238, 246–249
    coordination of evolutionary theory and natural-law ethics on, 307–313
    emergent complexity and, 186

evolutionary theory
    coordination with natural-law ethics, 313–319
    kin altruism or nepotism (*See* kin altruism)
    positions of, 297–307
    human dignity and, 197–198, 199, 205–206
    human nature and human flourishing, 313, 317
    moral ambiguity and, 250–252
    natural-law tradition on
        coordination with evolutionary theory, 313–319
        normative standards of, 307–313
    passion or *amor*, 228
    reciprocity, 302–303
    reductionist treatment of, 300, 316
    sociobiological views of, 298, 300, 316
Shannon, C. F., 68
sibling rivalry, 300
Simon, Thomas Yves, 179n78
Simpson, C. G., 193
sin, doctrine of, 154–157, 266
Singer, Peter, 236
situation ethics, 294
Skinner, B. F., 179
Smith, David Livingstone, 219n22
Sober, Elliott, 79, 139n39, 219–221
Sobrino, Jon, 177
social capacities, emergence of, 143–147
social context of scientific inquiry, 64–65
social Darwinism, 82, 85, 119, 191–193, 195, 302
social dimensions of Christian faith, 87
social justice and human dignity, 210
social passivity, biological determinism as encouraging, 162
sociobiology, 1
    adaptive view of morality, 250–255
    culture
        acceptance of impact of, 163
        explanations for, 63–65
    evolutionary accounts of religion and, 18–21, 28, 29
    in free will vs. determinism debate (*See under* freedom and responsibility)
    on human dignity and common descent, 193–194
    human nature as viewed by, 129–131
    on love and altruism, 221–223, 248–249
    on motivation, 222–223
    original sin vs. pessimism of, 156
    problem of evil in, 9–12
    self-transcendence in, 269
    on sex, marriage, and family, 298, 300, 316
soft vs. hard naturalism, 83
somatic markers, 142, 257, 262
Spencer, Herbert, 82, 191–192
Sperry, Roger, 137
Spohn, William, 92, 228n44
Stark, Rodney, 247

Stevenson, C. L., 251
Stoeger, William B., 90, 111, 119–123, 125–126, 207
Strawson, Peter F., 83
sublation, 186–187
survival of the fittest, 82
symbolic representation, 132, 135
Symons, Donald, 2
sympathy. *See* compassion

Tannehill, Robert, 230
Tatersall, Ian, 132n10, 132n11, 186
Taylor, Charles, 43, 134, 135n20, 142, 182
teleology
    contingency problem and, 119
    in free will vs. determinism debate, 183
    modern shift of theology and philosophy away from, 111
    place of, 15
    theological interpretation of evolution and, 86
tendencies. *See* proclivities, predispositions, or tendencies
terminology
    altruism, 221–223
    contingency, 111
    evolution (*See under* evolution and Christian ethics)
    of freedom vs. determinism debate (*See under* freedom and responsibility)
    narrative theology, 33
    naturalism, 82–83
    nature, 5n11
    science, 1n1
Tertullian, 71
Theissen, Gerd, 241
theistic science, 73–74
theological anthropology, 153–157
theological interpretation of evolution, 85–111
    Christian ethics, 90–93
    Christian faith, 86–90
    contingency problem, 123–127
    creation, doctrine of, 93–95
    divine action in the world, 107–110
    human nature and (*See under* human nature and human flourishing)
    in natural-law tradition (*See* natural-law tradition)
    origins of morality, Christian ethical appropriation of (*See under* origins of morality)
    primary and secondary causation, 103–107
    providence, doctrine of, 95–103
theological naturalism, 34
theology
    as interpretive enterprise, 2
    science, dialogue with, 1–7

theory
    evolution as, 77–82
    fact, relationship to, 79–81
    of mind, 133, 137
Thomas Aquinas and Thomism
    beatitude, notion of, 295
    on causality, 182
    on chance, 124
    on choice, 180
    on Christian faith and Christian ethics 90, 91
    on creation, 94
    on divine action in the world, 107–108
    human dignity and, 194, 195, 196–200
    on human nature and human flourishing, 148, 149, 150, 151
    on mutual indwelling, 242
    natural-law theory of (*See under* natural-law tradition)
    origins of morality and, 260, 263
    on primary and secondary causation, 103, 104, 105, 106–107
    on providence, 96, 98, 103
    on rational vs. irrational animals, 117
    religion, evolutionary accounts of, 28
    on self-love, 238
    on sex, marriage, and family, 307–309, 310
    theology and science, dialogue between, 5
Thornhill, Randy, 300
Tolstoy, Leo, 255
Toner, Jules, 310
Tooby, John, 2, 166
Trivers, Robert, 1, 216, 226
truth
    evolution of intelligence and human desire for, 138
    narrative theology on, 33–34, 36, 43–44
    unity of, 2, 3–4

UNICEF, 288–289, 290
unified center of self, 181–182
unilateralism of Christian love, 239n75, 241, 242
United Nations General Assembly, *Millennium Declaration*, 288
Universal Declaration of Human Rights, 188
Universal Declaration of the Human Genome and Human Rights, 188
universality of Christian love, 248–249

Van Till, Howard, 3, 80, 84–85, 126–127, 175–176, 207, 308
Vann, Gerald, 100
Veritatis Splendor, 45–46
violence and aggression, 300, 305–307
vitalism, 70

Ward, Keith, 127
Watchmaker theory, 283, 315
Warts, Fraser, 137n31
Weber, Max, 1, 25, 33
Whitehead, Alfred N., 100, 202
Whiten, Andrew, 133n14
Wilberforce, Samuel, 190
Wilkinson, Gerald S., 214n14
will, 180–182. *See also* freedom and responsibility
Williams, George C., 10–11, 15
Williams, Rowan, 131, 156, 281
Wilson, David Sloan, 21–23, 79, 219–221
Wilson, E. O.
  on free will vs. determinism debate, 166, 181
  on human dignity and common descent, 193
  on love and altruism, 226, 230
  natural-law tradition and, 276, 280, 282
  on origins of morality, 250–251, 254–255, 259
  other evolutionary accounts of religion and, 24, 27
  Placher's critique of scientific narrative and, 43
  reductionism of, 61–62, 64, 70, 74
  replacement theory of religion, 18–21, 29–30, 87
  as sociobiologist, 1, 4, 5
  theological interpretation of evolution and, 106
  truth, human desire for, 138
Wilson, Margo, 305, 306
Wright, John H., 96, 97, 98, 99, 100
Wright, Robert, 2, 161, 162, 164, 251

Yoder, John Howard, 295n72